# Medical Microbiology and Immunology

**Prof. S.M. REDDY**

Emeritus Professer
Sectional President Indian Science Congress Association
Department of Botany, Kakatiya University
Warangal-506009

**G. NARENDRA BABU**

Department of Microbiology,
Sri Raja Rajeshwari Government Degree and
Post Graduate College, Karimnagar, Telangana



SCIENTIFIC PUBLISHERS

Published by
**SCIENTIFIC PUBLISHERS (INDIA)**

Jodhpur –
5 A, New Pali Road
P.O. Box 91
Jodhpur - 342 001 INDIA

Delhi –
4806/24, Ansari Road
Daryaganj
New Delhi - 110 002 INDIA



ISBN: 978-93-86102-91-1

Visit the Scientific Publishers (India) website at
http://www.scientificpub.com

Printed in India

# Preface

The present book on medical microbiology and immunology is fourth in the series written to cater the beginners in microbiology. The syllabus mainly covers common core syllabus of B.Sc. Microbiology students of Andhra Pradesh and Telangana Universities and it also meets the requirements of pharmacy and basic Biotechnology of many Indian Universities. Initially, Microbiology mainly dealt with harmful activities especially as causative agents of diseases of man and animals, and popularly called as enemies of man. However, subsequent strides in the understanding of these organisms enlighted their other side beneficial activities and they got the name as an 'essential friends of man. Further, knowledge on these helped man to check effectively the disease-causing activities of microorganisms, which lead to the development of distinct discipline called as medical microbiology.

The present book deals with the diseases caused by the microorganisms and ways to protect man from them. The methods of detection and diagnosis of diseases caused by these organisms formed a separate chapter. The process of infection, and mechanism of disease, cause, epidemiology and methods of control formed the nucleus of the book.

Immunology constitutes second part of the book. It discusses basic concepts of immunity, organs of immunity, antigen and antibodies interaction. It also discusses autoimmunity and hypersensitivity.

Every effort is made to make the subject matter simple. The language written is lucid supported by neat diagrams. The necessary information is precised in tabular form. Question bank is given at the end of every chapter for self evaluation and to learn the subject perceptibly.

The work of this magnitude needs an unstinted cooperation from different persons who deserve our whole-hearted thanks and gratitude. We are fully conscious of the fact that science subject is never final and it changes with more and more studies. We earnestly request the students and teachers alike for their constructive criticism and valuable suggestion, which will help to make the book perfect and acceptable to one and all, and serve the student community for long time to come. We thank M/s Scientific Publishers, Jodhpur for their sustained interest in the publication of this book neatly and timely.

**Authors**

# Contents

# 1

# History of Medical Microbiology

Though the association of microorganisms with dreadful diseases was known ever since human beings existed, the causal agents were unknown. The ancient people attributed such sufferings to divine wrath and other supernatural forces, displeasure of Gods and Godesses. Later, other concepts such as environmental effects on the body constitution, poisonous materials, imbalances among four humors of human body viz., phlegm, blood, black bile and yellow bile (choler), and of faulty diet. It was also believed from very early times that diseases may result from invasion of the body by external contagion. However, all these were more speculations since microorganisms, the causal agents, could be seen only after the discovery of microscope in 19$^{th}$ century. Oliver Wendell Holmes (1848) in USA and Ignaz Semmelweis (1846) in Vienna have independently conduced that puerperal sepsis was transmitted by the contaminated hands of obsterician and medical student, and demonstrated that the disease could be prevented by washing the hands with antiseptic solution.

## 1.1. MICROORGANISMS AND DISEASE

Friedrich Henle in 1840 proposed criteria for providing evidence for the involvement of the microorganisms as a cause of human diseases. This theory is called as the germ theory of disease. Koch and Pasteur confirmed this theory in 1870 and 1880 respectively by series of experiments providing proof that microorganisms were responsible for causing anthrax, rabies, plague, cholera and tuberculosis. Pasteur has been called as father of modern medicine while Koch, as the founder of Diagnostic Microbiology. Subsequently, several diseases caused by different microorganisms were described.

While working on the prevention of wound infections, Joseph Lister, an English surgeon, gave convincing evidence that the microorganisms were the agents of human diseases. He developed a system of antiseptic surgery intended to prevent the microorganisms from entering into wounds. For this, the instruments were heat-sterilized and phenol was used in surgical dressing or sprayed over the surgical area. This approach was remarkably successful and it transformed the surgery significantly. It also provided strong indirect evidence that the microorganisms are involved in disease development because the phenol not only killed bacteria, but also prevented wound infection.

Later the German physician, Robert Koch, gave convincing and direct evidence about the role of bacteria in causing diseases. While working on the disease, Anthrax of cattle, he reported that the disease was caused by the bacterium, *Bacillus anthracis,* which he published in 1876. In his experiments, Koch infected a healthy mouse with a bacterial material from diseased animal and found that the mouse became ill. After successful demonstration of *B. anthracis* as a casual agent of anthrax disease in cows, Koch conducted series of experiment with spleen containing bacillus. He injected serum of spleen incubated with bacillus into mouse, which resulted in the development of anthrax. When the incubated beef serum was inoculated into the mouse, anthrax developed. Thus, by a series of experiments he gave convincing evidence about the role of *Bacillus anthracis* in the development of the disease anthrax. Then he proposed certain postulates about the relationship of the microorganisms and specific diseases, which are popularly known as Koch's Postulates that are summerized below:

1. A specific organism can always be found in association with a given disease.
2. The organism can be isolated and grown in pure culture in the laboratory.
3. The pure culture will produce the disease when inoculated into a healthy susceptible animal.
4. It is possible to recover the inoculated organism in pure culture from the experimentally-infected animal.

These postulates are being practiced even today before declaring any microorganism as pathogen.

Pasteur and Koch attracted many dedicated disciples who discovered causative agents of several bacterial infections, and enlarged the scope and content of microbiology in general and medical microbiology in particular. Important of them include Hansen, who discovered leprosy bacillus in 1874; Neisser, who discovered gonococcus in 1879; Ogston, who discovered the *Staphylococcus* in 1884; Nicolaier, who observed the tetanus bacillus in soil in 1884; Fraenkel, who discovered Pneumococcus in 1886; Bruce, who identified the causative agent of malaria fever in 1887; Schaudinn and Hofmann, who discovered spirochaete of syphilis in 1905.

Roux and Yersui (1999) discovered diphtheria toxin, and similar toxins were discovered in tetanus and other bacterial infections, which were specifically neutralized by their antitoxins. Thus, many infectious diseases were proved to be of bacterial origin in the beginning of the $20^{th}$ century.

## 1.2 DEVELOPMENT OF PURE CULTURE

During the course of his studies, Koch realized the need to isolate suspected bacterial pathogen in the pure form, which required a solid medium. For this, at first, he cultured the bacteria on cut surfaces of cooked potatoes, but this trial did not give satisfactory results because of the failure of sufficient bacterial growth on potatoes. Then he tried to solidify the liquid media by adding gelatin. Separate bacterial colonies developed after the gelatin surface was streaked with a bacterial sample when the gelatin medium hardened. Despite its advantages, gelatin was not an ideal solidifying agent because it was digested by many bacteria by their proteolytic activity and it melts when the temperature rises above 28°C. Later, on the suggestion of F.E. Hesse, instead of gelatin, he successfully used agar as a solidifying agent in the preparation of solid medium. This proved to be more ideal in the preparation of a solid medium. Agar is not a nutrient material and does not melt until 45°C. These developments made possible the isolation of many bacteria in pure culture that contained only one type of bacterium. This directly triggered rapid strides in all the branches of microbiology. Later on, meat extract and protein digest media was also developed by Koch, which was ideal substrate for the growth of pathogenic bacteria isolated from the human and animal bodies. These developments led to the discovery of nutrient broth and nutrient agar media, which are are extensively employed even today. Koch isolated *Mycobacterium tuberculosis,* which causes the disease tuberculosis in humans, in 1882 by employing pure culture technique and solid media. Then followed a golden age of about 30–50 y during which most of the human bacterial pathogens were discovered (Table 1).

The first clear demonstration of pathogenic role of fungus was made by Agostino and Bassi (1835) who showed that the muscardine diseases of silkworms were due to the invasion of their tissue by the fungus. However, characterization of the fungi responsible for systemic mycoses such as histoplasmosis and coccidiomycosis had to await investigations in Central and South America in the early years of the $20^{th}$ century. The techniques necessary for the systemic study of fungi in pure culture were refined by Raymond Sabouraud in 1890.

**Table 1:** Some important discoveries in Medical Microbiology

| Year | Name of the discoverer | Event in Medical Microbiology |
|---|---|---|
| 1492 | Cristopher Columbus | Initiated European-American contact, which led to transmission of European diseases to the Americans and *vice versa* |
| 1546 | G. Fracastoro | Suggested that invisible organisms are responsible for human diseases |
| 1835–1844 | Agostino Basse | Reported silkworm diseases caused by a fungus. Similarly, many diseases of human beings are caused by some fungi. |
| 1847–1850 | Semmellic | Discovered the use of antiseptics to prevent the disease spread |
| 1854 | John Snow | Recognized a link between the spread of cholera and drinking water supply |
| 1849 | Snow | Reported the cholera epidemics in London |
| 1867 | Joseph Lister | Developed antiseptic surgery |
| 1876–1877 | Robert Koch | Demonstrated that anthrax in cattle is caused by *Bacillus anthracis* |
| 1879 | Neisser | Reported gonorrhea caused by *Neisseri gonorrhoeae* |
| 1880 | Laveren | Discovered *Plasmodium*, a protozoa, as the causative agent of malaria disease |
| 1880 | Eberth | Reported typhoid four caused by salmonettion *typhi* |
| 1881 | Robert Koch | Cultured the bacteria on gelatin |
| 1881 | Louis Pasteur | Developed vaccine for anthrax |
| 1882 | Robert Koch | Reported *Mycobacterium tuberculi* as the causative agent of tuberculosis |
| 1883 | Klebs and Loefflar | Reported dyptheria caused by *Corynebacterium diphtherae* |
| 1883 | Nicolaier | Described titamus disease caused by *Clostridium titani* |
| 1884 | Robert Koch | Proposed Koch's postulates |
| 1884 | Graham | Developed gram stain |
| 1885 | Eschericha | Reported *Escherichia coli* causes diarrhoea |
| 1885 | Koch | Reported cholera caused by *Vibrea choleare* |
| 1886 | Frankel | Reported *Streptococcus pneumoniae* as causative agent of pneumonia disease |

| Year | Name of the discoverer | Event in Medical Microbiology |
|---|---|---|
| 1888 | Roux and Yersin | Toxin produced by *Corynebacterium diphtheria* is responsible for diphtheria disease |
| 1889 | Kitasato | Obtained pure culture of *Clostridium titani*, causative agent of titanus |
| 1890 | Von Behring and Kitasato | Developed antitoxin for diphtheria and titanus disease, and got Nobel prize in 1901 |
| 1892 | Ivanovsky | Demonstrated that tobacco mosaic disease was caused by a virus |
| 1892 | Welch and Nuttail | Reported that *Clostridium perfringens* causes gangrene |
| 1894 | Kitasato and Yersin | Discovered *Yersinia pestis* as a causative agent of plague |
| 1896 | Van Ermengen | Discovered *Clostridium botulinum* as a causative agent of the disease botulism, food poisoning |
| 1898 | Shiga | Described dysentry caused by *Shigella dysenteriae* |
| 1899 | Ronald Ross | Demonstrated that malaria disease is caused by the bite of female mosquito And got Nobel prize in 1902 |
| 1900 | Reed | Proved that yellow fever is transmitted by mosquito |
| 1905 | Schaudinn and Hoffman | Reported *Tryponema pallidum* as a causative agent of syphilis |
| 1906 | Wasserman | Developed complement fixation test for syphilis |
| 1906 | Bordet and Gengou | Whisping caugh caused by *Bordella pertussis* |
| 1909 | Ricketts, H. | Reported Rocky mountain spotted fever caused by *Rickettsia rickettsii,* which is transmitted by ticks |
| 1910 | Paul Ehrlich | Developed chemotherapeutic agent for syphilis |
| 1911 | Rous, F. P. | Discovered sarcoma virus in chicken |
| 1926 | Thomas Rivers | Established virology as a separate branch of science |
| 1929 | Alexander Fleming | Discovered penicillin, first antibiotic in the field of chemotherapy. Got Nobel prize in 1945 |

| Year | Name of the discoverer | Event in Medical Microbiology |
|---|---|---|
| 1935 | G. Domagk | Synthesized protocol for first sulpha drug to kill *Streptococcus*, got Nobel prize in 1939 |
| 1940 | H. Ruska | Used electron microscope to obtain first pictures of a virus |
| 1944 | Waksman | Discovered streptomycin, which was able to cure tuberculosis, got Nobel prize |
| 1949 | Ender, Weller and Robbins | Developed poliovirus in human tissue culture, got Nobel prize in 1954 |
| 1950 | Robert Hungate | Developed Roll-tube culture technique for culturing of anaerobic bacteria |
| 1952 | Park and Strominger | Gave mechanism of penicillin action |
| 1957 | Gajdusek | Discovered slow virus as a causal agent of wasting disease kuru, got Nobel prize in 1976 |
| 1957 | Issacs and Lindermann | Discovered interferon, which prevents viral infections |
| 1959 | Yalow | Developed the radioimmunoassay |
| 1959 | Sawada et al | Discoverd that the bacterial resistance is due to extra chromosomal genes i.e (plasmids) |
| 1967 | John Collier | Described the mechanisms of action of diphtheria toxins |
| 1976 | Bishop and Varmus | Identified oncogenes from Rous Sarcoma virus that can be found in the cells of human beings, got Nobel prize in 1989 |
| 1977 | Mc Dade and Sheppard | Discovered that *Legionella Pneumophilla* causes pulmonary disease in human beings |
| 1982 | Stanley Prusiner | Discovered prions as cause of scrapic disease in sheep, got Nobel prize in 1997 |
| 1983–1984 | Gallo and Montagnier | Isolated and identified HIV HIV (Human Immunodeficiency Virus), Azidothymidine (AZT) inhibits HIV |
| 1984 | Barry Marshall | Discovered that *Campylobacter pylori* causes ulcers in human beings |
| 1990 | | First human gene therapy started |
| 1992 | | First human trial of antisense therapy |

## 1.3 VIRUSES AS HUMAN PATHOGEN

The failure of Pasteur (1881) to recognize the organisms responsible for rabies was mainly due to its ultramicroscopic nature. This was suggested only when he could reproduce the disease in dogs and rabbits by the intracerebral injection of brain tissue or saliva from a fatal case. Then the discovery of viruses and their role in the disease was made possible when Carl Chamberland, one of Pasteur's associate, prepared a porcelain bacterial filter in 1884. The first viral pathogen to be studied was the tobacco mosaic disease virus. It was Loeffler and Frasch (1898), and Agiamarte and Lazear (1900) who demonstrated that foot and mouth disease of cattle and yellow fever of human beings respectively were due to bacterial filter-passing organisms. Walter Reed established that yellow fever was not only caused by virus, but also transmitted to humans by infected mosquito bite. Subsequently, it was developed into a different discipline of microbiology called virology.

Land Steiner and Popper (1909) showed that poliomyelitis was caused by a virus and it is transmitted the diseases experimentally to monkeys. Subsequent studies proved beyond doubt that smallpox, chickenpox, measles, influenza and common cold are caused by viruses. However, detailed studies on their morphology were possible only after the discovery of electron microscope by Ruska (1934) and subsequent refinement in electron microscopic techniques. Cultivation of virus was possible only in the animals or in human volunteers till the techniques of gowning them in cell cultures by Endus (1940) leading the way for large scale production of virus cultures for vaccine production. The application of tissue culture techniques enlarged the scope of virology considerably positive proof for malignancy T-cell leukaemia in humans was due to virus exhibited in 1980.

Knowledge on the microorganisms has transformed the diagnosis, prevention and cure of diseases. The conquest of epidemic and fatal infections has resulted in the graphic downfall of deaths of human beings. However, infections still remain to be the agents of deaths, especially in children. It is estimated that 16 million young children die each year due to the effects of infectious diarrhoea, measles, malaria, tetanus, diphtheria and whooping cough. Tragedy is that when it was learned to handle these diseases, many new diseases appeared and caused death. Tuberculosis, cholera, typhoid and leprosy are far from being controlled even today. Surprisingly, new agents of infectious diseases continued to be recognized (Table 2).

**Table 2:** Some newly-recognized infectious agents during 1980-2000

| Agent | Disease |
|---|---|
| *Balamuthia mandrillaris* | Granulomatous amoebic encephalitis |
| *Bartonella henselae* | Cat scratch disease |
| *Borellia burgdorferi* | Lyme disease |
| *Chlamydophila pneumoniae* | Pneumonia |
| *Cyclospora cayetanensis* | Diarrhoea |
| *Ehrlichia ciafeenis* | Human Ehrlichiosis |
| *Escherichia coli* 0157 | Haemolytic uraemic syndrome |
| *Helicobacter pylori* | Gastritis |
| Hepatitis C virus | Hepatitis |
| Hepatitis E virus | Hepatitis |
| Hepatitis G virus | Hepatitis |
| Human herpesvirus | Exanthema subuturn |
| Human herpesvirus | Kaposis sarcoma |
| HIV | Acquired Immunodeficiency Syndrome (AIDS) |
| Human T-cell | Adult T-cells leukaemia |
| Lymphotropic virus | Tropical spastic paraparenis |
| *Microspora* (various genera) | Diarrhoea |
| *Mobiluncus* spp | Bacteria/vaginosis |
| Nipah virus | Encephalitis |
| *T ropheryma whippelii* | Whipplei disease |

## 1.4 SOURCES AND SPREAD OF INFECTION

A few of the microorganisms that abound in nature produce disease in human beings and are called as pathogens. Some of the organisms associated with human beings (on the body and within body) are harmless for most of the time and are called as commensals. These organisms, on certain occasions, turn to be pathogenic and cause diseases to human beings. These organisms are called as opportunistic pathogens. The microorganisms use either enzymes or toxins to overcome the natural barriers of human beings and cause disease. The process of microbial invasion of the body is called infection. The diseases, which can readily be transmissible from person to person, are called infectious or contagious diseases. The sources of infection from which pathogenic microorganisms are disseminated to susceptible human beings are listed in Table 3.

**Table 3:** Sources of infection

| | |
|---|---|
| Environmental sources | Inanimate objects such as water, food, soil etc. Animals (bites and scratches, zoonoses), arthopod vectors |
| Person to person | Air-borne infection, direct contact (touching, kissing), maternal transmission, (intrauterine, transplacental), sexual transmission |
| Medical or surgical procedures | Instrumentations, operations, transfusion (blood and blood products), sexual transmission, transplants (tissues, organs) |
| Self infection | Endogenous sources (normal flora), shared needles, needle stick injuries. |

## 1.5 IMMUNITY AND IMMUNIZATION

During the early part of the 20$^{th}$ century, attempts were made to exploit the available immunological information by the development of vaccines and sera for the prophylaxis and treatment of infectious diseases. Toll Domaghls (1935) initiated scientific chemotherapy with the discovery of prontosil; till then antisera was the only specific therapeutic agent available for the management of infectious diseases.

It was known from ancient times that the persons who had suffered from a distinctive disease such as small pox or measles, developed resistance and they rarely contracted it a second time. Artificial immunization against small pox was practiced in various communities by the unsafe method of inoculating small pox exudates through skin (variolation). It was popularized in England by Lady Mary Eortley woman. She practiced variolation while in Turkey and even applied to her son in 1718. It was Edwar Jenna in 1796 who inoculated cow pox exudates to a healthy human being. Though the variolation and vaccination were vigorously opposed in certain parts of the world, vaccination has been instrumental in eradicating natural small pox, the last case of which occurred in Somalia in 1977.

Subsequently, live bacterial (Attenuated) vaccine (Pasteur, 1881) and killed bacterial vaccine (Salman and Smith, 1996), bacteral toxoid (inactivated toxin) were practiced successfully. Behring and Kitasato (1890) tried to elucidate the mechanism of immunity to the production of antibodies in response to infection by a pathogen. It gives protection to man from disease. Antibodies have a highly-specific affinity for the microbial substances i.e., antigen. Their antigens induce specific antibodies and provide protection to the human beings. The antibodies produced in blood are used in the laboratory for precise identification of the pathogen. The development of technique of perpetuating the growth of single antibody-producing cells i.e., monoclonal antibodies facilitated cheap and effective vaccination. Discovery of sulphur drugs (Gehard Domagk, 1935), antibiotics and development of chemotherapy helped to lower the human deaths due to infectious agents. Explosion of the knowledge in immunology further increased the confidence of human beings in controlling their diseases.

The era of chemotherapy was started in 1910 by the discovery of arsenic- containing salvarsan that could cure the disease caused by *Treponema*. Alexander Fleming (1929) made the accidental discovery that the fungus *Penicillium* produced a substance, which inhibits *Staphylococci.* Discovery was only due to the needs brought about by the Second World War that led to the isolation of the active substance – penicillin – and its subsequent mass production. Thus, it was the beginning of the antibiotic era, and other similar antibiotics were discovered in rapid succession with the availability of wide range of antibiotics with potent antibacterial activity. It was hoped that bacterial infections would be controlled within a short period. However, soon the development of drug resistance in bacteria presented serious difficulties.

With the discovery of a wide variety of antibiotics active against the whole spectrum of pathogenic bacteria and effective vaccines against most viral diseases, expectations were raised about the eventual elimination of all infectitious diseases. The global eradication of small pox inspired several of the similar campaigns against other major diseases of human beings. However, such hopes were dampened when new infectious diseases began to appear that were caused by either unknown microorganisms or known microbes producing novel manifestations. The climax came in 1981 when AIDS was identified in the USA and began its pandemic spread. Relentless vigilance appeared to be essential to protect the human beings from such microbes.

Apart from the obvious benefits such as specific methods of diagnoses, prevention and control of infectious diseases, Medical Microbiology has played a significant role in human welfare.

## 1.6 FUTURE

There has been a phenomenal decrease in the death rate from classic infections, curative effect due to increased resistance because of improved nutrition and living conditions, and subsequent immunization programmes and antimicrobial theory. These all helped to save life, but it is not safe to say that we have conquered infectious diseases; several challenges are being faced due to newly-emerged pathogenic microorganisms.

## REVIEW QUESTIONS

**I. Answer the following questions:**

1. Describe the studies made by Robert Koch.
2. Write an essay on the development of pure culture.
3. Give historical events that led to development of virology.
4. Write brief essay on the history of Medical Microbiology.
5. Discuss in brief the discovery and development of antibiotics in the treatment of infectious diseases.

**II. Write short notes on:**

1. Antibiotics
2. Importance of electron microscope
3. Agustine Barsi
4. Kochs postulates
5. Joseph Leister
6. Salvarsan
7. Emergence of immunization
8. Germ theory of disease

## 1.7 FURTHER READING

Balows A (eds) et al 1992 *The Prokarytes,* New York: Springer-Verlag

Baron EJ, Peterson L, and Finegold S 1995 *Bailey and Scott's Diagnostic Microbiology,* St. Louis: Mosby

Brock TD (ed) 1999 *Milestones in Microbiology,* Washington: American Society for Microbiology,

Cox FEG (ed) 1996 *Illustrated History of Tropical Diseases*, London: Wellcome Trust

Foster WD 1970 *Illustrated History of Medical Bacteriology and Immunology,* London: Cox and Waym

Murray PR 1996 *Pocket Guide to Clinical Microbiology*, Washington, DC: American Society for Microbiology

Murray PR et al 1995 *Manual of Clinical Microbiology,* Washington, DC: American Society for Microbiology.

# 2

# Normal Microbial Flora of Human Body

## 2.1 INTRODUCTION

Many different species of microorganisms such as bacteria, viruses, fungi and others inhabit the environment present around us. However, bacterial species predominate over other microorganisms. Apart from inhabiting the environment, they are also closely associated with healthy human being. Some of them live on the surface of the body while others are associated with different internal organs such as mouth, digestive system, respiratory system, urinary tract organs etc. (Fig. 1). As they are always present in and over the human body, they are called as natural, indigenous resident or normal flora or normal microbiota of man. All these microorganisms are intimately involved in the metabolic reactions of the human body (Fig. 2).

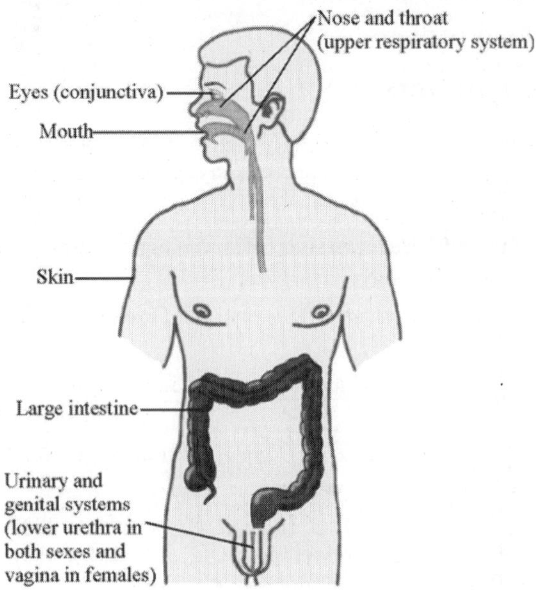


**Fig. 1** Location of normal microbiota on and in the human body

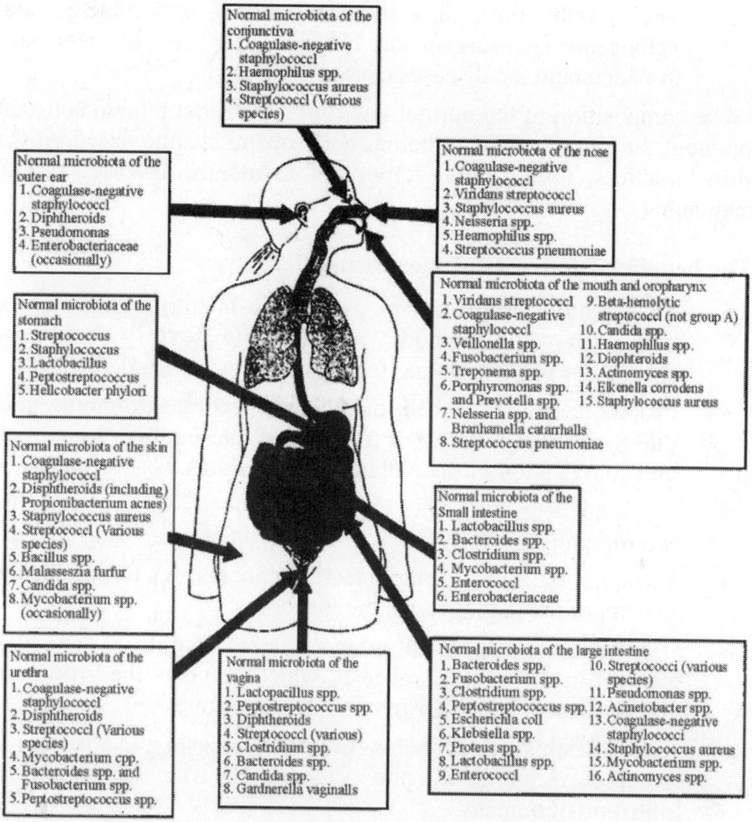


**Fig. 2** Normal flora of human beings

It is reported that human body consists of 1013 eukaryotic cells. It is interesting to note that the number of microorganisms associated with the healthy human body is ten times greater than the number of the body cells. They are able to live in different regions of the human body by adopting themselves to the continuously-occurring physical, chemical and biochemical changes. Some of them provide benefits to the human body while others derive benefits from human body. The former are called as mutualistic while the later as commensalistic microorganisms.

Knowledge about the activities of normal microbiota of the human body is essential to understand the following phenomena:

1. How are these organisms advantageous to the human body?
2. What are the disadvantages that the human body has to face in their absence, especially when antibiotics are used to control the diseases?

3. The normal flora also includes certain opportunistic and pathogenic microorganisms. In such case, it is also necessary to understand the diseases caused by them.

The composition of the normal microbiota does not remain constant throughout the life of a human being, but tends to change based on the healthy practices, food habits, activity of harmones, age etc., of the human being.

**1. The beneficial activities of the normal flora are:**

- They inhibit the growth of pathogenic microorganisms on or within the human body by secreting toxic / inhibitory substances and competing for the nutrients as detailed below:
- Propionic bacteria inhabiting the skin secrete propionic acid due to which acidic pH is maintained that inhibits the growth and colonization of the pathogenic organisms.
- Staphylococci of the mouth prevent the pathogenic microorganisms by producing toxic hydrogen peroxide ($H_2O_2$).
- Intestinal coliform bacteria secrete colicins, which inhibit the growth of the pathogens in the intestine.
- The vaginal lactobacilli produce lactic acid from glycogen present in the vaginal mucosa, which prevents the growth of the gonococci and other pyrogenic organisms.
- The intestinal flora helps in the synthesis of Vitamin B complex, Vitamin K and Vitamin E, which help in the nutrition of humans.
- Certain organisms associated with the digestive system help in the degradation and subsequent elimination of intestinal epithelial cells, mucins and carbohydrate fibres from the body.
- They also help in the maintenance of the body immunity and resistance against diseases.

**2. Harmful activities of normal microflora**: Microflora associated with the human body sometimes act adversely and causes health problems. Some of them are:

- Certain opportunistic microorganisms may become pathogenic under favourable conditions and cause diseases e.g., *Propionebacterium acne,* a bacterium of normal flora becomes pathogenic and causes acne vulgaris disease.
- They interfere in the disease diagnosis.
- Microflora of one system may cause diseases in other systems for e.g., *E. coli* of digestive system may cause infection in urinary tract.

## 2.2. STATUS OF MICROFLORA IN HUMAN BODY

Compared to other organisms, microorganisms in human body generally remain firmly attached to the skin and mucous membranes. This firm attachment is established due to an interaction between certain metabolites produced on the surface of both, the cells of microorganisms and the host cells. Filamentous structures called pili also help in this establishment. Certain microorganisms remain confined only to the specific regions of the body for e.g., *Staphylococcus salivarius* on tongue and *Streptococcus mutans* on the surface of teeth.

Because of desquamation of old epithelial cells on the skin and mucous membranes, old epithelial cells are removed and new epithelial cells are produced in their place, which occurs slowly in some areas and rapidly in other areas; for e.g., stomach, intestine and colon. Desquamation results in expulsion of weakly-attached and abnormal microflora. However, normal flora is not removed as they remain firmly-attached; even if they are removed they get re-established.

## 2.3. TYPES OF MICROFLORA

When compared to other types of microorganisms associated with the human body, generally bacteria predominate. Therefore, a detailed account of bacterial distribution in human body is given in this chapter. Certain fungi and protozoa are also present, but their number is relatively very less. As far as viruses are concerned, it is not clear whether they should be considered as normal flora or not; even though some of them live within the human body, e.g., Echoviruses and coxsackie viruses in the intestine.

**Table 4:** Normal microflora of human body

| Part of the body | Name of the microorganism |
|---|---|
| Skin | Anaerobic diptherioides- *Propionibacterarium acnes* |
| | *Staphylococus aurens* |
| | *Streptococus viridians* |
| | *Micrococci* |
| | *Streptococcus faecalis* |
| | Nonpathogenic *Bacillus* species |
| | *Escherichia coli* |
| | *Proteus* species |
| | Non-pathogenic mycobacteria |
| | Non-pathogenic corynebacteria |
| | *Candida albicans* |
| | Other fungi: *Cryptococcian pityrosporium* |
| | The pathogenic organisms such as *Salmonella typhi*, |

| Part of the body | Name of the microorganism |
|---|---|
| | *S. pyogenes* of bactericidal substances |
| Conjunctiva | Conjuctival sac is generally free from organisms because of lysozyme and flushing action of tears.<br>*Corynebacterium xerosis, Staphylococus epidermidis, Moraxella* sp and non-haemolytic streptococci are occasionally present is low number. |
| Mouth | Aerobes, non-haemolytic streptococci sps, *Neisseria* species, Micrococci, *Bacillus protes*, coliforms, *Mycoplasma*, *Candida* and *Peptococci*, P*eptostreptococci*, bacteriodes, *Fusobacterium*, *Actinomyces*, Non-pathogenic spirochates, *Lactobacillus, Veillonella* |
| Respiratory tract | The lower repiratory tract is normally sterile while upper respiratory tract is normally heavily colonized. |
| Nose | Non-pathogenic corynebacteria are constantly present. It is also a natural home for *Staphylococcus aurens.*<br>Species of *Proteus, Enterobacter, Pseudomonas aeruginosa, Candida albicans* and *Mycoplsma* species.<br>Anaerobes: Species of *Bacterioides, Bifidobacterium, Eubacterium*, *Leptobrichia* and *Clostridium*<br>Peptostreptococci and Lactobacilli: The flora increase progr- essively beyond duodenum to large intestine. The number of organisms in various parts of intestine are as follows:<br>Duodenum = $10^3$–$10^6$/g of content<br>Jejunum and proximal =$10^{5-}10^6$/g of content<br>Lower illeumand caecum = $10^8$–$10^6$/g of content<br>Colon and rectum = $10^3$–$10^6$/g of content |
| Faeces | Contain large number of bacteria, which are mostly dead. They constitute 15–30% of faecal matter. The number of live bacteria in faeces is about $10^{10}$/g and almost all (99.9%) are anaerobes. |
| Genitourinary tract | Female genital tract is more heavily colonized than males' because of its anatomy.<br>The normal vaginal secretions contain up to $10^8$ bacteria/ ml. Some of them are *Corynebacterium*<br>Aerobes: Species of *Micrococus, Dipththeria, Mycoplasma, Candida, Neisseria* and *Listeria, E. coli, Smegma* bacilli, *S. epidemidis*, *S. faecalis*, *Gardnerella vaginalis,* alpha-haemolytic streptococci<br>Anaerobes: Species of bacterioides, *Bifidobacterium, Eubacterium, Treponoma, Staphylococus epideridis, S. viridians, Neisseria catarrhatis (Branhamella catarrhalis), Moraxella lacunata* and *Haemophilus.* Species are other than those organisms present in the nose. |

| Part of the body | Name of the microorganism |
|---|---|
| Throat | *S. viridians* and *Branhamella catarrhialis* are commonly present.<br>Pneumonococci and Triendlanders bacilli are frequently present. Nasopharynx is a natural habitat of common pathogens causing infections of the nose, throat, bronchi and lungs.<br>*Pseudomonas aeruginose, E.coli* and *Proteus* sp may also present occasionally in normal healthy persons. |
| Gastrointestinal tract | It is similar to throat healthy stomach is practically sterile because of acidic pH.<br>The gastrointestinal tract of foetus in uterus is generally sterile, however it is colonized with organisms within 4–24 h. Lactobacilli.are found in breast-fed infants. The gram-negative bacilli dominate with the introduction of solid food. In adults, nearly 96–99% microorganisms are anaerobes while aerobes are upto 1–4%. |
| Intestine | It consists of both aerobes and anaerobes.<br>Aerobes: *E. coli, S. faecalli,* Clostridia, *Streptococcus, Peptococcus, Peptostreptococcus*<br>Flora changes with the deposition of glycogen in the vagina. At birth, intestine is sterile, but subsequently gets colonized by non-pathogenic micrococci and enterococci within 24 h. Subsequelntly, lactobacilli colonize within 2–30 hrs subsequently, lactobacilli, yeasts, and *E. coli* appears.<br>During pregnancy, there is a predominance of *Staphylococus epidermides.*<br>Smegma bacillus (*Mycobacterium smegmatis*) is present both in male and female. |
| External auditory meatus | It is profusely colonized with *Staphylococus epidermidis*, Diptherioirides (sometimes) and fast bacteria are present in the wax. |
| Blood and tissues | Commensal organisms from normal flora of the mouth, nasopharynx, intestinal tract and other sites may enter into the blood and tissues, but they are quickly eliminated by the defence mechanism. |

Most of the bacteria establish three types of relationships with human body viz., commensalism, mutualism and parasitism. Within each category, the association may be either ectosymbiotic or endosymbiotic. In ectosymbiosis, one organism remains outside the other. In endosymbiosis, one organism is present within the other. The following description consists of examples of both ecto and endosymbiotic relationships that exist with the human host (Table 4).

### 2.3.1 Microflora Associated with the Skin

An adult human is covered with approximately two square metres of skin on which about $10^{12}$ bacteria harbour. They live either on or in the skin either in the form of resident or transient microbiota. The former are present permanently and the later temporarily. Some of the microorganisms associated with the skin are: *Propionibacterium acne, Staphylococcus epidermidis, Staphylococcus viridians, Streptococcus faecalis, Eschereschia coli, Bacillus* species (non-pathogenic), *Proteus* species, anaerobic cocci, *Micrococci,* Non-pathogenic *Mycobacteria,* and Non-pathogenic Corynebacteria.

Anatomically, skin consists of two layers; an outer epidermis and inner dermis layer. The epidermal layer comprises of epithelial cells and keratinocytes. Most of the microorganisms cannot penetrate the epidermis because of the presence of thick and closely-arranged keratinocytes, which form a strong barrier against bacteria. Keratinocytes produce keratins, which are sclero proteins that are the important structural constituents of hair, nails and the outer skin cells. Moreover, continuous shedding of epithelial cells by desquamation prevents adherence of pathogenic bacteria to the surface of the skin.

#### Factors influencing the growth of bacteria in the skin

Skin surface or epidermis is not a favourable environment for colonization of bacteria because of the presence of certain skin factors, which are described below.

- **Dryness**: Lack of sufficient moisture in the skin leads to dryness, which prevent growth and colonization of many bacteria as they remain dormant or in an inactive state. Severe dryness also makes some microbiota die instantaneously. However, moisture is sufficiently high in certain parts of the body (scalp, ears, axillary areas, genitourinary and anal regions), which favours growth of the microbiota.
- **pH**: Skin pH varies between 3–5 and it is higher in theregions of the body having sufficient moisture. This acidic pH is due to the secretion of organic acids by normal staphylococci, from skin, which prevent growth and colonization of the microorganisms.
- **Inhibitory substances**: Sweat contains high concentration of sodium chloride, which makes the skin surface hyperosmotic causing osmotic stress that prevents the growth of microorganisms.
- **Sweat glands**: They release lysozyme, an enzyme, which lyses *Staphylococcus epidermidis* and other gram-positive bacteria

preventing their growth. Sweat glands also produce antimicrobial peptides called cathelicidins, which destroy bacterial invader by punching their plasma membranes or dissolving their membrane through some detergent action.

- **Oil glands**: The oil glands of the skin secrete complex lipids, which are partially degraded to unsaturated fatty acids such as oleic acid by certain gram-positive bacteria that prevents the growth and colonization of gram-negative bacteria. Some of the fatty acids are also highly volatile and are responsible for the formation of bad odours from the skin.

**Acne vulgaris:** It is a skin disease, which is more prominently seen on the facial skin. It is caused by the bacterium Propionibacterium acne, which is gram-positive, anaerobic and lipophilic. It is one of the constituent bacterium of the normal microbiota, but it becomes opportunistic and causes the disease. Acne vulgaris occurs during adolescence when endocrine system is active. Hormonal activity stimulates over-production of sebum, a fluid secreted by the oil glands. The sebum is normally eliminated along with sweat through the sweat pores. When these pores are blocked, sebum accumulates within the glands and provides suitable atmosphere for the growth of Propionibacterium acne. Sebum accumulation leads to inflammation of the oil glands resulting in the formation of pimples. If the disease is severe, a crop of pimples may be formed on the face making it ugly in appearance. P. acne produces lipase, which hydrolyses sebum triglycenides into free fatty acids. These fatty acids then enter into the dermis, and cause inflammation and irritation that also results in the formation of pimples. As this bacterium is sensitive to tetracycline, the disease is controlled by this antibiotic along with use of Vitamin A.

### 2.3.2 Normal Microflora of Eye

The microflora of eye includes the following organisms: Staphylococcus epidermidis, Staphylococcus aureus, Corynebacterium species, Streptococcus pneumoniae, Neisseria species, Moraxella species and Haemophillus parainfluenza.

A small number of the above-mentioned bacterial commensals are found in the conjuctiva of the eye at birth and throughout human life. Conjuctiva is continuously cleaned by tears. Most of the bacteria associated with conjuctiva are washed away along the rolling tears. This is the reason for the occurrence of small number of bacteria in the eyes. Moreover, lysozyme present in the tears also inhibits their growth.

### 2.3.3 Normal Micro Flora of the Respiratory Tract

***Upper respiratory tract***

The region from the nostrils to the middle part of the pharynx constitutes the upper respiratory tract. This region is covered with mucous membranes, which secrete sticky mucous. The mucous membranes are lined with constantly-vibrating cilia due to which mucous substances roll down the tract constantly.

External air is constantly inhaled during respiration. Many atmospheric microorganisms, especially bacteria enter the upper respiratory tract along with the inhaled air. However, majority of them get entangled in the rolling down mucous and are, therefore, pushed downwards. Hence, very less number of bacteria are present in this region.

The normal microbial flora of the upper respiratory tract includes: *Staphylococcus epidermidis, Staphylococcus aureus, Streptococcus pneumoniae* and species of *Staphylococcus, Corynebacterium, Neisseria, Branhamell, Haemophillous* and *Micrococous.*

Some of the above-mentioned bacteria remain firm by attaching to the mucous membranes due to which they are able to remain in this region. The other bacteria, which are not firmly attached move down, and finally reach the stomach where they are killed by the acid produced within the stomach and also by the action of the lysozyme enzyme.

**Lower respiratory tract**

The important parts of the lower respiratory tract are trachea, bronchi, pair of lungs and associated alveoli. Trachea and bronchi are tubular structure whose lumen is lined by delicate mucous membranes. Many contractile cilia are present on the upper surface of the mucous membranes, which help in rolling down of the mucous that finally reaches the alveoli. Along with the moving down mucous, bacteria also move down and reach the alveoli where majority of the bacteria are killed by the macrophages, which are associated with the alveoli. Hence, very few bacteria are present in the lower respiratory tract.

### 2.3.4 Normal Micro Flora of Mouth

There are about 108 microorganisms in the oral cavity, which include both aerobic and anaerobic bacteria. The normal bacterial flora includes species of *Staphylococcus, Neisseria, Veilonella, Actinomyces, Lactobacillus* and Yeasts.

The oral cavity remains suitable for the occurrence and growth of many microorganisms as sufficient quantities of water with dissolved salts is always present in the mouth along with the food particles. However, those microorganisms that resist mechanical removal by adhering to the surfaces such as the gum and teeth predominate. Others, which cannot attach to gum and teeth, are flushed down along with saliva to the stomach where they are killed by hydrochloric acid.

Apart from water, the saliva secreted in the mouth contains free amino acids, proteins, lipids, carbohydrates and inorganic salts, which are needed as nutrients by the bacteria. Therefore, the oral cavity forms a suitable region for the growth and multiplication of the bacteria.

### 2.3.5 Normal Micro Flora Associated with Teeth

Generally, aerobic and facultative anaerobic bacteria grow in the mouth of newly-born child till the teeth appear. Once the teeth start appearing, obligate anaerobic bacteria dominate the oral cavity because of prevalence of anaerobic conditions around the teeth. These bacteria include species of *Bacteroides, Fusobacterium, Streptococcus mutans, and Lactobacillus.* Among these bacteria, *Streptococcus mutans* grow more abundantly on the surface of the teeth and causes dental caries disease. During the course of attachment on the surface of the teeth, these bacteria secrete mucilaginous end-branched carbohydrates called glucans, which help in the firm attachment of the bacterial cells on the teeth surface. This glucan is produced from disaccharide sugars when its concentration is more due to the catalytic action of glycosyl transferase enzyme produced by the bacteria. Apart from glucans, fructose and glucose sugars are also formed during this reaction.

$$\text{N}-\text{Sucrose} \xrightarrow{\text{Glycosyl transferase}} \text{Glucose} + \text{n}-\text{Fructose}$$

The sugars, thus formed, are put to fermentation in the presence of *Streptococcus mutans* resulting in the formation of lactic acid on the surface of teeth. This acid gradually dissolves enamel layer present on the teeth surface, which leads to development of tooth decay. Once the process of tooth decay starts, plaques are formed on the tooth surface. Each plaque is formed by the accumulation of dissolved tooth debris, bacterial cells and certain organic chemicals, and contains $10^8$ bacterial cells. As the tooth-decay process continues, the environment becomes suitable for the growth and colonization of *Lactobacillus* and Actinomycetes on the tooth surface. This leads to further decay of the teeth, and finally results in the total decay of the affected teeth, which amounts to dental carries disease. When the total sets of teeth are formed in the oral cavity of a child, the composition of microbial florae becomes similar to the microbial florae of an adult. Along with bacteria, *Trichomonos tennax*, a protozoan, also colonizes on the surface of the teeth and in their crevises in those who do not practice oral hygiene.

### 2.3.6 Normal Microflora of External Ear

The normal flora of the external ear resembles those of the skin with coagluase negative staphylococci and *Corynebacterium* predominantly. Apart from these bacteria, certain fungi such as *Aspergillus, Alternaria, Penicillium, Candida and Saccharomyces* are also present on the external ear.

### 2.3.7 Normal Microflora Associated with Gastrointestinal Tract

Gastrointestinal tract consists of stomach, small intestine and large intestine (colon).

**Stomach**

Microorganisms are constantly moved from the oral cavity to the stomach. However, most of them are killed due to acidic pH (2–3) of gastric contents. The stomach normally contains less than 10 living bacteria per ml of gastric fluids. The normal bacterial flora includes species of *Peptostreptococcus, Staphylococcus, Lactobacillus, Streptococcus, Helicobacter* and yeasts such as *Candida* sp. Microorganisms survive in the stomach if they are moved rapidly or if they are resistant to acidic pH.

**Small intestine**

It consists of three anatomical parts - the duodenum, jejunum and ileum.

1. **The duodenum:** The first 25 cm part of the small intestine is called as duodenum. It harbours few microorganisms because of combined inhibitory action of stomach's acidic juice, bile and pancreatic secretions. Most commonly, bacteria found here are gram-positive cocci and rods.
2. **The jejunum:** It is the middle part of the small intestine. *Enterobacter faecalis,* lactobacilli, diphtheroids and the yeast *Candida albicans* are found in the jejunum.
3. **The ileum:** It is the last part of the small intestine. The microorganisms harbouring this part are similar to those present in the colon. As pH is alkalilne in the ileum, only anaerobic gram-negative bacteria and members of the family *Enterobacteriaceae* are present.

**Large intestine or colon**

It has the largest number of microbial flora than in the other parts of the body. It is estimated that $10^{12}$ microorganisms are present per gram of faeces. Over 400 different species have been isolated from the human faeces. It is reported that more than 3 X $10^{13}$ microorganisms are eliminated daily through human faeces.

Various physiological processes of colon help in moving the microorganisms out, which include peristaltic movements and desquamation of the surface epithelial cells, and continuous flow of the mucus that carries adhering microorganisms with it. To maintain homeostasis of microbial flora there occurs continuous replacement of those which are lost. The bacterial population in the human colon usually doubles once or twice a day. However, if the colon environment is disturbed due to stress, altitude changes, starvation, parasitic organism, diarrhoea and use of antibiotics, the normal microbial flora may change greatly. However, composition of the normal microbial flora depends upon one's dietary habits.

As stated earlier, obligate anaerobes and facultative anaerobes dominate the colon microbial flora, which include the following:

1. **Obligate anaerobes**
   a. Gram-negative organisms: *Bacteroides fragilis, B. melaminogericus, B. oralis, Fuso bacterium* and *Eschereschia coli*
   b. Gram-positive organisms: Species of *Bifidobacterium, Eubacterium* and *Lactobacillus*
2. **Facultative anaerobes:** *Eschereschia coli,* species of *Proteus, Klebsiella, Enterobacter, Peptostreptococci* and *Candida albicans*
3. **Protozoa:** *Trichomonos hominis, Entamoeba histolytica,* species of *Iodamoeba* and *Endolimax*

Most of the above-mentioned microorganisms grow and colonize the colon by feeding on other colon bacteria and are normally non-pathogenic. However, though *Entamoeba histolytica* is normally not pathogenic, it causes amoebic dysentry when colon environment becomes favourable for the disease.

The initial residents of the colon of breast-fed infants are the species of the gram-positive *Bifidobacterium* because human milk contains a disaccharide amino sugar, which acts as growth factor to these bacteria. On the other hand, in bottle-fed infants, gram-negative *Lactobacillus* predominates because bottle milk lacks the required growth factor. When the ingestion of solid food starts, these initial colonizers of the colon are eventually displaced by a typical gram-negative microbial flora and ultimately the composition of the adult's microbial flora is established.

### 2.3.8 Genitourinary Tract

It includes parts of urinary tract and genital systems. The urinary tract consists of kidneys, ureters and urinary bladder. Generally, parts of the urinary tract are usually free of microbial flora. However, *Staphylococcus epidermidis, Enterococcus faecalis* and *Corynebacterium* species are usually present in the distal portion of the urethra.

In contrast, the adult female genital-tract, because of its large surface area and mucus secretion, has a complex microbial flora that constantly changes with the female menstrual cycle. The important microorganisms include acid-tolerant lactobacilli, primarily *Lactobacillus acidophilus*. They ferment the glycogen produced by the vaginal epithelium forming lactic acid. As a result, the pH of the vagina and cervix is maintained between 4.4– 4.6.

## Review Questions

**I. Answer the following questions**

1. Give an account of beneficial and harmful activities of normal microbial flora.
2. Write an essay on the microbes associated with skin.
3. Describe the normal flora of respiratory tract.
4. Write an essay on the microbes associated with gastrointestinal tract.

**II. Write short notes on the following**

1. Desquamation
2. Opportunistic microorganisms
3. Acne vulgaris
4. Normal microbial flora of eye
5. Normal flora of mouth
6. Dental carrier
7. Normal flora of external ear
8. Microbial flora of the genitourinary tract

## 2.4 FURTHER READING

Broady TM, Larner J, and Minneman KP 1998 *Human Pharmacology: Molecular to Clinical*, St. Louis: Mosby

Balows A and Truper H 1992 *The Prokaryotes*, New York: Springer-Verlag

Murray P 1997 Human microbiota, in Topley and Wilson's Microbiology and Microbial Infections (eds) Balows et al

Senberg H and Amato RD 1995 Indigenous Penicillium and Pathogenic microorganisms of human, in Manual of Clinical Microbiology (eds) PR Murray et al, Washington, DC: American Society for Microbiology.

# 3

# Host Defence

## 3.1 DEFINITION OF DISEASE

A disease can be defined as any form or state of health in which, part or all of the host body is not capable of carrying on its normal functions due to the presence of an organism or its product. Any organism or agent that produces a disease is called a pathogen. When a pathogen is growing and multiplying within or on its host, the host is said to have an infection. The nature of the infection can vary widely with respect to severity, location, and number of pathogens involved in the cause of infection. The ability of a pathogen to cause disease is called as its pathogenicity. A primary pathogen is an organism that causes a disease in a healthy host by direct interaction. On the other hand, opportunistic pathogen, either usually free living or a part of the host's normal microbial flora, may act as a pathogen under certain circumstances such as under compromised immunity.

The term 'virulence' refers to the degree or intensity of pathogenicity. It is determined by three characteristics of the pathogen; invasiveness, infectivity and pathogenic potential. The ability of a pathogen to spread from the site of entry into adjacent or other tissue is called as invasiveness. The ability of an organism to establish an infection is called as infectivity. The degree or intensity of a disease caused by a pathogen is called as pathogenic potential. Toxigenicity is the pathogen's ability to produce toxin that will damage the host and produce disease. Virulence is often measured experimentally by determining the Lethal Dose 50 (LD50) or the Infectious Dose 50 (ID50). These values refer to the dose or number of pathogens that will either kill or infect 50% of an experimental group of hosts within a specified period.

## 3.2 NON-SPECIFIC DEFENCE MECHANISM

A microbial pathogen attacking a human body immediately encounters a vast variety of non-specific defence mechanism (Fig. 3). Although these mechanisms, individually, may not provide complete protection against invading pathogen, but collectively they counter the pathogen effectively. Some of the factors, which influence non-specific defence mechanisms, are direct and others are indirect. The direct factors include nutrition, physiology, fever, age, genetics and the indirect factors include personal hygiene, socioeconomic status, living conditions etc. At times they provide some type of defence to the host and at other times fail to prevent pathogenic organisms from causing infection in the human body. For example, susceptibility to infection increases when the host is either very young or very old. After the maternal immunity disappears in young children and before their own immunity matures, they are at greater risk of infection. Similarly, in elders there is a decline in immune system activity and decline in the functioning of many organs that reduce human defence. A vertebrate host possesses some specific physical and mechanical barriers in addition to the direct and indirect factors.

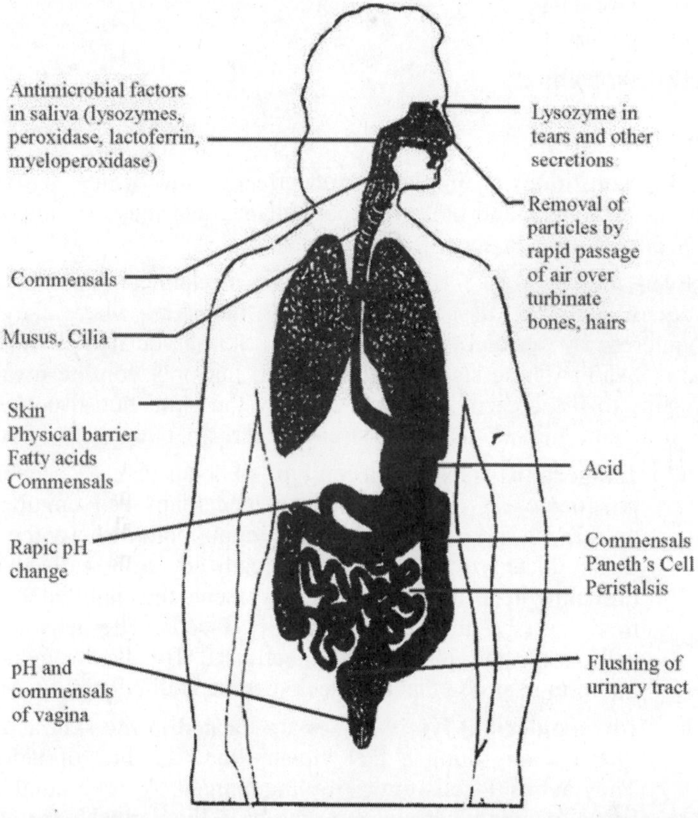


**Fig. 3** Host defences: Some non-specific (innate) host defence mechanisms that help to prevent entry of the microorganisms into the host's tissue

### 3.2.1 Mechanical Barriers

Mechanical barriers along with host's secretion constitute first line of defence against pathogens. Most important mechanical barriers are:

1. Skin
2. Mucous membranes
3. Gastrointestinal tract
4. Respiratory system
5. Genitourinary system
6. Inflammation
7. The Phagocytosis
8. The complement system
9. Cytokines
10. NK cells
11. Fever
12. Properdine

**Skin**

The skin constitutes an important and effective mechanical barrier to invading pathogens and other microorganisms, and plays an important role in conferring non-specific host resistance.

Even though skin forms an effective mechanical barrier, some pathogens enter the tissue present under the skin, where they are encountered by specialized cells called Skin-associated Lymphoid Tissue (SALT). These specialized cells will not only confine invading pathogens to the area of their entry below the skin, but also prevent them from entering into the blood stream. There are two types of SALT:

a. **Langerhans cells** are special type of dendritic cells capable of phagocytosing antigens. Once a langerhans cell engulfs the antigen, it migrates to nearby lymph node where the cell differentiates into another specialized cell called inter digitating dendritic cell. This cell presents the engulfed antigen to its surface and activates nearby T-cells. The activated T-cells can then interact with activated B-cells to induce a humoral response that produces specific antibodies (Fig. 4).

b. **Intraepidermal lymphocytes** are located in the skin and can interrupt any antigen that violates the first line of defence. They act as T-cells to destroy the antigen. A large number of tissue macrophages are also present in the dermal layer of the skin and phagocytose most pathogens that can enter the skin.

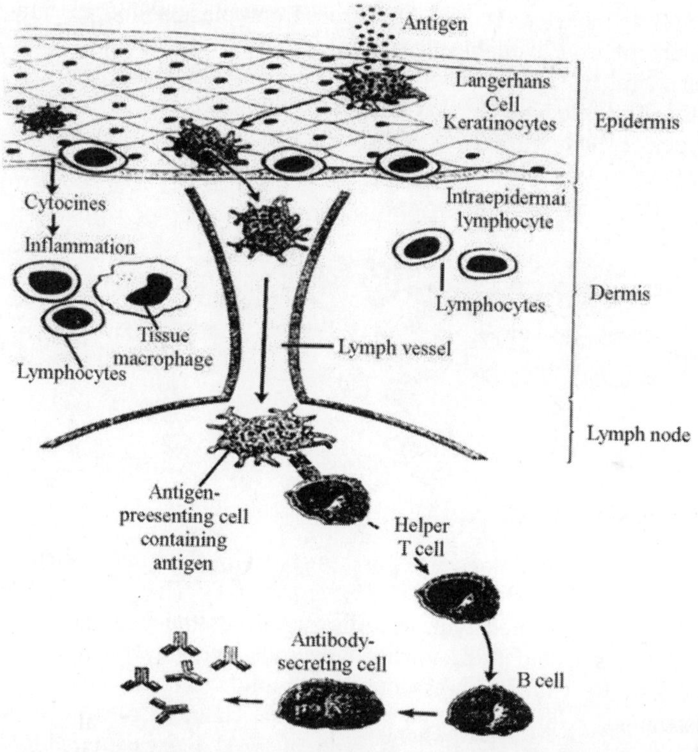


**Fig. 4** Langerhans cells action on antigen

## Mucous Membranes

Mucous membranes associated with many organs of human body such as digestive system, respiratory system, urinogenital system and eye prevent attack by the pathogens because of the layers of squamous epithellium and secretions of the mucous membranes. Thus, they not only resist penetration of the pathogen, but also trap many of them.

This mechanism constitutes a non-specific immunity. Cervical mucus, prostrate fluid and tears are toxic to many bacteria because they contain specific antimicrobial secretions, and hence prevent the growth and multiplication of many microorganisms. One such antimicrobial agent is lysozyme, which is an enzyme that lyses the bacteria by hydrolyzing B 1-4 glycosidic bond of peptidoglycon, a cell wall material of gram-positive bacteria (Fig. 5).

Mucous secretions also contain significant amounts of an immune protein, which not only prevents attachment of the pathogens on to the mucous membranes, but also reduces the non-binding substances lactoferrin that is released by the macrophages and Polymorphonuclear

(PMN) leukocytes. It separates iron from plasma due to which the amount of iron available to attacking pathogen is reduced, that limits their ability to multiply. Mucous membranes also produce lactoperoxidase enzyme that stimulates the production of toxic form of oxygen called superoxide radicals, which are poisonous to many microorganisms, both pathogenic and non-pathogenic.

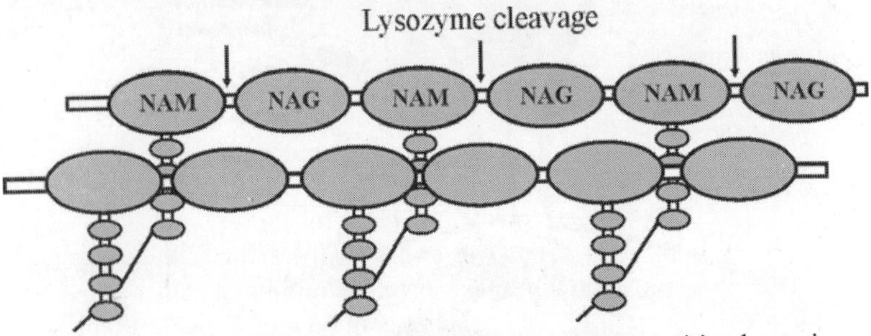


**Fig. 5** Action of lysozyme on the cell wall of gram-positive bacteria

Mucous membranes also possess specialized immune barriers called Mucosal-associated Lymphoid Tissue (MALT). There are several types of MALT associated with specific organs of the body. For example, MALT associated with tonsils, adenoids and payers patches in the intestine are called Gut-associated Lymphoid Tissue (GALT), MALT associated with respiratory system is called Bronchial-associated Lymphoid Tissue (BALT). A type of MALT is also reported from urinogenital system, but it does not have a specific name.

**Mechanism of MALT function:** The MALT carries its function in two basic mechanisms.

In the first mechanism, a special cell present in the epithelial cells, called M cell, contacts the antigen when it arrives at mucosal surface and releases it into a large pocket (present in M-cell) containing B-cells, T-cells and macrophages. The macrophages engulf the antigen or pathogen and destroy it (Fig. 6).

In the second mechanism, after M cell has phagocytosed, an antigen or pathogen transports it into a cluster of cells called an organized lymphoid follicle. The B-cells present in the follicle recognize the antigen and lead it into plasma cells, which leave the follicle and secrete secretory Immunoglobulin A (sIgA) antibody. The sIgA antibody is then transported into the lumen of the gut where it interacts with the antigen and destroys it (Fig. 7).

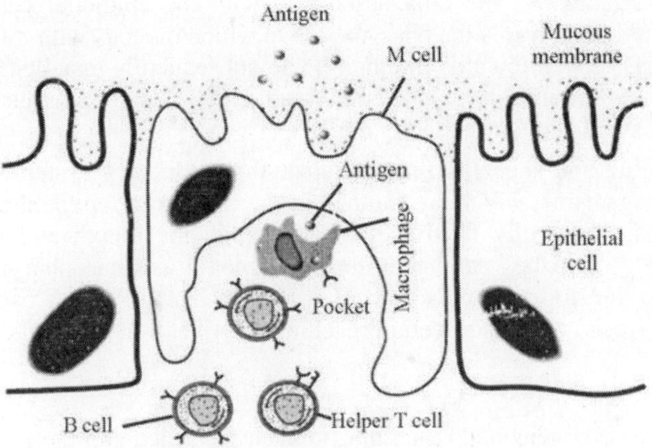


**Fig. 6** Action of MALT cells on antigens

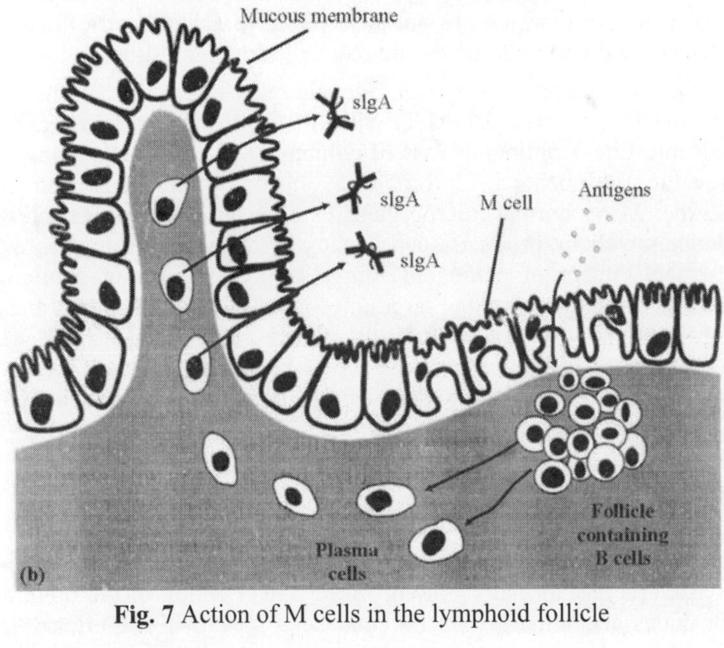


**Fig. 7** Action of M cells in the lymphoid follicle

## Respiratory System

Air is inhaled during breathing by a mammalian organism due to rhythmic movements of respiratory organs. Along with inhaled air, many microorganisms (8 microorganisms/ minute: 10,000/ day) may reach the upper and lower respiratory tracts. As the air flow is very turbulent in these tracts, pathogens are deposited on the moist, sticky mucosal surfaces. As the cilia associated with the epithelial cells of nasal cavity beat towards the pharynx due to which mucous with trapped pathogens is moved towards the mouth and subsequently expelled. Due to presence of humidity within the nasal cavity, many hygroscopic organisms swell and consequently die.

Coughing and sneezing reflexes clear the respiratory system from the microorganisms, which are trapped in the respiratory epithelium, by expelling air forcefully from the lungs through the mouth and nose. Salivation also washes pathogens from the mouth and nasopharyngeal region into the stomach acids. Alveolar macrophages kill most bacteria by phagocytosis, which succeed in reaching the alveoli of the lungs.

### Gastrointestinal Tract

Most of the pathogens that reach the stomach are killed and their toxins get inactivated (except protozoan cysts, toxins of *Clostridium* and *Staphylococcus*) by the acidic gastric juice with pH 2–3. However, some of the pathogens, which are not susceptible to acidic gastric juice, get protected by the food particles and reach the small intestine.

Various pancreatic enzymes, bile and other enzymes, present in the intestinal secretions and GALT, will kill the pathogens that reach the small intestine. Continuous loss of columnar epithelial cells along with peristaltic movements also promotes expulsion of pathogens from small intestine. Many normal microorganisms associated with large intestine produce metabolic products such as fatty acids, which not only prevent the establishment of pathogens, but also prevent unwanted microorganisms from colonizing the large intestine. The mucous membranes of gastrointestinal tract possesses specialized cells called paneth cells, which produce lysozyme and a group of toxic peptides called cryptins thatprevents the establishment of certain bacteria. However, action of cryptins is not clear.

### Genitourinary Tract

Kidneys, ureters, bladder and urethra constitute the urinary system. All these parts including the bladder normally remain sterile. However, a few bacteria are normally present in the apical region of the urethra of both males and females. Sterile conditions prevail in the urinary tract due to low pH and certain metabolic end products such as urea, uric acid, certain enzymes, mucin, hippuric acid, fatty acid indican, and urine that kill many pathogenic organisms. Only few organisms survive in the kidney medulla because of its hypertonic status. The lower urinary tract is usually flushed with urine and some mucous 4–10 times per day, which results in the elimination of pathogens.

The length of the urethra in males is about 20 cm and this lengthy urethra becomes a barrier in excluding pathogens from the urinary bladder, whereas in females it is short with 4 cm length and becomes a pivotal site for the colonization of the pathogens. Hence, the urinary tract infections are 14 times greater in females than in males.

The vagina has its own defence mechanisms. Under the influence of the hormone estrogen, the vaginal epithelium produces increased amounts of glycogen. The glycogen is converted into lactic acid by the activity of *Lactobacillus acidophilus,* making the vaginal fluid acidic (pH 3–5). This prevents the growth and survival of many pathogenic bacteria. Cervical mucosa also exhibits antimicrobial activity.

The conjuctiva of an eye lines the interior surface of the eyelids, and the exposed surface of the eyeball contains mucous-secreting epithelial cells. It is kept moist by the continuous secretion of tears from the lacrimal glands. As the tears contain large amount of lysozyme, lactoferrin and the antibody sIgA), it provides chemical as well as mechanical protection against the pathogens.

## Inflammation

Inflammation is a non-specific defence reaction caused in response to tissue injury, which may be due to invasion of a pathogen or wound formation by mechanical reasons. The signs of inflammation include redness, swelling, warmth and pain, and altered function at the site of inflammation.

There are two types of inflammation:

**1. Acute Inflammation:** Acute inflammation is a very complex process. It starts when the cells of injured tissue release inflammatory chemicals that activate the inner lung or endothelium of nearby capillaries. Secretions, which are cell adhesion molecules, are displayed on the activated endothelial cells within the capillaries. They help in randomly attracting and attaching neutrophils to the endothelial cells and allow them to roll down slowly along the endothelium. As the neutrophils roll down along the endothelium, they come across the inflammatory chemicals, which act as activating agents, and in turn, activate adhesion receptors, present on the neutrophils. The activating agents are called integrins, due to which neutrophils get attached to the endothelial adhesion molecules. This results in sticking of neutrophils to the endothelium and prevention of rolling. Quick changes take place in the shape of the neutrophils due to which they come out of the endothelial wall. This process is called as diapedesis. The neutrophils then enter into the interstitial tissue fluid, from where they migrate to the site of the tissue injury. There they attack the pathogen or other causes of tissue damage. Chemotactic factors or chemotaxis, released by the bacteria or mast cells play an important role in attracting neutrophils and other leukocytes to the site of injury. Based on the severity and nature of the tissue damage, other types of leukocytes such as lymphocytes, monocytes and macrophages may follow neutrophils, and reach the site of injury (Fig. 8).

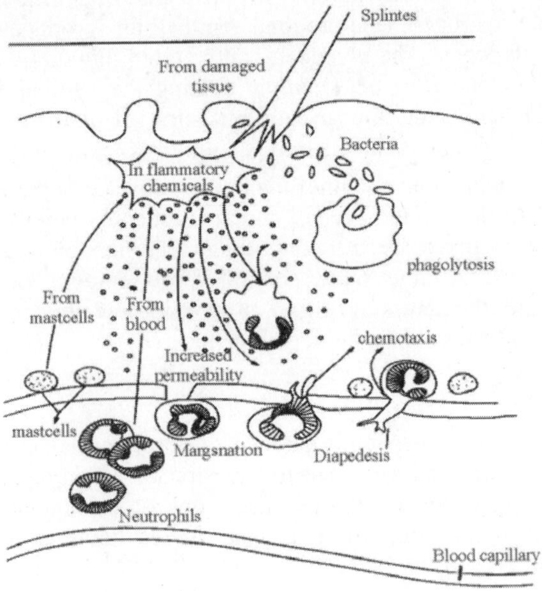


**Fig. 8** Physiological events of the acute inflammatory response

When this process of killing the pathogens and other causes is going on simultaneously, the processes of tissue swelling operate. The swelling process begins with the release of inflammatory chemicals from the injured tissue, which induces acidity in the surrounding extracellular fluid. Decrease in pH stimulates extracellular enzymes called Kallikrein, which split other long-chain products called bradykinin. Bradykinin binds to the receptors on the capillary due to which junction between the cells opens, making the fluid and infection-fighting leukocytes to leave the capillary, and enter the infected tissue. Simultaneously, some other bradykinin binds to mast cells present in the connective tissue associated with smallest blood vessels. This activates mast cells, and results in a massive flow of calcium ions, which leads to degranulation and release of histamine. Histamine, in turn, makes the intercellular junctions in the capillary wall wider. Hence, more fluid, leukocytes, kallikrein and bradykinin processors move out to cause a swelling. Bradykinin then binds to nearby capillary cells and stimulates the production of prostaglandins, $PGE_2$ and $PGF_2$, to promote tissue swelling in the infected site. A pain impulse also starts in the wound area when prostaglandins bind to the free nerve endings.

Phospholipase A2 releases arachidonic acid after a change in mast cell plasma membrane permeability associated with activation. Arachidonic acid utilization by the activity of cyclooxygenase or lipoxygenase pathway leads to the formation of certain metabolites, which include prostaglandin $E_2$ and $F_2$, thromboxine, $A_2$, slow reacting substance and leukotrienes (LTC4 + LTD4). All of them play an important role in various phenomena associated with inflammatory response (Fig. 9).

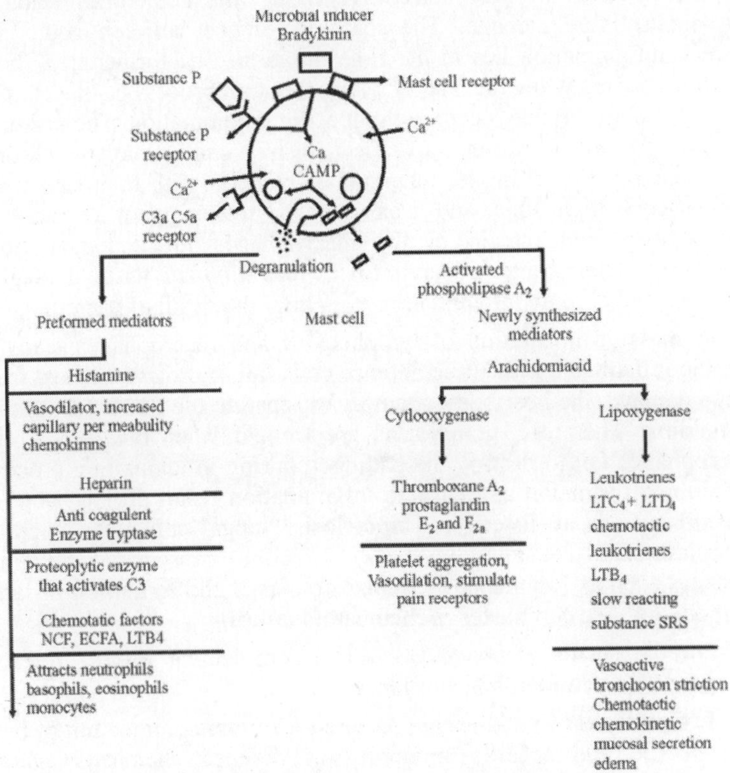


**Fig. 9** Biochemical events of inflammation

The invading pathogen is neutralized and subsequently eliminated by a series of events during acute inflammation. They are:

- Phagocytes collect in the swollen area and phagocytose the pathogen.
- A fibrin clot often formed may inhibit the spread of the pathogen and there by it is confined to the site of infection.

- The inflammatory chemicals stimulate the bone marrow to release more neutrophils and increase in the rate of production of granulocytes.
- The rise in temperature stimulates the inflammatory response and may inhibit the growth of the pathogen.
- More antimicrobial factors and leukocytes move into the inflamed area that destroys the pathogen, due to dilatation of capillaries, and increase the blood flow.

**2. Chronic Inflammation:** It is a very slow process, which results in the formation of new connective tissue and sometimes leads to permanent tissue damage. The main difference between acute and chronic inflammation lies in the time duration. The former is a short-duration process while the later is a long duration process lasting for two weeks or longer. It may occur without acute inflammation. The presence of bacteria by a variety of mechanisms can stimulate chronic inflammation. For example, tuberculosis bacteria with their cell walls having very high lipid and wax content, makes them resistant to phagocytosis and remain in the macrophages for a longer time. Moreover, certain bacteria secrete toxins that stimulate tissue-damaging reactions and chronic inflammation even after the death of bacteria.

A massive infiltration of lymphocytes and macrophages leads to chronic inflammation. If these defence cells fail to protect the host from tissue damage, the host body attempts to separate the site by forming a granuloma. Therefore, granulomas are formed when neutrophils and macrophages fail to destroy the pathogen during inflammation process. Granuloma formation and chronic inflammation result during bacterial infections such as listeriosis, brucellosis; fungal infections such as histoplasmosis, coccidiodomycoses; infection caused by helminth parasites such as leishmaniasis, schizostomiasis, and formation of large antibody-antigen complexes in rheumatoid arthritis.

For the details of phagocytosis and complement system, refer to **Chapter 6** on antimicrobial substances.

**1. Cytokines:** Cytokines are required for immunoregulation of both non-specific and specific immunity as defence chemicals against infections caused by bacteria and their products, viruses, parasites and cancer cells.

Cytokines are soluble proteins or Glycoproteins (GPs) released by one cell population that act as intercellular mediators or signalling molecules for providing protection against invading pathogens. Their production is stimulated by non-specific stimuli such as bacterial, viral or parasitic infection, or cancer or inflammation or the reaction between a T-cell and antigen. Sometimes, cytokines can also stimulate the production of other cytokines.

Cytokines function in three different ways:

a. An autocrine function: Cytokines affect the same cell from which they have originated.

b. A paracrine function: Cytokines affect the nearby cells.

c. An endocrine function: Cytokines move along circulatory system and affect the target cell.

**Biological function**

As far as biological function is concerned, cytokines are involved in all the aspects of infection. Biological activity of cytokines starts when they act as ligands and bind to specific, high-affinity receptors called Cell-associated Differentiation Agents (CDA) present on the surface of the target cell. The affinity of cytokine receptors for their ligands is very high. Therefore, they are effective even at a lowest concentration ($10^{-10}$–$10^{-15}$ m). When cytokines bind to the receptors of target cells, specific intracellular signalling-pathways are activated, which induce genes that encode proteins essential for the inhibition of suitable target cell. These proteins include cell-cell adhesion receptors, proteases, nitric oxide synthase, lipid synthesizing enzymes and other cytokines (Fig. 10).

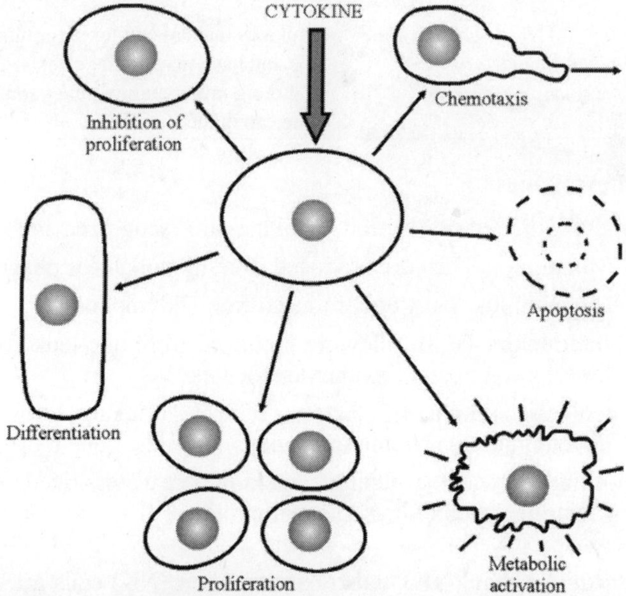


**Fig. 10** Range of biological actions that cytokines have on eukaryotic cells

Cytokines activate cell proliferation and differentiation, but inhibit cell division and cause programmed cell-death or apoptosis. Chemokines are one type of cytokines, which stimulate chemotaxis and chemokinesis (cell movement), and thus play an important role in the acute inflammatory response. Important cytokines and their functions are given in Table 5.

**Table 5:** Important cytokines and their functions

| Type | Examples | Functions |
|---|---|---|
| Chemokines | IL-8, RANTLES, MIP (Macrophage Inflammatory Protein) | Cytokines that are chemotactic and chemokinetic for leukocytes, stimulate cell migration and attack phagocytic cells and lymphocytes. Chemokines play a central role in the inflammatory response. |
| Haemato-poietins | EPO (Erythropoietin) various colony-stimulating factor | Cytokines that stimulate and regulate the growth and differentiation process are involved in cell formation (haematopoiesis). |
| Interleukine | IL-1 to IL-18 | Cytokines produced by lymphocytes and monocytes that regulate the growth and differentiation of other cells, primarily lymphocytes, and haematopoietic stem cells. They often also have other biological effects. |
| Tumour Necrosis Factor (TNF) type as ligand | TNF-a, TNF-B, F as ligand | Cytokines that are cytotoxic for tumour cells and have many other effects such as promoting inflammation, fever and shock, some can induce apoptosis. |

## Types of cytokones

Based on the source of production, cytokines are recognized as follows:

- Monokines: They are produced from mononuclear phagocytes.
- Lymphokins: They are produced from T-lymphocytes.
- Interleukins (ILs): They are produced from one leukocyte and they exhibit action on other leukocytes.
- Colony stimulating factors: They stimulate growth and differentiation of immature leukocytes in the bone morrow.

According to recent grouping, cytokines are categorized into the following families, chemokines, haematoprotiens, ILs, and members of TNFs.

**1. Natural Killer (NK) cells:** Natural Killer (NK) cells are a small group of large non-phagocytic granular lymphocytes. They destroy malignant cells and cell infected with pathogens. They recognize their target cells by one of the two methods.

**Table 6:** Nonopsonic modes of regulation and signalling by phagocytes

| Type of interaction | Bacterial ligand (and example) | Phagocytic receptor and example |
|---|---|---|
| Lectin-carbohydrate | Lectin-Type-I fimbrae<br>Polysaccharide-capsule | Glysoprotein (integrins)<br>Lectin (man/Glc N-Acetylglucosamine receptor) |
| Protem-protein | Arginine-glycine-aspartic acid<br>Arginine-glycine-aspartic acid (RGD) containing proteins (filamentous Haemagglutinin (HA)) | Retinal ganglion cell receptor (integrins) |
| Hydrophobic protein | Glycolipid (lipoteichoic acid) | Lipid receptors (integrins) |
| Signalling | Proteolycans (gram-positive bacteria)<br>Lipopolysaccharide (LPS) (gram-negative bacteria, heat shock proteins<br>Flagellin<br>Bacterial cytosinetrip-hosphate deoxynucleotide (CPG) DNA | Toll-like Receptor (TLR-2)<br>TLR-4<br>TLR-5<br>TLR-9 |

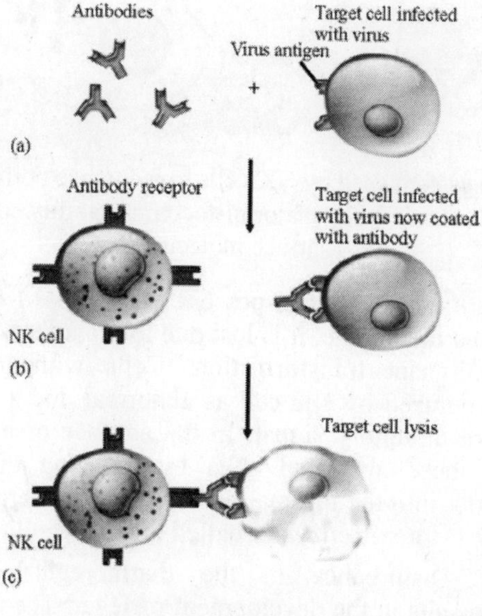


**Fig. 11** Antibody-dependent cell-mediated cytotoxicity

The NK cells possess crystallisable fragment (Fc) receptors, like many cells, that bind Immunoglobulin G (IgG) antibody. These Fc receptors link NK cell to IgG-coated target cells, which they kill by a process called Antibody Dependent Cell-mediated Cytotoxicity (ADCC) (Fig. 11).

The NK cell possesses killer activating receptors and killer inhibitory receptors. The former receptors recognize a number of different molecules present on the surface of all nucleated cells while the later receptors recognize a surface marker called Major Histocompatibility Complex (MHC) class I molecule, which is also present on all nucleated cells. If the killer activator receptor binds to the surface of a nucleated cell, a killer signal is released to NK cells due to which the target cell is killed. If the killer inhibitory receptor binds to the MHC class I molecule present on all nucleated cells, no killing takes place (Fig. 12).

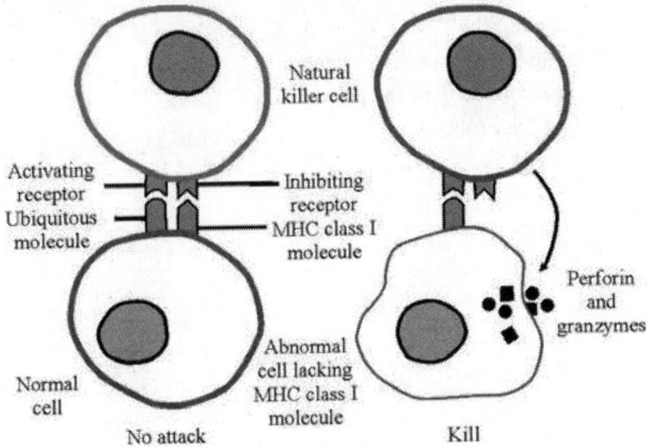


**Fig. 12** The system used by NK cells to recognize normal cells and abnormal cells that lack the major histocompactability complex Class I surface molecule

Although all nucleated cells possess MHC class I molecule on the surface of plasma membrane, it is lost due to the infection by pathogens or from a malignant transformation. Cells without MHC class-I molecule is recognized by NK cell as abnormal due to which NK cell does not receive inhibitory signal. In the absence of inhibitory signal, NK cells kill these abnormal cells by inserting the pore-forming proteins, perforin, into the plasma membrane of the target cell and then injecting it with cytotoxic enzymes called granzymes.

**2. Fever:** Disturbance in the thermoregulatory activity of hypothalamus results in the development of fever. The temperature of a normal human body is 98.6°F or 37°C with plus or minus difference of

1°C. When the body temperature increases above the normal range, it is called as fever. The most common cause of fever in humans is an infection caused either by bacteria or viruses or bacterial toxins. However, the endogeneous cause in the formation of pyrogenes also stimulates fever. The most common types of pyrogenes associated with fever are IL I, II, 6 and TNF that are produced by host macrophages in response to the invasion of pathogen. Pyrogenes, after their release, reach hypothalamus along with circulating blood where they stimulate neurons to secrete prostaglandins. Prostaglandins reset the thermoregulatory activity of hypothalamus at a higher temperature and then the temperature-regulating reflex mechanisms act to bring the body temperature above normal (Fig. 13).

The fever stimulated by a pathogen enhances the host defences by three complimentary processes.

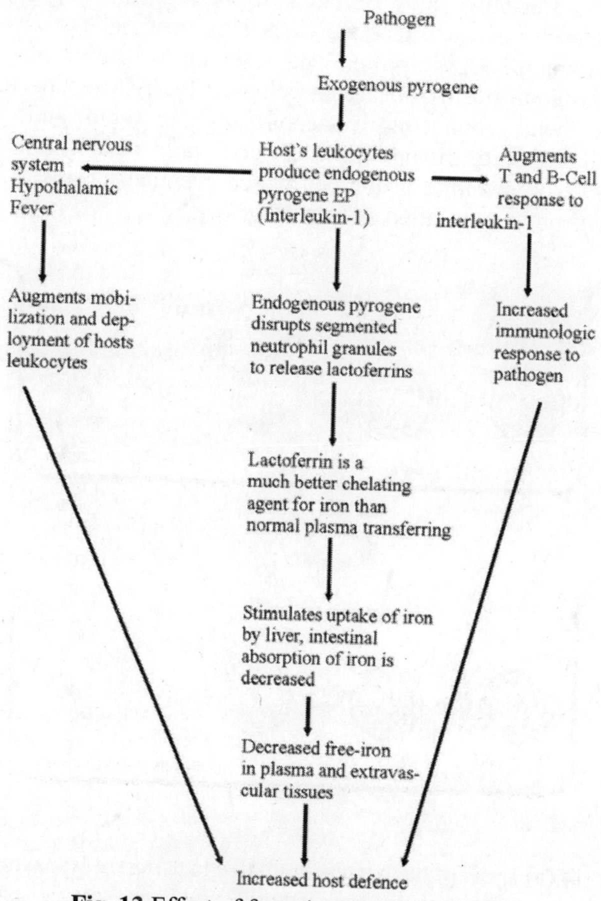


**Fig. 13** Effect of fever in mammalian host

It stimulates leukocytes hence, they can destroy the pathogens.

a. It enhances growth inhibition of pathogen by decreasing the availability of iron to them (Microbiostatis)

b. It augments the specific activity of the immune system.

Some hosts, it is reported, redistribute iron during fever so that iron is not made available to pathogens. This phenomenon is called hypoferremia, which results in inhibition of growth of pathogens. It is also reported that virulence of the pathogens increase when more amount of iron is made available to them, a phenomenon called hyperferremia. For example, gonococci population increases during menstruation period due to increased concentration of free iron available to the bacteria.

**1. Lysozyme:** Lysozyme is found in animal secretions such as saliva, tears and other body fluids, and probably functions as a major line of defence against infections caused by bacteria. Lysozyme, also called as muramidase, is a protein and functions as an enzyme. It breaks most of the gram-positive bacterial cell wall by hydrolyzing the β 1-4 glycosidic bond connecting N-acetyl muramic acid and N-acetyl glucosamine. Due to disruption of the cell wall, water enters the cell, which results in swelling of the cell which eventually leads to bursting of cell. This process is called as lysis (Fig. 14).

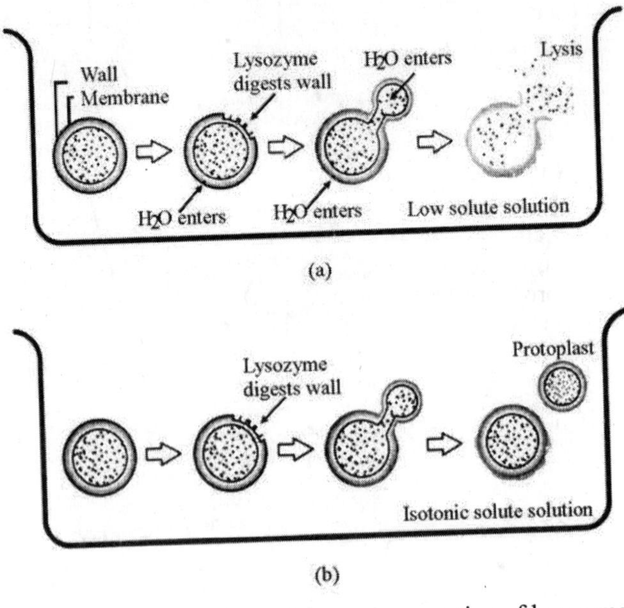


**Fig. 14** (a) Lysis of the bacterial cell due to action of lysozyme and (b) Release of protoplast from bacterial cell in an isotonic solution

If the proper concentration of a solute that does not penetrate the cell, such as sucrose, is added to the medium, the solute concentration outside and inside the cell becomes equal (isotonic). In such conditions, even if lysozyme distrupts peptidoglycan cell wall, water does not enter the cell and lysis does not occur. Instead a protoplast is formed (Fig. 14b). However, if such glucose-stabilized cells are placed in water, lysis occurs immediately. Such protoplasts are often called as spheroplasts. There is a slight difference between these two. Protoplasts are generally naked without any traces of cell wall whereas spheroplasts contain pieces of cell wall here and there on their surface.

**2. Complement System:** Complement is a series of proteins that are activated by the interactions of antigen-antibody complexes. Several distinct complement proteins may react in a sequence to cause lysis of the bacterial cells. These proteins also enhance phagocyte recognition and destruction of antigens. Thus, complement plays an important role in the immune response.

### Activation of complement

Complement consists of a group of proteins; some of them have enzymatic action. These proteins are activated in a sequential manner by antigen and antibody complexes on the surface of the bacterial cell. They cause damage to cell membrane and lysis or leakage of cellular contents. Complement proteins are present in large number in human plasma. An important function of an antibody is to recognize an attacking antigen or pathogenic cells and activate the complement system for attack. Thus, many antibodies, which are specific for a specific antigen, can employ complement proteins.

The individual proteins of complement system are termed as $C_1$, $C_2$, $C_3$ and upto $C_9$. Activation of complement occurs only with IgG and IgM antibody classes. When IgG or IgM antibodies bind on cell surfaces of an antigen, the antibodies fix or bind complement proteins. The complement proteins react in a sequential manner with activation of one complement protein followed by the activation of the next and so on. The entire process of complement system activation occurs as follows (Fig. 15):

- It starts with binding of an antibody with an antigen.
- Binding of $C_1$ protein to the antibody-antigen complex leading to $C_4$-$C_2$ binding at an adjacent membrane site, and activation and binding of $C_3$ protein.
- Membrane-bound $C_3$ protein catalyses formation of $C_5$-$C_6$-$C_7$ complex at a second membrane site, and $C_8$, $C_9$ are then deposited with the $C_5$-$C_6$-$C_7$ complex resulting in membrane damage and cell lysis.

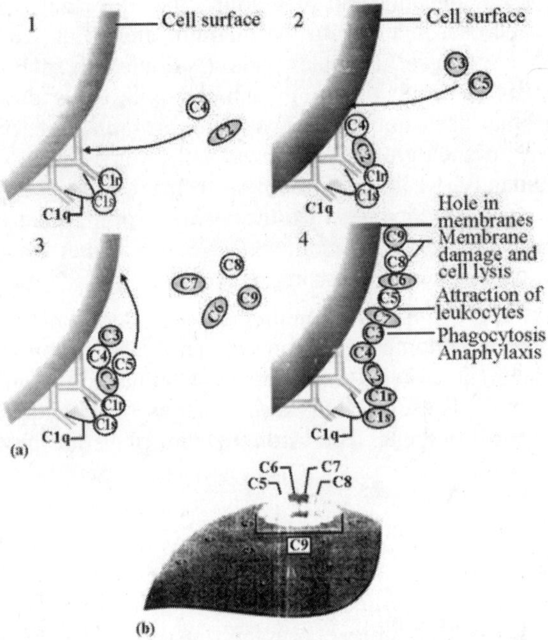


**Fig. 15.** Complement system: a) The sequence, orientation, and activity of the various components as they interact to lyse a cell. Panel 1: Binding of the antibody recognition unit C1q and other C1 proteins; panel 2: the C4–C2 complex; panel 3: the C4–C2–C3–C5 complex – after activation, the C5 complex – after activation, the C5 unit travels to an adjacent membrane site; panel 4: binding of C6, C7, C8, and C9, resulting in membrane damage. b) Three-dimensional view of the hole formed by complement components C5 through C9

Some of the by-products of complement activation are powerful chemoattractants called anaphylatoxins, because they cause inflammatory reactions at the site of the complement deposition. Reactions involving $C_3$ protein resulting chemotactic attraction and activation of phagocytes resulting in increased phagocytosis. Reactions involving $C_5$ proteins lead to T-cell attraction and cytokine release.

Most of the gram-negative bacterial cells are put to lysis because of the catalytic action of complement system on the antibodies. However, gram-positive bacteria are not killed by specific antibodies even in the presence of complement system. Gram-positive bacteria however, can be killed by opsonization.

**1. Phagocytosis:** The process of recognization, ingestion and killing of pathogenic microorganisms, which cause diseases in humans and other vertebrate animals, by phagocytic cells or phagocytes, is called phagocytosis. Monocytes, macrophages, neutrophils and dendritic cells are involved in killing the pathogen.

**Recognition of pathogenic microorganisms:** Phagocytes adapt two types of molecular mechanisms for the recognition of pathogenic microorganisms:

a. **Opsonin-dependent Recognition:** It is also called opsonization. In this process, pathogenic microorganisms or other particles are coated by serum components such as antibodies, mannose binding proteins, and complement-binding $C_3$ b proteins, due to which they are activated for recognition followed by ingestion by the phagocytes. The serum components act as a bridge between the pathogenic microorganisms and the phagocyte in the process. Serum components bind to the surface of the pathogenic microorganisms at one end and to specific receptors on the phagocytic surface at the other end (Fig. 16).

| Phagocytic cell | Degree of binding | Opsonin |
|---|---|---|
| (a) 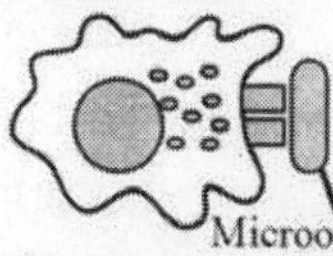  Attachment by nonspecific receptors | ± | - |
| (b) 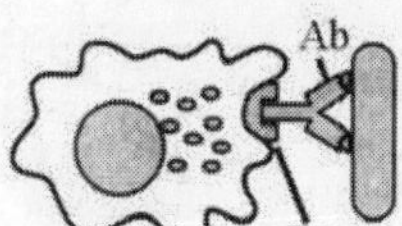  | + | Antibody |
| (c) 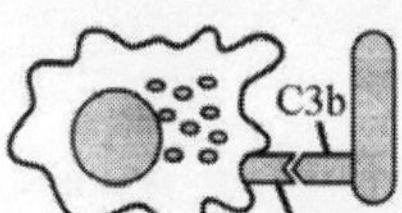  | ++ | Complement C3b |
| (d) 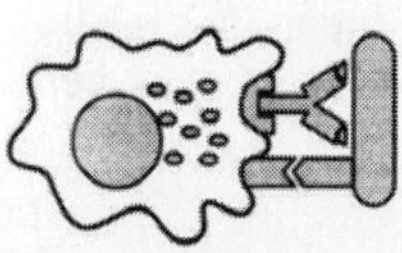 | ++++ | Antibody and complement C3b |

**Fig. 16** Mechanism of binding of serum components to the surface of pathogenic microorganism and specific receptors on the pathogenic surface

**b. Opsonin-independent Recognition:** In this process, opsonins are not used instead other non-specific or specific receptors present on the phagocytes, which are capable of recognizing the structures associated on the surface of pathogenic microorganisms, are employed. Three main types of recognition and one type of signalling mechanism have been identified in phagocytes under opsonin-independent recognition process as indicated in Table 6.

- In lectin phagocytosis, lectin, which is a protein, is involved in the recognition of pathogens. Recognition occurs between lectins present on the surface of one cell and carbohydrates present on the opposite cell.
- In protein-protein interactions, interaction between Arginine-Glycine-Aspargin peptide sequence of pathogenic microorganisms and phagocytic cells leads to the recognition of pathogens.
- In hydrophobic interactions, hydrophobic reactions between pathogen and phagocyte establish recognition and phagocytosis (Fig. 17).

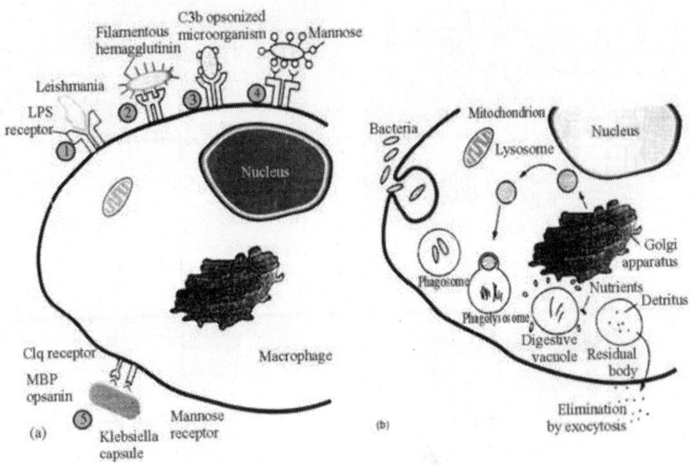


**Fig. 17** Different events of phagocytosis

- In the signalling mechanism, the pathogen is recognized based on the type of arrangement of certain chemical molecules on the cell surface, which is called as non-self. Hence, recognition of the pathogen is based on the detection of molecular patterns, which are the products of normal microbial physiology. These molecular patterns are called as Pathogen-associated Molecular Patterns (PAMPs), which are specific to specific microorganisms. These are different in different microorganisms and are not produced by the host. The famliar examples of PAMPs are the LPS on gram-negative bacteria and the peptidoglycan of gram-positive bacteria. These and other PAMPs are recognized by receptors present on phagocytic cells called as Patterns Recognition Receptors (PRPs). Since, PAMPs are produced only by the pathogenic microorganisms and they are perceived by the phagocytic cells of the innate immune system as molecular signals of infection, allows innate immune system to distinguish self from microbial non-self.

### 3.2.2 Classes of PRPs

Several structurally-distinct classes and functionally-distinct classes of PRPs have evolved in phagocytic cells to recognize PAMPs, and to stimulate several host defensive pathways. Among them, secreted PRPs and Toll-like Receptors (TLRs) are very important.

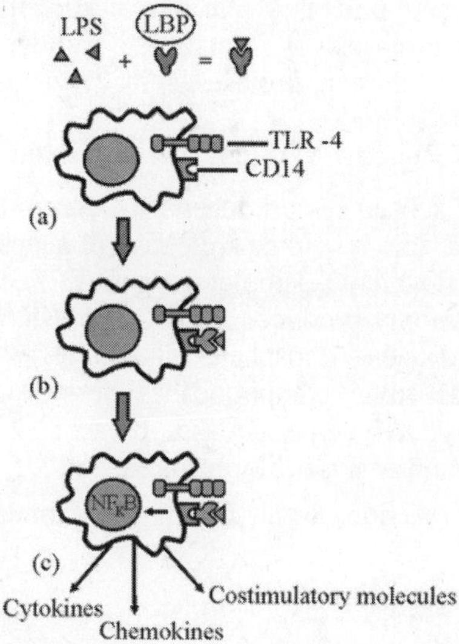


**Fig. 18**. Toll-like Receptor 4 on a macrophage signals the presence of a pathogen: a) Bacterial LPS from a gram-negative bacterial pathogen binds to LPS-binding protein found in blood plasma and other body fluids, b) The LPS-LBP complex binds to the receptor protein CD14 on the surface of the macrophage, and c) The bound LPS-LBP interacts with the Toll-like receptor 4 resulting in a signal (arrow in the cytoplasm) to the nucleus. This activates the transcripton factor NFkB, which in turn, activates genes that produce cytokines, chemokines, and costimulatory molecules that play a role in defence against the gram-negative bacterial pathogen.

Secreted PRPs bind to the microbial cells, which facilitate the destruction of microbial cells either by complement system or by phagocytosis.

The TLRs function exclusively as signalling receptors. They do not recognize and bind pathogens directly, but are involved in signalling the appropriate response to different classes of pathogens such as bacteria, viruses or fungi (Fig. 18). There are nine types of TLRs in mammals. The TLR-4 in mammals signals the presence of LPS and heat-shock proteins. The TLR-9 signals the dinucleotide CpG motif present on DNA released by dying bacteria. The TLR-2 signals the presence of peptidoglycan of gram-positive bacteria.

The ingested pathogens are enclosed in a vacuole called phagosome after phagocytosis. The phagosome fuses with another vacuole called lysosome due to which a new vacuole called phagolysosome is formed. The ingested pathogens are transported in phagolysosome. A variety of hydrolases are produced in the phagolysosome by the activity of lysosome, which include lysozyme, phospholipase-A, ribonuclease, deoxyribonuclease and proteases. An acidic vacuolar pH favours the activity of these hydrolases. The entrapped pathogen is killed by the combined action of all these hydrolases.

### 3.2.3 Formation of Reactive Oxygen Intermediates (ROIs)

The ROIs are toxic substances produced by oxygen-dependent enzymes that are present in the lysosomes of macrophages and neutrophils. Superoxide radical ($O_2$), $H_2O_2$, singlet oxygen ($O_2$) and hydroxyl radical (OH) are the commonly-produced ROIs. The Nicotinamide Adenine Ddinucleotide Hydrogen Phosphate ($NADPH_2$) required for the production of these toxic compounds is derived from the pentose phosphate pathway. Myeloperoxidase present in the neutorphils catalyses the formation of hypochlorous acid.

The following reactions occur during the formation of these toxic substances.

1. Hydroxyl radical formation

$$O_2^- + H_2O_2 \longrightarrow OH^+ + OH^- + O_2$$

2. Hypochlorous acid formation

$$H_2O_2 + cl^- \xrightarrow{\text{myeloperoxidase}} HOCl + OH^-$$

3. Singlet oxygen formation

$$clo^- + H_2O_2 \longrightarrow {}^1O_2 + cl^- + H_2O$$

4. $H_2O_2$ formation

$$2O_2 + 2H^- \xrightarrow{\text{superoxide dismutase}} H_2O_2 + O_2$$

5. Superoxide formation

$$NADPH + 2O2 \xrightarrow{\text{NADPH oxidase}} 2O2 + HT + NADP$$

These reactions occur from the respiratory burst that accompanies the increased oxygen consumption and ATP generation needed for phagocytosis. Lysosome fusion is not necessary for the respiratory burst, because these reactions occur immediately after the formation of phagosome. These toxic oxygen compounds are effective in killing the pathogenic microorganisms.

### 3.2.4 Reactive Nitrogen Intermediates (RNIS)

The RNIs have been recently shown to be formed in macrophages, neutrophils and mast cells. These include nitric oxide (NO), nitrite ($NO_2$) and nitrate ($NO_3$). They are cytotoxic agents and may be either released from cells or generated from vacuoles of a cell. Among all these oxides of nitrogen, nitric oxide is most effective toxic substance and is produced from the amino acid arginine in macrophages. Nitric oxide functions by inhibiting the cellular respiration by reacting with iron in electron transport chain. The pathogens destructed by RNIs include Herpes Simplex Virus (HSV), the opportunistic fungus *Cryptococcus neoformans,* the protozoa *Toxoplasma gondii Leishmania major*, the metazoan pathogen *Schistosoma mansom.*

A variety of other microbicidal substances such as several cationic proteins, the bactericidal permeability-increasing proteins and a group of broad spectrum antimicrobial peptides called defensins are produced in neutrophil granules. Four types of defensins are produced in humans and are referred as Human Neutrophil Proteins (HNPs) 1, 2, 3 and 4. These defensins are synthesized by myeloid precursor cells during their formation in the bone marrow and are then stored in the cytoplasmic granules of mature cells. They are then released as extracellular secretions or delivered to phagocytic vacuoles. These defensins act against bacteria and fungi by making cell membranes more permeable by the formation of membrane channels through which some efflux occur. They kill viruses by neutralization of viral envelop. They are not effective against non-enveloped viruses. These defensins are effective against a variety of gram-negative and gram-positive bacteria, yeasts, molds and certain viruses.

### 3.2.5 Properdin

It is one of the protein factors of complement system, which is universally present in the blood plasma of all the humans. It plays an important role in the innate, non-specific immune defence against intravascular invasion by bacteria and some fungi. The complement system consists of complement proteins $C_1$–$C_9$, factor-B, factor-D, factor-H, factor-I, $C_4$ b binding protein, $C_1$ INH complex, S protein and properdin protein. It is involved in an alternative complement pathway, one of the pathways of the complement activation.

The alternative pathway begins with the cleavage of $C_3$ protein into $C_3a$ and $C_3b$ by a blood enzyme. The $C_3b$ becomes stable when it binds to LPS receptor present on the surface of gram-negative bacterial cell walls, to aggregates of IgA or Immunoglobulin E (IgE) or to some endotoxins. A protein in blood called factor B adsorbs to bound $C_3b$ and is cleaved into two fragments by factor D, leading to the formation of active enzyme $C_3b$ Bb. This comlex is also called as $C_3$ convertase because it claver more $C_3$ to $C_3a$ and $C_3b$. $C_3b$ Bb is further stabilized by a second blood protein called properdin, which allows another addition of $C_3b$ forming $C_5$ convertase, which is represented as $C_3b$ Bb 3b. The convertase then cleaves $C_5$ into $C_5a$ and $C_5b$. The complement proteins $C_6$ and $C_7$ rapidly bind to $C_5b$ and form $C_5b$ b7 complex. This complex becomes stable once it binds to the cell membrane of a pathogen. The complement proteins $C_8$ and $C_9$ then bind forming the membrane complex that creates a pore in the membrane of the pathogenic cell. If the cell is eukaryotic, $Na^+$ and $H_2O$ enter through the pore causing cell lysis.

In case of gram-negative bacteria, lysozyme enters through the pore formed in the outer membrane; digests the peptidoglycan cell wall leading to weakening of the cell wall and subsequent lysis of the cell.

Thus, properdin is one of the proteins of the complement system, which is involved in the formation of membrane attack complex, and it thereby helps in the lysis of pathogenic cells, which represent innate and non-specific defence mechanism adopted by the host.

## Review Questions

### I. Answer the following questions

1. Why does the skin constitute as the first line of defence against pathogenic microorganisms?
2. How do mucous membranes resist invasion of microbes in the host?
3. How do M cells function in MALT?

4. Describe the different antimicrobial defence mechanisms that operate within the respiratory system of humans.
5. Why is the genitourinary tract sterile?
6. What major events occur during an inflammatory reaction and how do they contribute to pathogen destruction?
7. How does a chronic inflammation differ from an acute inflammation?
8. How can fever be beneficial to the host?
9. How do cytokines play a role as mediators of the natural immunity?
10. What are the major functions of a complement?

**II. Write short notes on the following**

1. Degranulation of mast cells
2. Opsonic phagocytosis
3. Nonopsonic phagocytosis
4. Reactive oxygen intermediates
5. Reactive nitrogen intermediate
6. Defensins
7. Natural killer cells
8. Non-specific defence mechanism
9. Lysozyme
10. Properdin

## 3.3 FURTHER READING

Dipersio JR et al 1996 Spread of serious disease producing M3 clones of Group A Streptococcus among family members and health care workers, *Clinical Infectious Diseases* **22** 490–495

Murray PR 1996 *ASM Pocket Guide to Clinical Microbiology*, Washington, DC: American Society for Microbiology

Chapin KP Clinical microscopy, in *Manual of Clinical Microbiology* (eds) Murray PR et al, Washington, DC: American Society for Microbiology

Mandell GL, Douglar RG Jr, and Bennet JE 1990 *Principle and practice of infectious diseases,* New York, St. Louis: Mosby

# 4

# Host-Pathogen Interactions

## 4.1 INTRODUCTION

An intensive struggle between a host and a pathogen occurs, before the pathogen establishes and causes a disease. In the competition, host tries to eliminate the pathogen while pathogen tries to establish by overcoming the barriers. If the pathogen succeeds in this struggle, it results in suffering from the disease. If the host is successful, the pathogen is eliminated and it helps the host to be healthy. All infectious diseases begin at some surface of the host, whether it is the external surface such as skin and conjunctiva of an eye or an internal surface such as the mucous membranes of the respiratory tract, intestine or urogenital tract. Many pathogens can selectively attack to a particular host surface.

In most infectious diseases, the pathogenic microorganism penetrates the body surface and gains access to the internal tissue. In some kinds of infections, the pathogen may remain localized, growing near its point of entry into the body. In other instances, it may be transported to other parts of the body. Some pathogens may cause generalized infections, in which the microorganisms become widely distributed and grow throughout the body. Some pathogens are capable of growing within the cells of the host causing severe disruption of normal physiological processes. In other infections, the pathogen may grow extracellularly. Within and outside, the damage to the body cells results due to the production of poisonous substances (toxins) by the microorganism.

A host must eradicate the pathogenic microorganism if it has to recover from an infection. However, as a group, pathogenic microorganisms exhibit a vast array of weapons called virulence factors that can combat various defence mechanisms of the host. Each species of pathogen possesses only one or few of these factors. Thus, an infection represents a battle between the defence mounted by the host and the particular type of virulence factor produced by the pathogen. Often the infection proves fatal to the host. On the other hand, microbe has the advantage of the battle, a disease results, if the disease causing pathogen is not so severe as to kill the host, killing the host it diminishes the pathogen's chance of survival, consequently host microbe interactions that result in chronic form (Table 7). Long-lasting infections are regarded as being more highly-evolved than infections that are short-lived or acute. The common procedure adapted to establish the disease causing capacity of the pathogen (Koch's postulates) in illustrated in Figure 19.

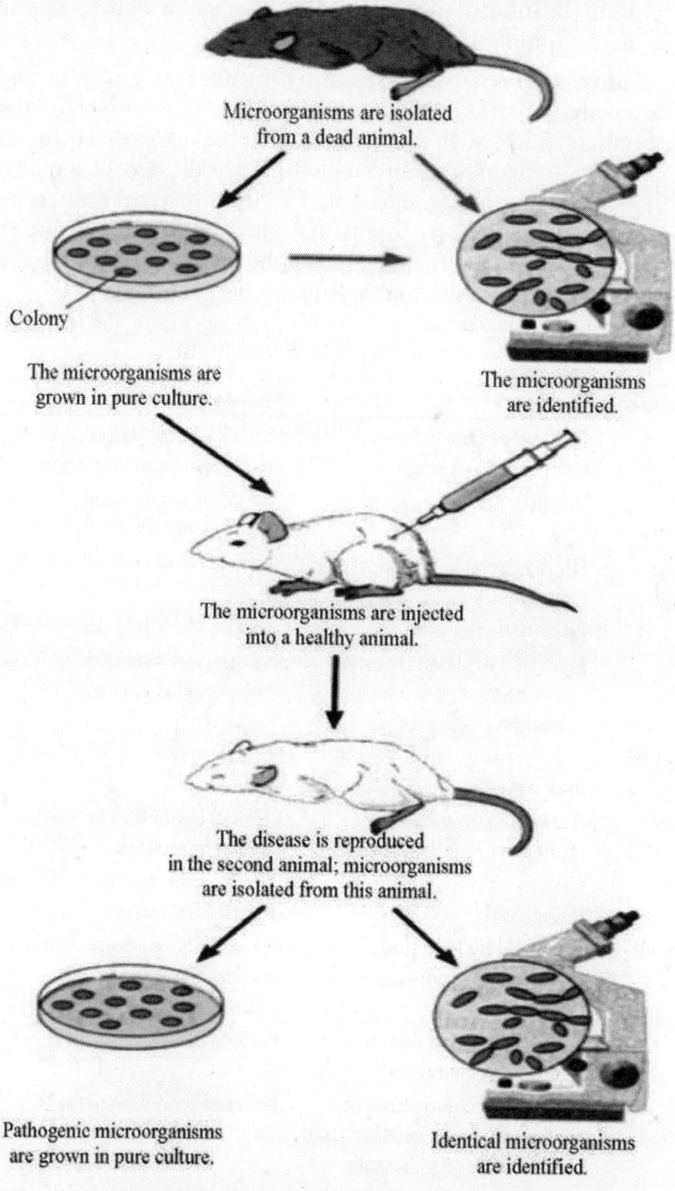


**Fig.19** Demonstration of Koch Postulates

## 4.2 BACTERIAL TOXINS

Based on the role of the bacteria in causing a disease, two distinct categories of the diseases can be recognized:

1. **Infections:** A disease caused directly by the growth and multiplication of the bacterial pathogen, which causes tissue disintegration is called as an infection.
2. **Intoxications:** A disease caused by the entry of a preformed metabolite of a pathogen in to the host's body is called as intoxication. Such preformed metabolites are called toxins (Latin–toxicum–poison), which induce disease in the absence of pathogen that produced it. The term 'toxemia' refers to the condition caused by toxins that entered into the host's blood. Toxins produced by bacteria can be categorized into two main types - a. Exotoxins and b. Endotoxins (Table 7).

**Table 7:** Characteristics of exotoxins and endotoxins

| Characteristic | Exotoxins | Endotoxins |
|---|---|---|
| Chemical composition | Protein, often with two components (A and B) | The LPS complex on outer membrane, lipid A portion is toxic |
| Disease examples | Botulism, diphtheria, tetanus | Gram-negative infections, meningococcemia |
| Effect on host | Highly variable between different toxins | Similar for all endotoxins |
| Fever | Usually do not produce fever | Produce fever by release of IL-1 |
| Genetics | Frequently carried by extra-chromosomal genes such as plasmids | Synthesized directly by chromosomal genes |
| Heat stability | Most are heat sensitive and inactivated at 60–80°C | Heat stable |
| Immune response | Antitoxins provide host immunity; highly antigenic | Limited antibodies produced; weakly immunogenic |
| Location | Usually excreted outside the living cell | Part of outer membrane of gram-negative bacteria |
| Production | Produced by both gram-positive and gram-negative bacteria | Found only in gram-negative bacteria; released on bacterial death and some liberated during growth |
| Toxicity | Highly toxic and fatal in microgram quantities | Moderate toxicity |
| Toxoid production | Converted to antigenic, non-toxic toxoids; toxoids are used to immunize (e.g., tetanus toxoid) | Toxoids cannot be made |

### 4.2.1 Exotoxins

Exotoxins are heat labile and soluble proteins that are released from the cells of the pathogen into the surroundings where the pathogen grows. They spread from the region of infection to other body tissues or cells in which they cause their effects.

**Important Characters of Extotoxins**

- They are synthesized by specific bacteria that contain plamids or exotoxin genes containing prophages.
- They are associated with specific diseases and possess specific mechanism of action.
- They are proteins, which get degraded at 60–80°C
- They are easily inactivated by formaldehyde, iodine and other chemicals resulting in the formation of immunogenic toxoids.
- They are the most known lethal chemicals.
- They cause high immune responses and induce the production of neutralizing antibodies called as antitoxins.
- They do not cause fever in the host directly.
- These are named based on the name of the disease they cause e.g., botulinum, diphtheria toxin, etc.

**Classification of Exotoxins**

Exotoxins are classified into four categories based on their structure and physiological activities:

1. AB toxin: It contains two parts designated as 'A' and 'B'. 'A' possesses enzyme activity and causes toxicity while 'B' binds to the receptors of host cell.
2. A second type also called as AB toxin, affects a specific host site such as nervous tissue (neurotoxins), the intestine (enterotoxins), and general tissue (cytotoxins) by acting either extracellularly or intracellularly on the host cells. It is also called as specific host-site exotoxin.
3. A third type, which does not contain A and B parts, and functions by degrading the host's cell membranes e.g., leukocidins, phospholipases and haemolysins, also called as membrane-disrupting exotoxin.
4. A fourth type possesses antigenic properties and acts by stimulating T-helper cells to release cytokines. These are also called as super antigenic exotoxin.

Some details of the structure and function of these four types of exotoxins are described here.

**AB Exotoxin**

As described above, this exotoxin contains two parts, fragment A and fragment B. Fragment A has enzyme activity, but lacks cell-binding and cell-entry capacity while fragment B binds and interacts with specific receptors on the target cell or tissue such as gangliosides GM, for cholera toxin, GT, or GD for tetanus toxin and GD for botulinum toxin. It is nontoxic and biologically inactive.

The general properties of some of AB exotoxins are given in Table 8.

**Table 8:** Properties of some AB model bacterial exotoxins

| Toxin | Organism | Genetic control | Subunit structure | Target cell receptor | Enzymatic activity | Biological Effects |
|---|---|---|---|---|---|---|
| Anthrax toxins | *B.anthracis* | Plasmid | Three separate proteins (EF, LF, PA) | Unknown, probably GP | EF is a calmodulin-dependent adenylate cyclase; LF enzyme activity is unknown | EF+PA: increase in target cell cAMP level, localized edema; LF+PA: death of target cells and experimental animals |
| Bordetella Adenylate cyclase toxin | *Bordetella* spp. | Chromosomal | A-B | Unknown, probably glycolipid | Calmodulin-activated cyclase | Increase in target cell cyclicAMP level; modified cell function or cell death |
| Botulinum toxin | *C.botulinum* | Phage | A-B | Possibly ganglioside ($GD_{1b}$) | Zinc-dependent endopeptidase cleavage of synaptobrevin | Decrease in peripheral, presynaptic acetylcholine release; flaccid paralysis |
| Cholera toxin | *V.cholera* | Chromosomal | A-5B | Ganglioside ($GM_1$) | ADP ribosylation of adenylate cyclase regulatory protein $G_s$ | Activation of adenylate cyclase, increase in cAMP level; secretory diarrhoea |
| Diphtheria toxin | *C. diphtheriae* | Phage | A-B | Heparin-binding, EGF-like growth | ADP ribosylation of elongation | Inhibition of protein synthesis; cell |

| Toxin | Organism | Genetic control | Subunit structure | Target cell receptor | Enzymatic activity | Biological Effects |
|---|---|---|---|---|---|---|
| | | | | factor precursor | factor 2 | death |
| Heat-stable entero-toxins[1] | *E. coli* | Phasmid | | Similar or identical to cholera toxin | | |
| Pertussis toxin | *B. pertussis* | Chromo-somal | A-5B | Unknown, probably Glyco-protein | ADP ribosylation of signal-transducing G proteins | Block of signal transduction mediated by target G proteins |
| Pseudo-monas exotoxin A | *P. aerug-inosa* | Chromo-somal | A-B | $\alpha_2$-Macro-globulin/ LDL receptor | Similar or identical to diphtheria toxin | |
| Shiga toxin | *S. dyse-nteriae* | Chromo-somal | A-5B | Globotriaos ylceramide ($Gb_3$) | RNA N-glycosidase | Inhibition of protein synthesis, cell death |
| Shiga-like toxin 1 | *Shigella* spp., *E.coli* | Phage | | Similar or identical to shiga toxin Ganglioside (GT) and or GD 1b | | |
| Tetanus toxin | *C. tetani* | Plasmid | A-B | Ganglio-side ($GT_1$ and/ or $GD_{1b}$) | Zinc-depe-ndent endo-peptidase cleavage of synapto-brevin | Decrease in neuro-transmitter release from inhibitory neurons; spastic paralysis |

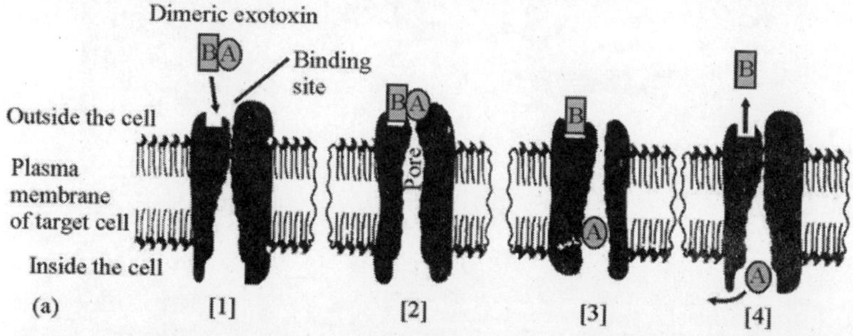


**Fig. 20** Formation of a pore in the plasma memebrane of a host cell by AB exotoxin

**Mechanism of action:** Entry of A subunit is facilitated by initial insertion of B subunit into the plasma membrane and creates a channel or pore through which A subunit enters the cell (Fig. 20).

The mechanism of action of AB toxin is quite complex for diphtheria toxin as shown in the Fig. 21. This toxin binds to cell surface receptor by the B fragment and is taken into the cell through the formation of a clathrin-coated vesicle. The toxin then enters the vesicle membrane and is cleaved into two parts, one of which, the A fragment, travels into the cytosol. The A fragment is an enzyme that catalyses the addition of an ADP-ribose group to the eukaryotic elongation factor $EF_2$ that aids in the traslocation during protein synthesis. The substrate for this reaction is the coenzyme nicotinamide-adenine dinucleotide (NAD).

$$NAD^{+} + EF2 \longrightarrow ADP\ ribosyl - EF2 + nicotinamide$$

The modified EF2 protein cannot participate in the elongation cycle of protein synthesis and the cell dies because it can no longer synthesize proteins.

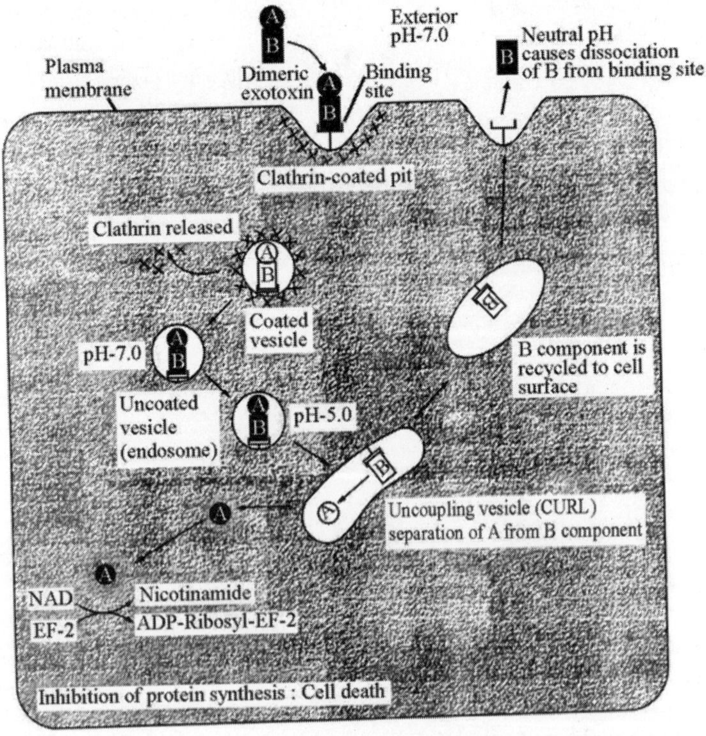


**Fig. 21** Transport mechanism of diphtheria exotoxin into the cell by endocytosis mechanism

## Specific Host-site Exotoxins

These are the second type of exotoxins categorized on the basis of its specific site of action. Some of the bacteria that produce these toxins are given in Table 8. Neurotoxins are usually ingested as preformed toxin that affect the nervous system and indirectly cause enteric symptoms. Examples include staphylococcus enterotoxin B, *Bacillus cereus* emetic toxin and botulinum toxin.

The classical enterotoxin of cholera has been studied in detail. It has a direct effect on the intestinal mucosa and causes profuse fluid secretion.

The B subunit of this toxin is made of five parts arranged as a donut-shaped ring. The B subunit ring anchors itself to the epithelial cells of the plasma membrane and then inserts A subunit into the cell. The A subunit activates tissue adenylate cyclase to increase intestinal cyclic adenosine monophosphate (cAMP) concentration. High concentration of CAMP provokes the movement of massive quantities of water and electrolytes across the intestinal cells into the lumen of the intestine. The genes for the enterotoxigencity are present on the chromosome of *Vibrio cholera.*

Cytotoxins have specific toxic action on the cells or tissues of special organs and are named according to the type of the cells or tissues or organs for which they are specific, such as nephrotoxins (kidney), hepatotoxin (liver) and cardiotoxin (heart).

## Membrane-distrupting Exotoxins

It is a third type of exotoxin, which breaks the host cells by disrupting the plasma membrane. It consists of two subtypes:

**1. Proteinaceous exotoxin:** It is a protein that binds to the cholesterol part of the host-cell plasma membrane, inserts itself into the membrane and forms a channel or pore, through which cytoplasmic contents leak out (Fig. 21). In another mechanism, the cell swells and finally bursts due to the entry of water into the cell by endosmosis. Two examples of this type of membrane-disrupting toxins are described below.

a. **Leukocidins:** These are the membrane-disrupting exotoxins that are produced by pneumococci, streptococci and staphylococci. They kill the leukocytes, which leads to decrease in the host resistance.

b. **Haemolysins:** These are also produced by certain pathogenic bacteria. They also form pores or channels in the plasma membrane of erythrocytes, through which haemoglobin and ions are released out of the erythrocytes. This results in the death of the erythrocytes. Streptolysin-O (SLO) is a haemolysin produced by *Streptococcus pyogenes* that is inactivated by oxygen hence; O is included in its name. SLO causes beta haemolysis of erythrocytes on agar plate incubated anaerobically. A zone of clearance around the bacterial colony growing on the blood agar is called beta haemolysis. Streptolysin-S (SLS) is also produced by *S. pyogenes*, which is insoluble and binds to the bacterial cell. It is oxygen stable and causes beta hemolysis, in aerobically-incubated blood-agar plates. SLO and SLS act as leukocidins and kill leukocytes. It should be noted that the haemolysins attack not only the plasma membranes of erythrocytes and leukocytes, but also of many other cells'.

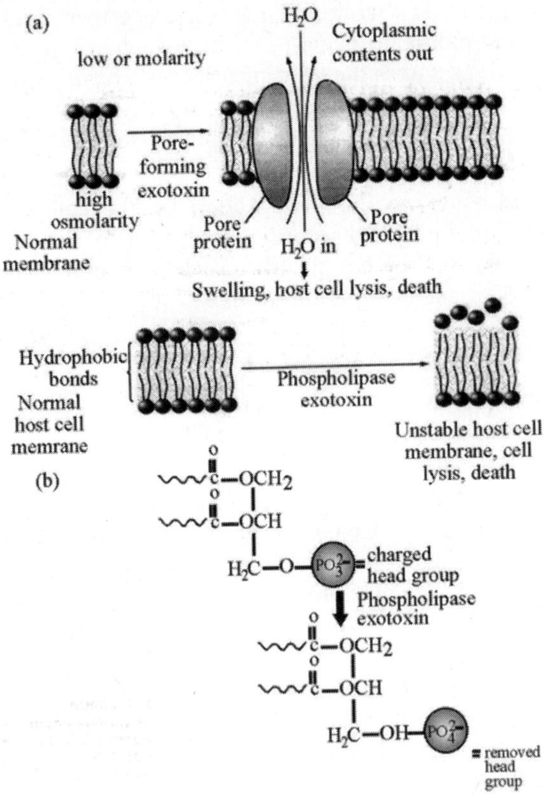


**Fig. 22** Two subtypes of membrane-disrupting exotoxin: (a) A channel - formimg (pore-forming) type of exotoxin that inserts itself into the normal host cell membrane and makes an open channel (pore) and (b) Phospholipid-hydrolising phospholipase exotoxin destroys membrane integrity

**2. Phospholipases:** These are the second subtype toxin of membrane-disrupting exotoxins. They remove the charges of head group from the lipid portion of the phospholipids in the host-cell plasma-membrane (Fig. 22). This destabilizes the plasma membrane and cell lyses, and dies. One example of the pathogenesis caused by phospholipases is the disease gas gangrene caused by *Clostridium perfringens*, whose α-toxin almost totally destroys the local population of the white blood cells that are drawn in by inflammation to fight the infection.

**3. Super Antigens:** These represent fourth type of exotoxins. These are the antigens, which stimulate drastic immune response and hence, are called as super antigens. They are the bacterial proteins, which non-specifically stimulate T-cells to proliferate by interacting with MHC products on antigen-presenting cells and the T Cell Receptors (TCRs) outside the antigen-binding site. This binding allows many different T-cells with different TCRs to become activated (Fig. 23).

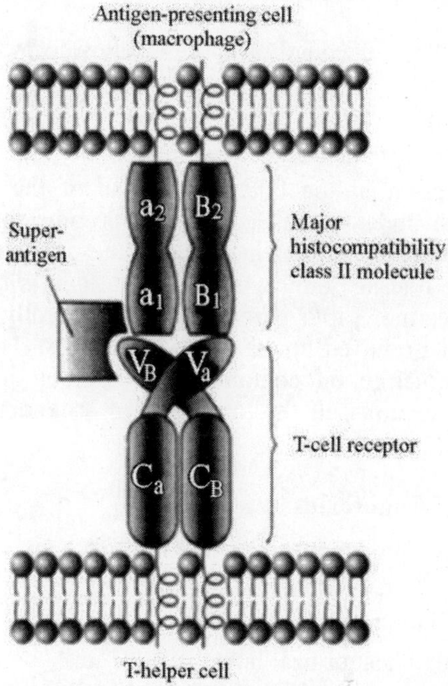


**Fig. 23** Super antigens: The most prevalent antigen receptor of T-cells is the TCR. Super antigens stimulate T-cells bearing a particular $V_\beta$ element of the TCR, regardless of the nature of the $V_\alpha$ receptor element. They bind directly to the outer surface of the MHC class II molecule without being processed by the antigen cell. Thus, a super antigen can activate many more T-cells than a normal antigen does.

Good examples of the super antigens are Staphylococcal enterotoxins that cause food poisoning and the toxin that causes toxic shock symptoms. They cause symptoms by stimulating the release of massive quantities of cytokines from CD4 type of T-cells. These cytokines have two effects on the host:

- Systematic toxicity
- Suppression of the specific immune response

Both of these effects contribute to the microbial pathogenicity. Super antigens are formed in such diseases as rheumatic fever, arthritis, Kawasaki syndrome, atopic dermatitis and one type of psoriasis.

**Role of Exotoxins in Disease**

Humans are exposed to exotoxins in three different ways:

1. Ingestion of preformed exotoxin
2. Colonization of a mucosal surface followed by production of exotoxin
3. Colonization of wound or abscess followed by local exotoxin production

### 4.2.2 Endotoxins

The LPSs present in the outer membrane of the gram-negative bacterial cell wall, under certain conditions, become toxic to specific hosts. They are called as endotoxins because they remain bound to the bacterium and are released only when the bacterium is lysed or during bacterial multiplication. Lipid part of a LPS, generally, acts as toxic component and is referred to as lipid-A. Lipid-A is not a single macromolecular structure, but contains complex group of lipid residues. This lipid-A part exhibits all the properties of an endotoxin of gram-negative bacteria.

**Characteristics of Endotoxins**

- Generally, similar in action though they come from different sources
- Weakly immunogenic
- Sensitive to heat or heat labile
- Exhibit toxicity at high doses (milligram to kilogram quantity)
- Cause general systematic effects such as fever shock, blood coagulation, inflammation, intestinal haemorrhage, general weakness and fibrinolysis – breakdown of fibrin, which is the major protein component of the blood clots.

Endotoxins do not act directly, but act indirectly by involving the host molecules and systems. For example, endotoxin can initially activate blood clotting factor XII (Hageman factor), which in turn, activates four humoral systems; - coagulation, complement, fibrinolytic and kininogen systems.

Gram-negative bacterial endotoxins also indirectly induce fever in the host by stimulating macrophages to release endogenous pyrogens that reset hypothalamic thermostat. The cytokine interleukin-1 is an important endogenous pyrogen. Other cytokines produced by macrophages include TNF, which also causes fever.

Recent reports indicate that LPS affects macrophages and monocytes by binding to special plasma protein called LPS-binding protein. The LPS and LPS-binding protein complex then attaches to the receptors on monocytes, macrophages and other cells. This stimulates several events including the production of cytokines Il-1, Il-6 and TNF. Cytokine Il-1 and TNF induce fever. These cytokines also promote other endotoxin effects such as complement activation, coagulation, prostaglandins formation etc.

## 4.3 VIRULENCE

The ability of a microbial species to cause disease is called pathogenicity. Different strains of pathogenic species differ with regard to their degree of pathogenicity or virulence. For example, some strains are highly virulent because only few bacterial cells are required to cause a disease while other strains may be less virulent because large numbers of cells are required to cause a disease. Some strains may be incapable of causing diseases even when large number of cells are inoculated into a host and they are called as avirulent.

The virulence of a pathogen is generally measured by determining the $LD_{50}$ for a particular type of laboratory animal. $LD_{50}$ is defined as the number of bacterial cells that are required to kill 50% of the experimental organism (e.g., a strains $LD_{50}$ is 10 and other strain is $LD_{50}$ is 1,000 reveals that strain X is 100 times more toxic).

The $LD_{50}$ can be determined more precisely than other end points such as the dose that kills 100% of the organisms ($LD_{100}$ dose sometimes also termed Minimal Lethal Ddose (MLD) because the rate of change in mortality verses change in dose is greatest around the point of 50% mortality.

Virulence is determined by three characteristics of the pathogen:

1. **Invasiveness:** The ability of the pathogenic organism to spread to adjacent or other tissues is called as invasiveness.

2. **Infectivity:** The ability of the pathogenic organism to establish a focal point of infection is called as infectivity.
3. **Pathogenic Potential:** The degree of disease that a pathogenic organism causes in a host is called as pathogenic potential.

The ability of a pathogen to produce toxins is called as toxigenicity. Toxins are metabolites produced by a pathogen that will damage the host and cause a disease.

## 4.4 ATTENUATION

Virulent strains of many pathogens when repeatedly cultured on laboratory media or grown in vivo in hosts other than their normal hosts, may lose their virulence and are called as avirulent strains. Such virulent strains are called attenuated strains and are widely used in the vaccination to confer immunity or disease resistance in respective hosts. The process of generation of virulent a strain of the pathogen is called as attenuation.

### Review Questions

**I. Answer the following questions:**

1. Describe host-pathogen interaction.
2. What are exotoxins? Write an essay on the properties and functions of AB exotoxins.
3. Give an account of specific host-site exotoxins.
4. What are the membrane disrupting exotoxins?
5. What are endotoxins? Describe their properties and functions.
6. What are the two general types of exotoxins? List examples of each type and routes through which exotoxin can infect a person least.
7. Name five examples of organisms that use each route.

**II. Write short notes on the following:**

1. Infection
2. Intoxication
3. Characters of exotoxins
4. Super antigen
5. Virulence
6. Pathogenic potential
7. Attenuation

## 4.5 FURTHER READING

Ellner PD and New HC *Understanding infectious diseases*, St. Louis: Mosby

Emond RT, Rowland HAK, and Weloby P 1995 *Color atlas of infectious diseases,* St. Louis: Mosby

Mandell GL, Douglar RG Jr, and Bennet JE 1990 *Principle and Practice of Infectious Diseases*, St. Louis: Mosby

Richman DD, Whitley RJ, and Hayden FG 1997 *Clinical Virology,* St. Louis: Mosby

Shulman ST, Phair JP, and Sommers HM 1992 *The Biologic and Clinical Basis of Infectious Diseases*, Philadephia: W.B. Saunders.

# 5

# General Principles of Diagnostic Microbiology

## 5.1 INTRODUCTION

Pathogens, particularly bacteria and fungi, coexist on the surface of or within the host. These pathogens must be properly identified for the diagnosis of infectious diseases. The diagnostic microbiologist identifies agents and organisms that are responsible for the infectious diseases, based on morphological, biochemical, immunological and molecular methods. Advances in diagnostic technology made it possible for rapid diagnosis of many infectious diseases. Molecular methods facilitate identification of the pathogens based on highly-specific genomic and biochemical characteristics. Once isolated and identified, the pathogens can be subjected to antimicrobial sensitivity tests. Time is a significant factor in the diagnostic process, especially in life-threatening situations. All these analytical methods used in the diagnosis of infectious diseases and their causal organisms will help in enforcing patients' well being and health.

## 5.2 SPECIMENS

Rapid isolation and identification of the pathogens from the clinical specimens is one of the important parts of diagnostic microbiology. The main purpose of diagnostic microbiology laboratory is to provide the doctor with information regarding the presence or absence of the pathogen responsible for the disease. Diagnostic microbiology makes use of the information obtained from microbial biochemistry and physiology, immunology, molecular biology, genomics and the host-parasite relationship, in the identification of disease-causing organism and the type of disease.

In diagnostic microbiology, a clinical specimen represents a portion or quantity of human material that is tested, examined or studied to determine the presence or absence of a particular pathogen. Other important aspects of the specimen include:

- The specimen should be forwarded properly to the laboratory.

- Attention should be given to specimen collection to avoid contamination from the many varieties of microorganisms, indigenous to the skin and mucous membranes.
- It is better to collect the specimen before antibiotics are administered to the patients.
- The specimen selected should adequately represent diseased area, and may include additional sites to isolate and identify the potential agents of the particular disease process.
- Adequate quantity of specimen should be collected to allow a variety of diagnostic testing.

## 5.3 COLLECTION

Several methods are available for the collection of specimen by using aseptic technique, in which specific procedures are followed to prevent unwanted microorganisms from contaminating the clinical specimen. The following methods are designed to enforce the collection of only proper specimen that can be sent to the laboratory:

**1. Intubation:** It is a process of inserting a tube into a body canal or hollow organs (e.g., stomach). Most commonly used intubation tube is the Levin tube. In this method, a long sterile tube is attached to the syringe and the tube is either swallowed by the patient or passed through the nostril into the patient's stomach. Specimens are then withdrawn periodically into the sterile syringe.

**2. Needle aspiration:** This method is used for the collection of specimen (anaerobic bacteria) aseptically from Cerebrospinal Fluid (CSF), pus and blood. Stringent antiseptic techniques are used to avoid skin contamination. Anticoagulants such as heparin or sodium citrate are included within the specimen bottle or tube to prevent blood from clotting and entrapping microorganisms.

**3. Swab:** A sterile swab is used for the collection of specimen from throat or anterior nose. A sterile swab is a polystyrene applicator with rayon, calcium alginate or dacron tip. Many commercially-manufactured swabs contain a transport medium intended to preserve a variety of microorganisms and to prevent multiplication of rapidly-growing members of the microbial population (Fig. 24). However, the use of a swab for the collection of specimen from other parts of the body, except throat and nose, is generally discouraged. This is because the swabs are associated with general risk of contamination with surface micro-organisms and possess narrow-volume capacity ($<0.1$ ml).

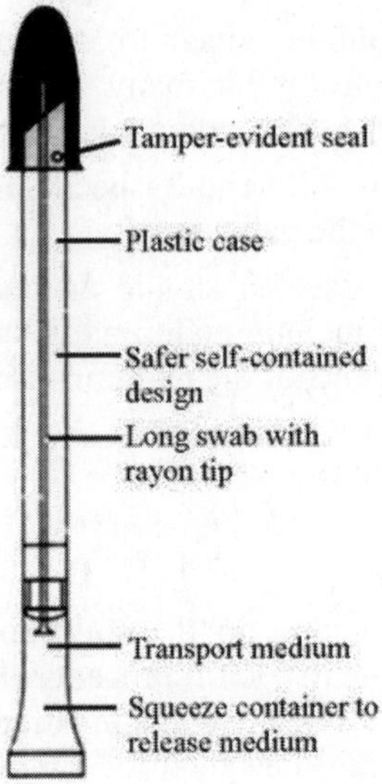


**Fig. 24** A sterile swab with a specific transport medium

**1. Use of catheter:** A catheter is a tubular instrument used for either withdrawing or introducing fluids from or into a body cavity. However, a catheter is generally used for withdrawing urine specimen from bladder, to detect urinary tract infections caused by bacteria and also from newborns and neonates who cannot give a voluntary urine specimen (Fig. 25). There are three types of catheters:

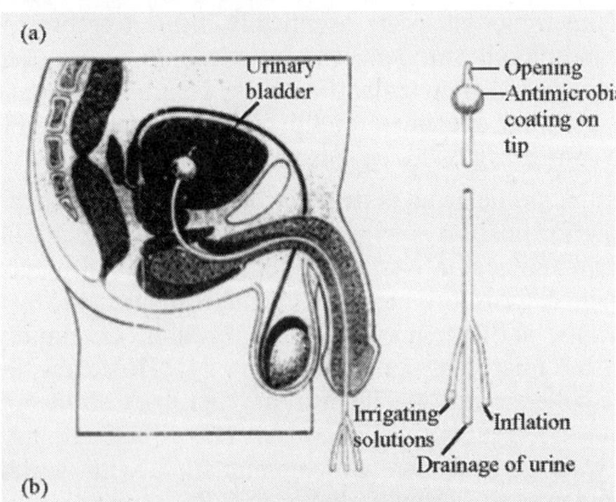


**Fig. 25** A foley catheter attached with three separate tubes within the round shaft for drawing urine, inflation and sanding irrigating solutions into urinary bladder

a. Hard catheter used when the urethra is either narrow or with strictures.
b. French catheter is a soft tube used to collect single specimen sample.
c. Foley catheter is used when multiple urine samples are necessary over a prolonged period.

Clean catch method is generally used for the collection of urine from a patient. The opening of urethra is cleaned, the catheter is inserted through the urethra upto the neck of the bladder and small container is used for urine collection. The urine collection is generally made in the morning because the urine contains more microorganisms as it remains in the bladder overnight. The first voided urine is not collected because it contains those transient microorganisms normally present in the basal part of the urethra. The midstream part of the urine is preferentially collected as it most likely contains those microorganisms seen in the bladder.

**2. Collection of sputum:** Sputum is the most common specimen collected in suspected cases of lower respiratory tract infection such as bronchitis. Sputum is the mucuous secretion expelled from the lungs, bronchi and trachea through the mouth. It differs from saliva, as the later is the transparent and thin fluid secreted from the salivary glands. Sputum is collected in a specially-made sputum cup (Fig. 26).

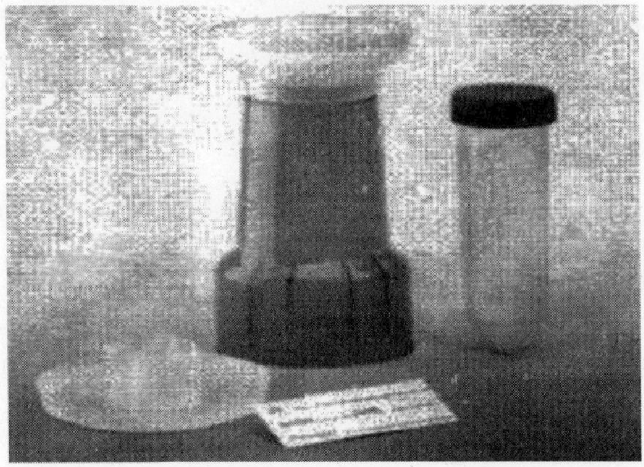

**Fig. 26** A sputum cup for the collection of sputum from a patient

## 5.4 TRANSPORT

Transporting the specimen to the diagnostic laboratory as soon as it is collected from the patient is of prime importance. Several changes may occur in the specimen if transportation is delayed and some laboratories refuse the delayed specimen.

Transportation of specimen is accomplished in several ways based on the type of the specimen collected. For example, certain specimens should be transported in a special medium that not only preserves the microorganisms, but also helps in preventing the contamination with microorganisms foreign to the body.

The CSF from the patient suspected of meningitis, should be collected immediately by skilled personnel in strigent aseptic conditions. The CSF is collected by lumbar puncture and the sample is transported to the laboratory within 15 min.

Specimens are kept in ice before transport for the isolation of viruses, and can be kept at 4°C upto 72 h. If the sample is to be stored for more than 72 h, it should be frozen at -72°C.

Urine specimens are to be transported to the laboratory within one hour after collection. If it is not possible to send the urine sample within the scheduled time, it must be refrigerated immediately.

Stool specimens for culture must be transported in special buffered preservatives.

Special treatment is necessary for the specimens with anaerobic microorganisms. The material is drawn with a needle and syringe, then the needle is removed and the syringe is capped with a seal. The specimen, thus collected, must be transported to the laboratory within 10 min, or else the specimen must be injected immediately into an anaerobic transport vial containing a medium and an indicator (such as resazurin). This indicator shows that the interior of the vial is anaerobic at the time of introduction of the specimen.

## 5.5 PROCESSING

Further processing of the collected specimen is done to facilitate definite identification of the pathogens and the infections caused by them based on the following (Fig. 27):

- Microscopic examination of the specimen
- Immunological tests to detect antibodies or microbial agents
- Molecular methods for identification of pathogens based on highly-specific genomic properties
- Study of the biochemical properties of the isolated pathogenic organisms

### 5.5.1 Microscopic Examination of the Specimen

Different types of microscopes are used for the identification of microorganisms, especially bacteria. The specimen is processed in the form of wet-mount, which is heat-fixed or chemically-fixed for the purpose of microscopic observation. Many pathogenic microorganisms can be observed and identified with ordinary bright field microscope. However, microscopes with high resolutions such as phase contrast or dark field microscopes are required for the identification of specific pathogens. Dark field microscope is employed for the identification of spirochetes present in the skin lesions, which develop during early syphilis or in blood specimen of patients suffering from early leptospirosis. Fluorescent microscope is used for the identification of acid fast microorganisms such as *Mycobacterium tuberculosis* after they are stained with fluorochromes such as auramine-rhodamine.

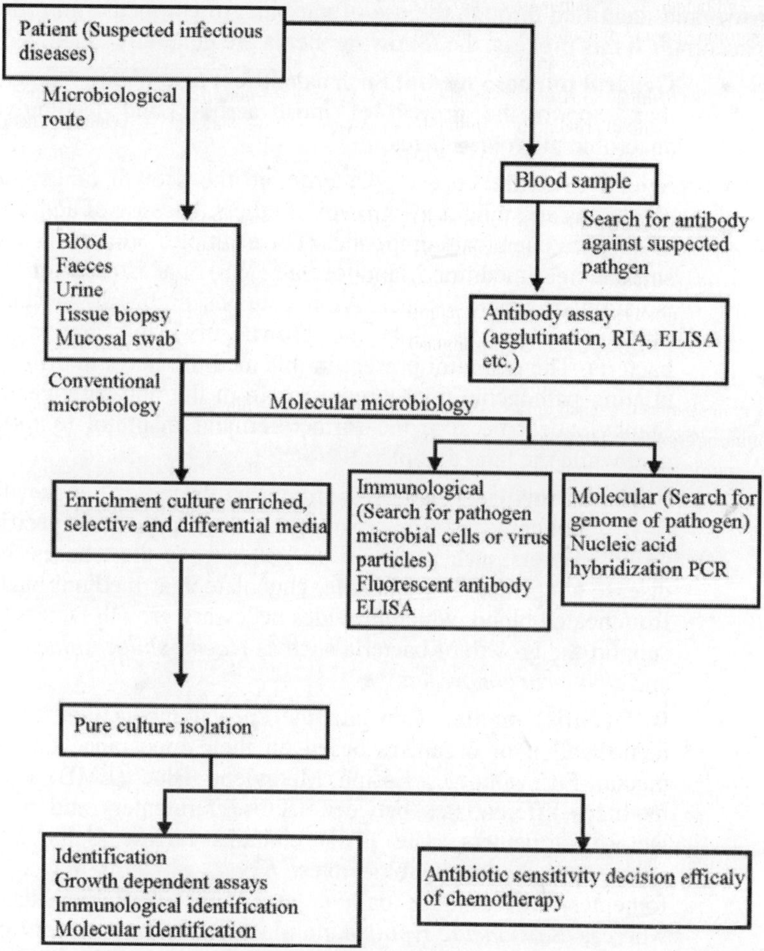


**Fig. 27** Clinical and diagnostic methods used for isolation and identification of infectious pathogens

Many stains are used to examine the specimens of specific microorganisms. Gram stain and acid fast stain are the two commonly and widely-used stains. These stains are based on the chemical composition of the cell walls; therefore, they are not useful in the identification of bacteria without cell walls.

### 5.5.2. Culture Media

Selected cultured media are used to isolate pathogenic microorganisms from natural samples. It is an important part of clinical microbiology. Most pathogenic microorganisms of clinical importance can be isolated, grown and identified through the use of various growth media. In order to accomplish this process, the following media are generally employed:

- **General purpose media:** Such as nutrient agar and blood agar that support the growth of most aerobic and facultative anaerobic microorganisms.
- **Selective media:** The media promote the growth of certain pathogens and inhibit the growth of others, because of addition of specific chemical compounds. For example, salmonella and shigella agar medium, Mannitol Salt (MS) agar is used for the isolation of Staphylococci. As it contains high salt concentration (7.5%), it inhibits the growth of many groups of bacteria. The mannitol present in this medium helps in differentiating pathogenic microorganisms from the non-pathogenic. Staphylococci because the former ferment mannitol to form acid while the later do not.
- **Enriched media:** This is a medium that promotes the growth of metabolically-fastidious pathogens on addition of specific growth factors such as *Neisseria gonorrhoea* that causes the disease gonorrhea. For example, chocolate agar medium made from heated blood, which provides necessary growth factors to support the growth of bacteria such as *Haemophilus influenzae* and *Neisseria gonorrhea*.
- **Differential media:** They are the special media that allow identification of organism based on their appearance on the media. For example, Eosine Methylene Blue (EMB) agar medium differentiates between lactose fermenters and non-lactose fermenters. The EMB contains lactose, salts, and eosine and methylene blue dyes. *E.coli*, which is a lactose fomenter, will produce dark colony with a metallic sheen, whereas *Salmohella typhi*, a non-lactose fermenter, appears colourless. Another differential medium is the MacConkey agar medium used for the selection and recovery of *Enterobacteriaceae* and related gram-negative rods. The bile

salts and crystal violet in this medium inhibit the growth of gram-positive bacteria, and some fastidious gram-negative bacteria. As lactose is the sole carbohydrate, lactose fermenting bacteria produce colonies with various shades of red colour, whereas non-lactose fermenters produce colourless colonies (Table 9).

**Table 9:** Colony characteristics of frequently-isolated gram-negative rods cultured on various clinically-useful media[a]

| Organism | Agar media | | | |
|---|---|---|---|---|
| | **Eosin Methyelene Blue** | **Mac Conkey Agar** | **Salmonella Shigella Agar** | **Bile Salts** |
| *Escherichia coli* | Dark centre with greenish metallic sheen | Red or pink | Red to pink | Mostly inhibited |
| *Enterobacter* | Similar to *E.coli*, but colonies are larger | Red or pink | White or beige | Mucoid colonies with silver sheen |
| *Klebsiella* | Large, mucoid, brownish | Pink | Red to pink | Mostly inhibited |
| *Proteus* | Translucent, colourless | Transparent, colourless | Black center, clear periphery | Green |
| *Pseudomonas* | Translucent, colourless to gold | Transparent, colourless | Mostly inhibited | No growth |
| *Salmonella* | Translucent, colourless to gold | Translucent, colourless | Opaque | Black to dark green |
| *Shigella* | Translucent, colourless to gold | Transparent, colourless | Opaque | Brown or inhibited |

### 5.5.3 Culture

#### Blood culture

In standard blood culture procedure, 10 ml of blood is drawn aseptically from a vein and injected into two blood-culture bottles, containing an anticoagulant and an all-purpose culture medium. One bottle is incubated aerobically and the other anaerobically at 35°C temperature, and examined directly upto 5 d in an automated system. Most clinically-important bacteria grow within this period. In certain blood culture methods, a chemical is used that lyse the red and white blood cells due to which intracellular pathogens are released, that might be otherwise overlooked. Microorganisms in blood cultures are commonly detected by visual inspection with regard to turbidity, microscopic examination and subculturing. Growth of the pathogens in these culture methods is detected by continuously monitoring carbon dioxide ($CO_2$) production and turbidity by taking readings for every 10 min (Fig. 28).

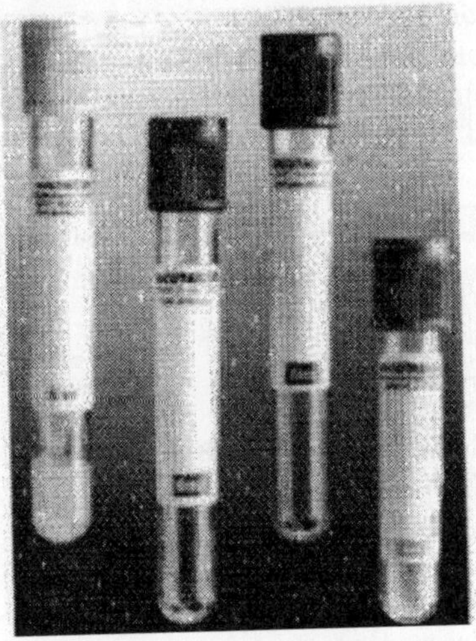

**Fig. 28** Stanely Vacutainer tube

Certain amounts of skin contaminations are unavoidable when blood is drawn from a vein. This contamination is indicated if certain organisms that are commonly found on the skin are isolated. They include *Staphylococcus epidermidis*, *Corynebacteria* and *Propionibacteria*, which sometimes also cause certain diseases such as bacterial endocarditis. Bacteraemia is the presence of bacteria in the blood, which indicates a systematic disease, because the condition of bacteriaemia is not seen in healthy humans. The most common pathogens found in blood include *Pseudomonos aeruginosa*, enteric bacteria such as *E.coli* and *Klebsiella peumoniae*, and the gram-positive cocci such as *Staphylococus aureus* and *Streptococcus pyogenes*.

Another blood infection is the septicaemia, which results from the growth of virulent organisms entering the blood from a site of infection; theymultiply and travel to various body tissues to initiate new infections. Septicaemia is indicated by the presence of severe symptoms in the patient such as fever, chills followed by prostration. Sometimes it may result in septic shock, which is a life-threatening systemic condition characterized by severe reduction in blood pressure and failure of several organs such as heart, kidney and lungs. In many disease conditions, blood culture provides the only immediate way of isolation and identification of the causal agent, and diagnosis.

## Urine culture

Two media are used to culture urinary tract pathogens:

1. Blood agar as non-selective general medium
2. Selective medium for enteric bacteria such as MacConkey or EMB agar

These specialized media permit initial differentiation of lactose fermenters and non-lactose fermenters. However, these media inhibit the growth of gram-positive organisms such as *Staphylococcus* sp (common skin contaminants). An experienced clinical microbiologist may make a tentative identification of a pathogen from a sample by observing the colour and morphology of the colonies of the suspected pathogen grown on these media (Table 10). However, such identification must be confirmed by more detailed tests.

**Table 10:** Sensitivity of immunodiagnostic assay

| **Assay** | **Sensitivity (µg antibody/ml)** |
|---|---|
| Precipitation reaction | |
| • In fluids | 24–60 |
| • In gels (double immnodiffusion) | 24–160 |
| Agglutation reaction | |
| • Direct | 0.4 |
| • Passive | 0.08 |
| Radioimmuno Assay (RIA) | 0.0008–0.008 |
| Enzyme-linked Immunosorbant Assay (ELISA) | 0.0008–0.008 |
| Immunofluorescence | 8.0 |

If no bacterial growth occurs despite persistant urinary tract infection symptoms, the pathogen can be identified by direct culturing of urine.

Urinary tract infections are very common and most of them may be due to bacteria of the normal flora. Therefore, considerable care must be taken in the bacteriological analysis of urine. The most common urinary tract pathogens belong to enteric bacteria such as *E.coli*, which accounts about 90% of the urinary tract infections. Other pathogens include *Klebsiella, Enterobacter, Proteus, Pseudomonos, Neissesia gonorrhea* (responsible for the disease gonorrhoea, does not grow in urine, but grows in the urethral epithelium and is diagnosed by Venereal Disease Research Laboratory (VDRL) tests.

Direct microscopic examination of the urine is also undertaken to assess presence of abnormal number of bacteria in the urine. This is done by using commercially-available dipstick tests. For example, one type of dipstick test monitors the reduction of nitrate by detecting the reduction product, nitrite. A positive test is indicated by a colour change from light yellow to colourless at a particular region on the dipstick. Significant production of nitrite occurs only when large number of (more than $10^5$/ml) enteric organisms are present. This method is used as an instantaneous check for urinary tract infections. Along with nitrate reduction test, other dipstick tests are also carried out simultaneously for esterases and proteases. These dipstick tests are followed by urine culture.

### Faecal cultures

Freshly-collected faecal samples are placed in a bottle containing phosphate buffer for transport to the laboratory. Pus or blood containing stools, water-borne or food-borne diseases, should be inoculated into a variety of selective media for the identification of disease-causing organisms. Intestinal parasites such as *Entamoeba histolytica* are identified microscopically in the stool sample rather than by the culture methods. Faecal acidity increases during storage. Therefore, any delay between sampling and sample processing must be avoided, which is important for the isolation of *Shigella* and *Salmohella* species, both of which are sensitive to acidic pH.

### Wounds and abscesses

Wounds and abscesses are generally formed due to traumatic injuries such as animal or human bites, burns, cuts or the penetration of foreign objects. Sampling from these injuries must be carefully done to recover the relevant pathogen.

**Table 11:** Enriched and selective media for primary isolation of pathogenic microorganisms

| Specimen | Blood agar | Enteric agar | CA | MTM | ANA |
|---|---|---|---|---|---|
| Fluids, chest, abdomen in pericardium | + | + | + | - | + |
| Faeces, rectal shoake | + | + | + | - | - |
| Surgical tissue bioprocess; lung, lymphnodes, throat swabs, sputum, tonsil, nasopharynx | + | + | - | - | + |
| Genitourinary swabs, urethra, vaegina, cervix | + | + | + | + | - |
| Urine | + | + | _ | - | - |
| Blood | + | - | - | - | + |
| Swabs, wounds, abscesses, exudates | + | + | + | - | + |

A variety of pathogens can be associated with wound infections. Some of them may be anaerobic; therefore such samples must be transported from the site of collection under anaerobic conditions. *Staphylococcus aureus* is the common pathogen present in purulent discharges, along with enteric bacteria, *Pseudomonas aeruginosa* and anaerobes such as *Bacteroides and Clostridium*. Blood agar medium, several selective media are used for culturing and identification of these bacteria (Table 11). Blood agar containing additional nutrients along with reducing agents is employed for the isolation and identification of anaerobic bacteria. Smears from wound specimens are directly examined by microscopy.

Sampling, for abscesses and other purulent wounds, is usually done by aspirating the pus with a sterile syringe and needle after disinfecting the skin surface. Internal purulent lesions are commonly sampled by biopsy from the tissues removed by surgery.

**Genital specimen culture**

Gonorrheal infection is commonly a sexually-transmitted disease caused by *Neisseria gonorrhoea*. This organism colonizes mucosal surfaces of urethra, uterine cervix, conjunctiva, throat and anal canal (Fig. 29). A purulent urethral discharge is the classical symptom of gonorrhea in males. Sampling for this infection can be done in two ways. In the first method, sampling is done when there is no discharge from urethra. A sterile piece of cotton swab is inserted into the anterior part of the urethra and is left there for few seconds to allow the cotton swab to absorb any exudates, and then removed for culturing. In the second method, a sample of the first early-morning urine of an infected individual containing viable cells of the organism is collected into a sterile bottle. Samples are usually obtained by swab from the cervix or the urethra of a female, suspected to be infected with gonorrheal disease.

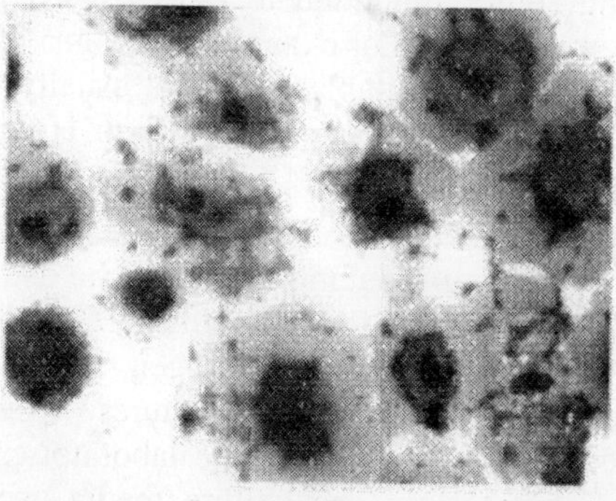

**Fig. 29** Cells of *N. gonorrhoeae* within human polymorphonuclear leukocytes from cervical smear with diplococci

Samples, thus collected, are cultured. When identifying the pathogen, culture methods are more sensitive than microscopic analysis. Non-selective enrichment media containing heat-lysed blood, generally called chocolate agar medium, is commonly employed for the isolation of gonorrheal bacterium. The heated blood interacts with the medium components absorbing compounds that are toxic to *N. gonorrhene*. Modified thayar-martin agar selective medium is used for the primary isolation of the bacterium. Vancomycin, nystatin, trimethoprim and colistin inhibit the growth of normal flora, but not the *N. gonorrhea*. Incubation is done in humid environment, for 24–48 h, in an atmosphere containing $CO_2$, which is required for the growth of *N. gonorrhea*. The colonies that are formed on chocolate agar are immediately tested by oxidase test, because all *Neisseria* are oxidase positive, which confirms the presence of this pathogen and the disease gonorrhea. However, definitive identification requires determination of carbohydrate utilization pattern and immunological or nucleic-acid probe tests.

**Cultures of anaerobic organisms**

Samples of anaerobic pathogens are collected by section or biopsy and are immediately placed in a tube containing oxygen-free gas with small amount of a dilute salt solution with a reducing agent such as thioglycolate and a redox indicator dye resazurin. The dye becomes pink when oxidized and colourless when reduced, indicating the presence or absence of oxygen in the tube. In the absence of suitable anaerobic transport device, the syringe itself can be used to transport the specimen, and then the needle is inserted into a sterile inhaler stopper. Hence, no air enters the syringe.

Agar plates are placed in a sealed jar for anaerobic incubation and the jar is made anaerobic (without $O_2$) by either replacing the air in the jar with an oxygen-free gas mixture, usually a mixture of nitrogen or $CO_2$, or by adding some compound to the enclosed jar that removes oxygen from the air of the jar. For example, hydrogen is generated in the jar in the presence of a suitable catalyst, usually palladium. The hydrogen is combined with free oxygen to form $H_2O$; thus, removing the contaminating oxygen (Fig. 30).

In another method, anaerobic conditions are provided with the use of culture media containing reducing agents or the use of anoxic glove boxes. Glove boxes are large gas-impermeable bags filled with an oxygen-free gas such as nitrogen or hydrogen, and fitted with an air-lock for inserting and removing the cultures. However, they are generally not used in chemical diagnostic laboratories, but frequently used in research laboratories. The culture media used for anaerobic microorganisms are basically similar to those used for aerobes except that these media contain a reducing agent, either cystein or thioglycolate, and a redox indicator such as resazurin.

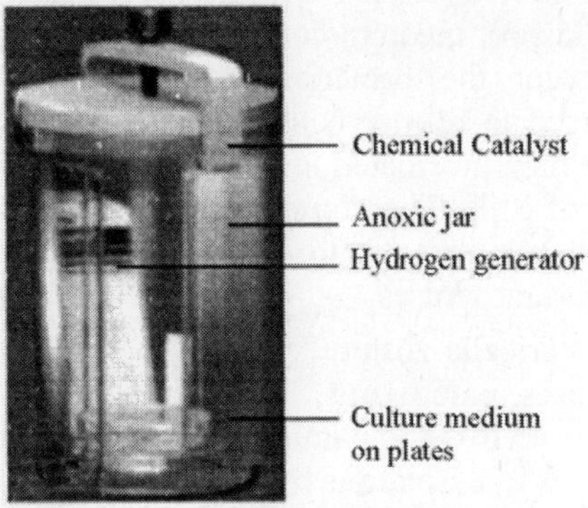


**Fig. 30** Sealed jar for incubating cultures under anoxic conditions

Obligate anaerobic bacteria also cause human diseases and are completely overlooked in chemical diagnosis unless special isolation culture media are used. Specimen collection, handling and processing requires special attention to recover obligate anaerobic organisms. There are several sites in the body such as oral cavity and the intestinal tract that are generally oxygen-free, and support the growth of anaerobic organisms, which may become a part of normal flora. However, other parts of the body can also become oxygen-free as a result of tissue injury or trauma, due to which blood supply is reduced resulting in oxygen-free conditions. Though most of the anaerobic bacteria in the human body constitute normal flora, they may become opportunistic pathogens and cause infection under certain circumstances such as injury or trauma.

### 5.5.4 Viral Isolation using Tissue Culture Methods

Viruses are obligate intracellular ultramicroscopic pathogens, which do not grow on animate culture media, but require living cells for their growth. Various types of cells in culture are maintained in many diagnostic virus laboratories wherein the chemical specimens can be inoculated. The living cells in culture may undergo morphological changes, when the viruses replicate in them. These changes include swelling due to changes in membrane permeability and shrinkage due to the cell death, which are visible under light microscope. They are referred to as cytopathic effect. The inoculated living cells are examined at intervals to determine whether cytopathic effect has developed or not. The occurrence of these changes indicates that the clinical specimen contains the virus.

A variety of samples can be used for viral isolation such as eye swab, CSF, throat swab, urine, faeces, nasopharyngeal aspirate etc. As viruses will not survive on dry swabs, the swabs with viruses are dropped into a transport medium consisting of an isotonic fluid and an antibiotic to prevent the bacterial growth. In the laboratory, the container is reversed to release cells from the swab into the transport medium, which is then inoculated into living cells again. Not all viruses grow in all types of cells. Therefore, most laboratories maintain two or three different cell culture types to increase the chances of isolation of the clinically-important viruses.

The HSV, Varicella-zoster Viruses (VZVs), Cytomegalovirus (CMV), adenoviruses, enterovirus, Respiratory Syncytial Virus (RSV), rhenovirus, influenza virus and parainfluenza virus are routinely isolated using cell culture. Viral isolation by cell culture is much slower than bacterial isolation in inanimate media. The fastest growing virus, HSV, may produce cytopathic effect after 24 h, but most viruses require several days and weeks to produce cytopathic effect. Some viruses cannot be grown using standard laboratory cell culture e.g., Hepatitis B Viruses (HBVs), HIV, EBV etc.

Various modifications of cell cultures have been devised to speed up the detection of certain slow-growing viruses. These include:

- **Haemadsorption:** Some viruses e.g., influenza viruses, possess HA, which are expressed on the surface of the infected cells. Thus, the red cells added to the cell culture tubes get adhered to the cell sheet. This enables to detect the presence of a virus before cytopathic effect has developed.
- **The Detection of Early Antigen Fluorescent Foot (DEAFF) Test:** This test is used to accelerate the diagnosis of CMV and adenovirus infections. Virus-infected cells express antigens (as they are expressed early in the virus replication cycle) in their cell nucleus before any cytopathic effect develops. These antigens can be demonstrated by staining the cells with suitable fluorescent-labelled monoclonal antibodies directed against the early antigens.

## 5.6 SEROLOGICAL TESTS

*In vitro* studies of antigen-antibody reactions are called as serology, which is the basis for all diagnostic immunology tests. The principles of antigen-antibody reactions lie in the specific interaction of determinants on the antigen with the variable regions of the antibody molecule.

### 5.6.1 Specificity and Sensitivity

For diagnostic purposes, the usefulness of a serological test depends upon the test's specificity and sensitivity.

The ability of an antibody preparation to recognize a single antigen is called as specificity. An optimum level of specificity requires that the antibody is specific for a single antigen, will not cross react with any other antigen and hence, will not provide false positive results.

The lowest amount of an antigen that can be detected by an antibody is called as sensitivity. The highest level of sensitivity requires that the antibody in a test should be capable of identifying a single antigen molecule. Highest sensitivity prevents false negative reactions. The sensitivity of some of the common tests is shown in Table 10. The amount of antigen detected by each test system is proportional to the amount of antibody used. For example, immune precipitation reactions require a large amount of the antibody and generally detect 0.1–1.0 mg quantities of antigen. This shows that precipitation tests are less-sensitive serological tests. On the other hand, ELISA test require 1,00,000 times less antibody and can detect one million times less antigens (0.1–1.0 mg quantity) than precipitation antigens. Hence, ELISA test is among the most sensitive tests in serological methods.

A variety of serological tests are used in the identification of antigens and pathogens, such as:

**1. Neutralization Test:** It is a process of interaction between antibody and antigen, to prevent the antigen to reduce or eliminate its pathogenic activity. These reactions can occur both *in vitro* and *in vivo*.

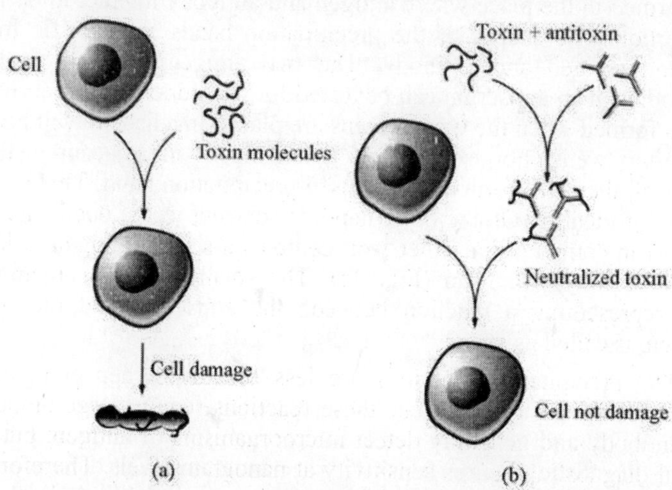


**Fig. 31** Neutralization of an exotoxin by antibody

For instance, neutralization of pathogen toxins by specific antibody can occur when toxin molecules and antibody molecules combine in such a way that the active part of the toxin is blocked (Fig. 31). Such neutralization reactions occur in case of many bacterial toxins. An antiserum containing antibody that neutralizes a toxin is called as antitoxin. Neutralization reactions can also occur when viruses are bound by specific antibodies. For example, antibodies directed against HA and Neuraminidase (NA) proteins of influenza viruses prevent the adsorption of the viruses to specific receptors on the host cells *in vitro*.

**2. Precipitation Test:** It is a process of interaction between a soluble antibody with a soluble antigen to form an insoluble complex or precipitation. Antibody molecules generally possess two antigen-binding sites and hence, it is possible for each site to combine with a separate antigen molecule. If the antigen also contains more than one determinant, a precipitate may develop consisting of aggregates of antibody and antigen molecules. As they are easily observed in vitro, precipitation reactions are very informative serological tests, especially for the quantitative measurement of the antibody concentrations. Maximum precipitation occurs only when there are optimum proportions of the two reacting substances. The presence of either antigen or antibody in excess quantities results in the formation of only small, soluble immune complexes (Fig. 32).

Precipitation reactions conducted in agar gel are called as immunodiffusion test. They are carried out to study the specificity of antigen-antibody reactions. Both the antigen and antibody diffuse outward from separate wells made in agar gel, and precipitation bands are formed in the place where antigen and antibody interact in optimum proportion. The shapes of the precipitation bands are specific for the reacting antigen and antibody. The two antigens reacting with the antibodies in an antiserum can be tested for relatedness by observing the bands formed when the two antigens are placed in adjacent well near the antiserum wells. For example, if two antigens in adjacent wells are identical, they will form a single fused precipitation band. This is called as line of identity. On the other hand, if adjacent wells; one contains an antigen in common, but other well contains a second antigen, a line of partial identity will form (Fig. 32). The small extended precipitation line, representing a reaction between the antiserum and the second antigen, is called as spur.

The precipitation reactions are less sensitive when compared to other serological tests because these reactions require large amount of the antibody and generally detect microorganisms of antigen, but most useful diagnostic requires sensitivity at nanogram levels. Therefore, the precipitation tests are not done in most diagnostic laboratories but are normally used in research laboratories.

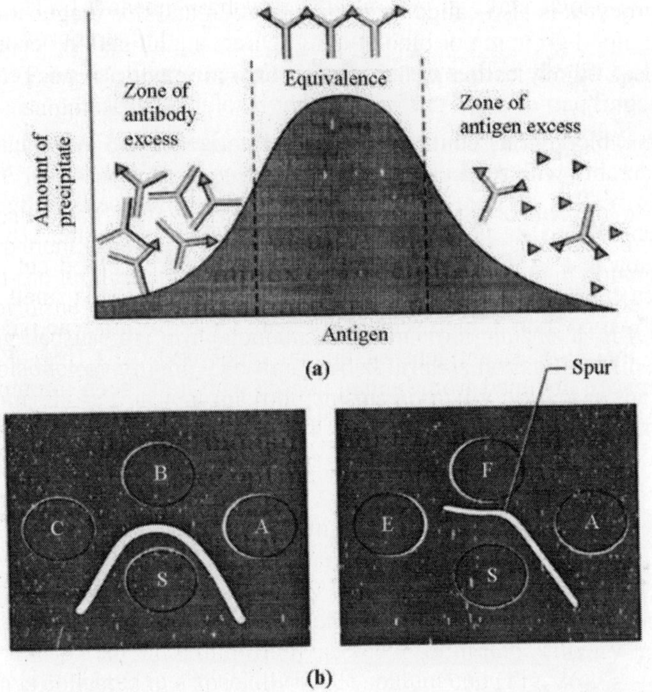


**Fig. 32** Precipitation reaction between soluble antigen and antibody: (a) Shows the extent of precipitation between antigen and antibody with respect to concentration, (b) Precepitation process called immunodilation wells indicates antibodies of *Proteus mirabilus;* wells A, B and C indicate soluble extracts of *Proteus mirabilus.* A line of identity is observed in the wells on the left and the right antigen, E does not react and antigen A shows partial identity with antigen F (Spur)

**3. Agglutination:** Although agglutination tests are not as sensitive as other serological tests, they are 100 times more sensitive than precipitation tests. A visible clumping of particulate antigen when mixed with a specific antibody is called as agglutination (Table 12). Agglutination tests are more frequently used in most of the diagnostic and clinical laboratories because they are simpler to perform, highly specific, rapid and inexpensive. Standardized tests are available for the identification of blood group antigens, and also for the identification of pathogens and their products.

There are two types of agglutination tests:

**a. Direct Agglutination:** The most widely used direct agglutination process is employed for the classification of antigens found on the surface of the red blood cells called erythrocytes. Agglutination of errythrocytes is also called haemagglutination and this diagnostic test is called blood grouping or blood typing. Direct agglutination occurs when soluble antibody causes clumping by interacting with an antigen that is an integral part of a cell surface or other insoluble particle.

Red blood cells contain a variety of antigens, and individuals vary considerably with respect to the antigens present on their cells. A, B and O (also called Rh) are the major human red blood cells antigen. The antigens A and B, and the antibodies are the bases for the ABO blood grouping test. Antibodies specific for particular red blood cell surface antigen cause red blood cells to form visible dump when a small drop of blood is mixed on a microscope slide with an antiserum reactive against either the A or B antigens of human red blood cells (Fig. 33). The antisera are obtained from human donors who have been immunized to A or B antigens by natural or artificial means.

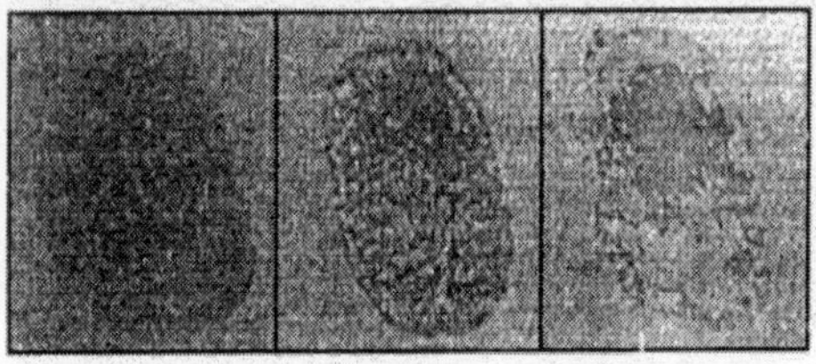

**Fig. 33** Direct agglutination of human red blood cells for ABO typing

Type A individuals make antibodies to group B antigens while type B individuals make antibodies to group A antigens. Type AB individuals have either A or B antigens, whereas type O individuals have antibodies to both A and B antigens. Antibodies against A and B antigens are generally called as natural antibodies (Fig. 34).

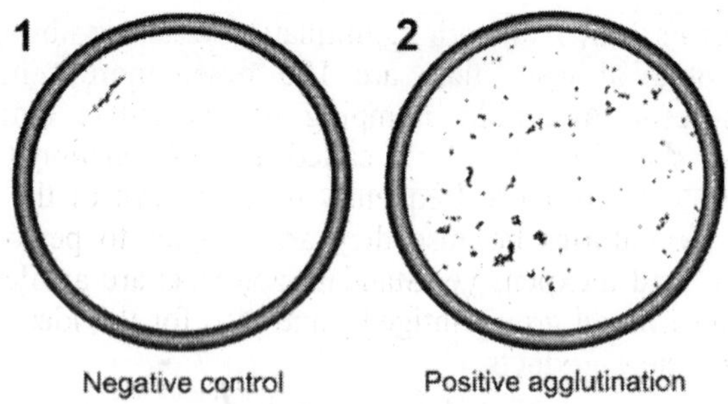


**Fig. 34** Latex bead agglutination test for *Staphylococcus aureus*

Direct agglutination reactions are also useful in the diagnosis of certain diseases. For example, Widal test is used for the diagnosis of typhoid. In this reaction, the clumping of typhoid bacilli occurs when the bacilli are mixed with a serum containing typhoid antibodies from an individual having typhoid fever.

**b. Passive Agglutination:** It is the agglutination process of soluble antigens or antibodies that have been absorbed or chemically coupled to cells or insoluble particles such as latex beads or charcoal particles. The insolubalized antigen or antibody can then be detected by agglutination reaction. The cells or particles serve as carriers of antigen or antibody. Passive agglutination reactions are about five times more sensitive than the direct agglutination tests.

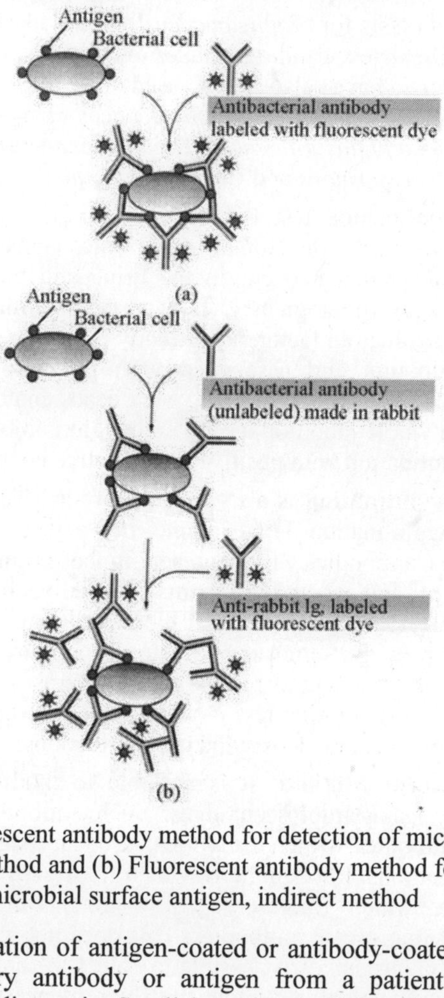


**Fig. 35** (a) Fluorescent antibody method for detection of microbial surface antigen, direct method and (b) Fluorescent antibody method for detection of microbial surface antigen, indirect method

The agglutination of antigen-coated or antibody-coated latex beads by complementary antibody or antigen from a patient is a typical method of rapid diagnosis. Small latex beads (0.8 μm), coated with a specific antigen or antibody, are mixed with patient's serum on a microscope slide and incubated for a short period. If patient's antibody binds the antigen on the head surface, the milky-white latex suspension will become visibly clumped indicating a positive agglutination reaction.

Latex agglutination is also used to detect the bacterial surface-antigens by mixing a small amount of bacterial colony with antibody-coated latex beads. For example, a commercially-available suspension of latex beads containing antibodies to protein A and clumping factor on the surface of *Staphylococcus aureus* will help to accurately identify the clinical isolates of *Staphylococcus aureus* (Fig. 35). Unlike traditional growth-dependent tests for *Staphylococcus aureus,* identification of this bacterium by the latex band test takes only 30 s. The latex band agglutination tests have also been used for the identification of pathogenic bacteria such as *Staphylococcus pyogenes, Neisseria gonorrhoeae, Hemophilus influenzae, Campylobacter sp* and fungi such as *Cryptococcus neoformans* and *Candida albicans.*

Passive agglutination test is also used in the pregnancy test, in which elevation levels of Human Chorionic Gonadotropin (HCG) hormone are detected that occur in the urine and blood of a female during early stages of pregnancy. It is also used for specific serum antibodies for rheumatoid factor, an antibody directed against the body's own immunoglobulin and associated with the autoimmune disease called rheumatoid arthritis. In this test, latex beads coated with human Ig are mixed with whole blood or serum, and agglutination is compared to that of the agglutination with positive and negative control sera.

**1. Haemagglutination** is also seen in certain viruses and is called as viral haemagglutination. For example, if a person has viral disease, such as measles, antibodies will be present in the serum to react with the measles viruses and neutralize them. Normally, haemagglutination occurs when measles viruses and red blood cells are mixed, but is not seen when the person's serum is added to the mixture. This shows that serum antibodies have neutralized the measles viruses and is considered a positive test. This positive test is widely used to diagnose influenza, measles, mumps, mononucleosis and other infections.

**a. Fluorescent Method:** It is possible to modify the antibodies chemically by using fluorescent dyes. Such antibodies are called as fluorescent antibodies, which are widely used in diagnostic as well as research applications. Rhodamine B, which fluoresces red or fluorescent isothiocyanate, which fluoresces yellow-green can be used in the preparation of fluorescent antibodies. This does not alter the specificity of the antibody, but makes it possible to detect the antibody bound to a cell or tissue surface-antigen under the fluorescence microscope. Cells with bound fluorescent antibodies emit a bright fluorescent colour, usually red or orange or yellow-green based on the dye used. Fluorescent antibodies are used in diagnostic microbiology because they permit the identification of a microorganism directly in a patient's specimen, and avoid the isolation and culturing of the organism.

There are two distinct fluorescent antibody processes available, the direct and the indirect staining methods. In the direct method, the antibody attached to the surface-antigen exhibits fluorescence. In the indirect method, the presence of a non-fluorescent antibody on the cell surface is detected by the use of fluorescent antibody attached to the non-fluorescent antibody (Fig. 36). This is done by immunizing one animal species e.g., a goat, with antibodies from a second animal species e.g., a rabbit, and then conjugating the fluorescent dye to the goat antibodies. The fluorescent goat anti-rabbit antibodies can then be used to detect the presence of any rabbit immunoglobulin that is bound to the cells.

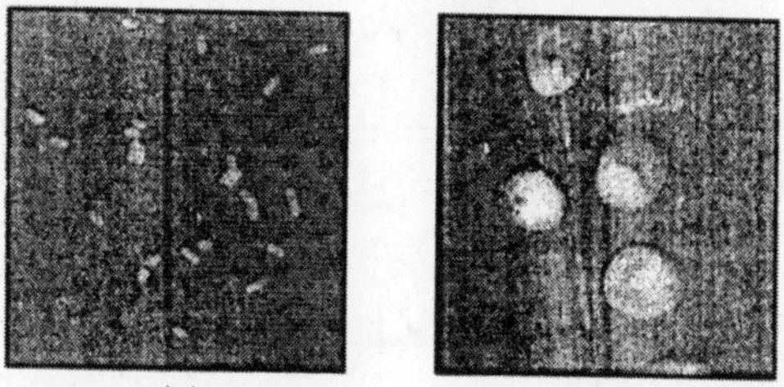

(a) (b)

**Fig. 36** (a) Immunofluorescent cells of *Lagionella pneumophila* and (b) Detection of virus infected by immunofluorescence

In a typical clinical test, a specimen containing a suspected pathogen is allowed to react with a specific fluorescent antibody and observed with a fluorescent microscope. If the pathogen contains a surface antigen, it reacts with the antibody resulting in fluorescence of the pathogenic cells (Fig. 35).

Fluorescent antibodies can be applied directly to infected host tissue, permitting diagnosis long before the primary isolation techniques yield a suspected pathogen. For example, in diagnosing legionellosis disease, a positive identification can be made by staining biopsied lung tissue with fluorescent antibodies prepared against cell walls of *Laegionella pneumophilia*. Similarly, a fluorescent antibody against the capsule of *Bacillus anthracis* can be used in the microscopic diagnosis of the anthrax disease (Fig. 36).

Fluorescent antibody reactions can also be used in diagnosis of viral infections and in a variety of non-infectious diseases. For example, to identify cell types that express a particular antigen, such as on malignant cells, fluorescent antibodies may be very valuable in following the course of the disease.

**b. ELISA:** ELISA has become one of the most widely used serological tests for the detection of antibody or antigen. This test involves linking of various labelled enzymes to either antigens or antibodies. There are two basic methods available:

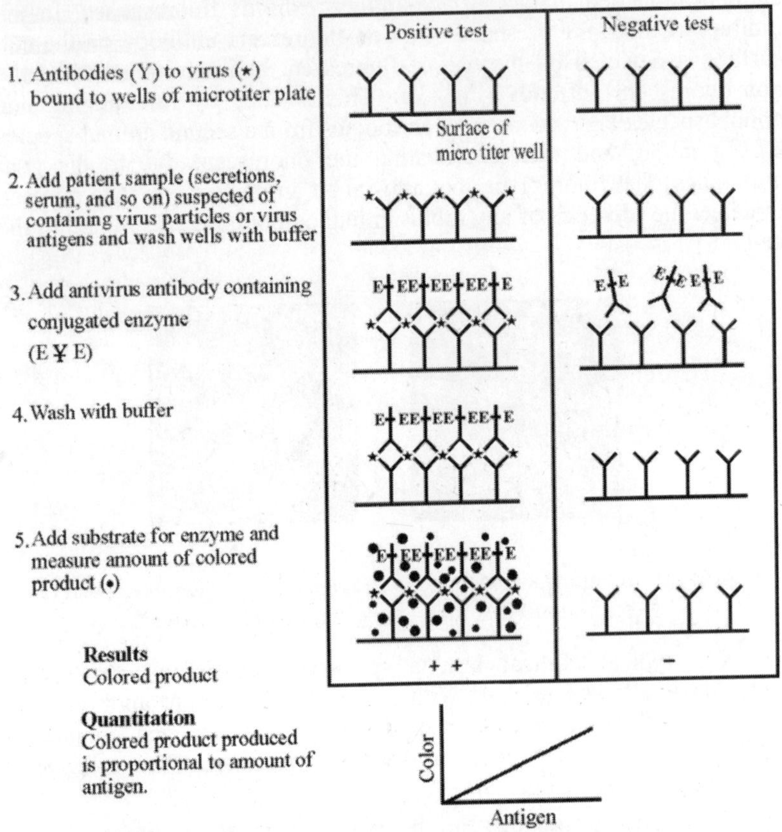


**Fig. 37** Detection of viruses by a direct ELISA test

i. **Double Antibody Sandwich Assay:** It is used for the detection of antigens. In this process, specific antibody is placed in wells of a microtiter plate. The antibody is absorbed on to the wells, coating and sensitizing the plate. A test antigen is then added to each well. If the antigen reacts with the antibody, antigen is retained when the well is washed to remove unbound antigens. An antibody-enzyme conjugate specific for the antigen is then added to each well. The final complex formed contains an outer antibody-enzyme, middle antigen and inner antibody, which is called as layered sandwich. A substrate that the enzyme will convert into coloured is then added and resulting product is quantitatively measured by optical density scanning of the plate. If the antigen has reacted with the absorbed antibodies in the first step, the ELISA test is positive. If the antigen is not recognized by the absorbed antibodies, the ELISA test is negative because the unattached antigens have been washed away and no antibody-enzyme is bound (Fig. 37).

This assay is currently being used for the detection of Helicobacter pylori infections, Brucellosis, Salmonellosis and cholera. Many other antigens can also be detected by this process. The ELISA kits are available in the market that can test for many different food allergens.

ii. **Indirect Immunosorbent Assay:** By this process, it is possible to detect antibodies rather than antigens. Antigen in appropriate buffer is incubated in the wells of a microtiter plate and is absorbed on the walls of the wells. Unbound antigens are washed away. The test antiserum is then added, and if specific antibody is present, it binds to the antigen. Unbound antibody is washed away. Instead of microtiter plate, latex beads can also be used in this method. The test sample is incubated with a suspension of latex beads containing the desired antigen on their surface. After antibody-antigen complex is formed, the latex beads are trapped on the filter and unbound antibody is washed away.

An antibody that has been covalently attached to an enzyme, such as horse radish peroxidase, is added next into the wells of microtiter plate. The antibody-enzyme complex binds to the test antibody, and after unbound antibody-enzyme complex is washed away, the attached complex is visualized by the addition of a chromogen, which is a colourless substrate. A coloured product is formed when enzyme part of the complex acts on the chromogen. The amount of test antibody is measured by optical density scanning of the plate (Fig. 38).

The indirect immunoassay is presently being used to test the antibodies to HIV, which causes Acquired Immunodeficiency Syndrome (AIDS) in human, and Rubella virus.

**1. Radioimmunoassay (RIA):** This process employs radioisotopes instead of enzymes as antibody conjugates. The iodine 12S isotope is commonly used in this method as a detection agent because proteins can readily be iodinated without disrupting immune specificity. The RIA is used clinically to quantify rare serum proteins such as human growth hormons, glucagon, vasopressin, testosterone and insulin present in humans in extremely small amounts. It has also become an important tool in medical research and clinical practice e.g., in cardiology, blood banking, diagnosis of allergies and endocrinology (Fig. 39).

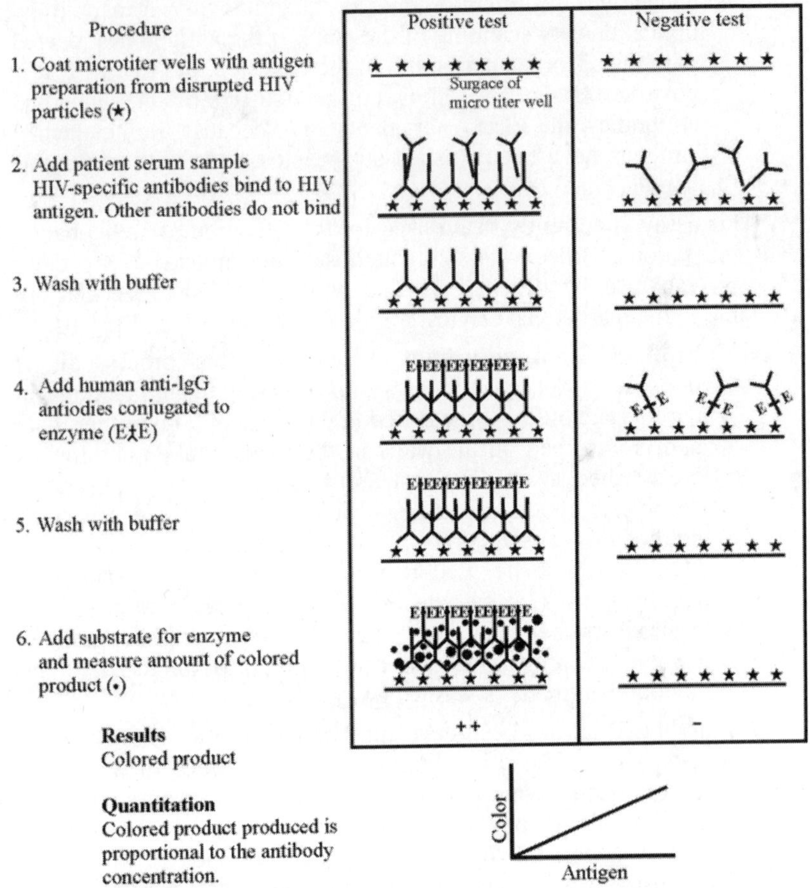


**Fig. 38** Indirect ELISA test for detecting antibodies to HIV

In most cases, direct RIA is employed, which is a two-step process. In the first step, radioactive antigen-specific antibodies are added to a series of microtiter wells containing a known concentration of pure antigen, such as a hormone, which is first bound to the wells. The radioactivity in each of these standard wells is then measured. These data establish a standard curve for antigen concentration. In the second step, the antigen sample from a patient is allowed to bind to another well, and radioactive antibodies are added and measured. The amount of radioactivity bound by the patient's sample is then compared to a standard plot generated from the binding data obtained using the pure antigen and the concentration of antigen in the patient's serum is interpolated from the standard plot.

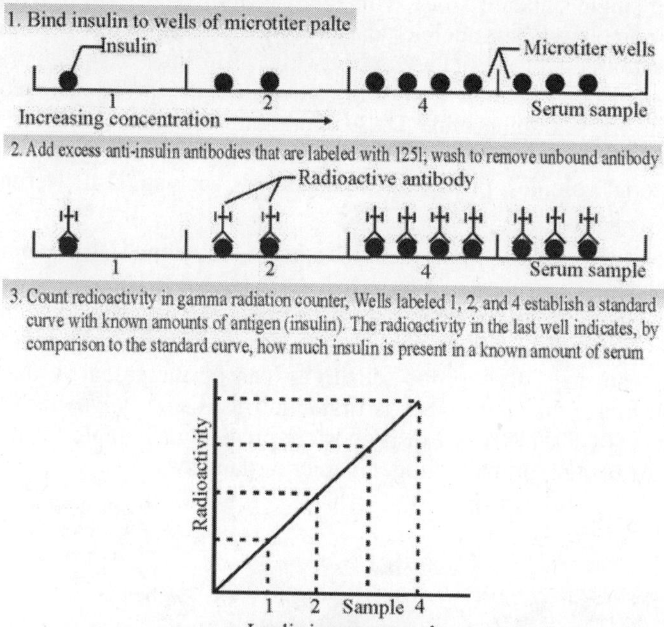


**Fig. 39** Radioimmunoassay (RIA): Using RIA to detect insulin levels in human serum

Though RIA has the same sensitivity range as ELISA and can be performed rapidly, it has certain disadvantages.

- The instruments used for RIA are quite specialized and costly.
- It generates considerable amount of radioactive waste, and the radioactive decay time, that is, half life of the radioisotope, may limit the useful life of the test kits. Therefore, RIA is used only when the results of ELISA are not accurate.

## 5.7 MOLECULAR METHODS

It is now possible to analyze the molecular characteristics of pathogenic microorganisms in the diagnostic laboratory with the application of new molecular methods, in which analyses of proteins and nucleic acids can be applied for their accurate identification. The most widely employed molecular methods are:

**1. Nucleic Acid Probes:** It is a recent technology used in clinical microbiology. Nucleic-acid based methods are used for the detection and identification of pathogens. The genetic composition of a microorganism is determined in DNA probe technology. This is used as DNA probe, whose DNA sequence is predetermined. It is used in the form of single standard DNA. This DNA is allowed to bind or hybridize with a complementary nucleic-acid sequence present in test specimen to form a double-stranded DNA hybrid (Fig. 40). Thus, a single-stranded DNA sequence obtained from one microorganism (probe) is used as a template for searching other DNA containing the same sequence. This hybridization process is presently applied to purified DNA preparations of bacterial colonies, or to clinical specimens such as tissue, serum and pus.

Recently, DNA probes have been developed that bind to complementary strands of ribosomal RNA (rRNA). These DNA-rRNA hydrids are more sensitive than conventional DNA probes, which give results in less than 2 h and require the presence of fewer pathogenic microorganisms. DNA probe sensitivity can be increased by over one million-fold if the target DNA is first amplified using Polymerase Chain Reaction (PCR). DNA-rRNA probes are presently available for most of the clinically-important pathogenic microorganisms.

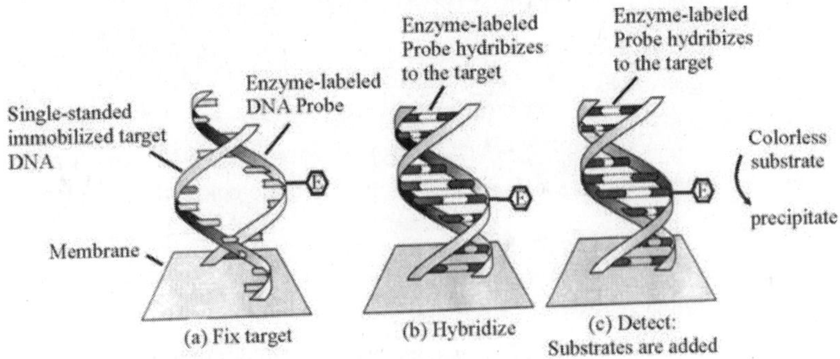


**Fig. 40** Basic steps in a DNA probe hybridization

**2. Gas-Liquid Chromatography:** Cellular fatty acids, products from pyrolysis (a chemical change affected by heat) of whole bacterial cells and specific microbial metabolites are analyzed in gas-liquid chromatography. These compounds are obtained from microorganisms, especially bacteria, growing on a medium by extraction with an organic solvent such as ether. The extract is then fed into gas-liquid chromatography. Common bacteria isolated from clinical specimens can be identified depending upon the pattern of production of fatty acids.

The reliability and accuracy of gas-liquid chromatography has been improved considerably in these days by the introduction of High Performance Liquid Chromatography (HPLC), use of mass spectro-scopy, nuclear magnetic resonance spectroscopy and associated analytical techniques. These techniques have recently been used to discover specific chemical markers of various infectious disease agents by direct analysis of the body fluids such as blood.

**3. Plasmid Fingerprinting:** Plasmids are self-replicating extra-chromosomal DNA molecules present in bacteria. It is possible to identify microbial isolate of the same or similar strain by plasmid fingerprinting because related strains contain the same molecular weight, same phenotype and same number of plasmids. Contrary to this, microbial isolates that are phenotypically-distinct possess different plasmid fingerprint. Plasmid fingerprinting of *E.coli, Salmonella, Campylobacter* and *Pseudomonos* strains and species have shown to be having plasmid fingerprinting biotyping, antibiotic resistance pattern, serotyping and phage typing.

Plasmid fingerprinting process starts with culturing the bacterial strains in broth or on agar plates. Once sufficient bacterial growth occurs, the cells are harvested and lysed with a detergent. The plasmid DNA is separated from chromosomal DNA by differential centrifugation. The plasmid DNA is then separated by agarose gel electrophoresis. The gel is stained with ethidium bromide, which binds to DNA allowing it to glow under UV light. The plasmid DNA bands are then located (Fig. 41).

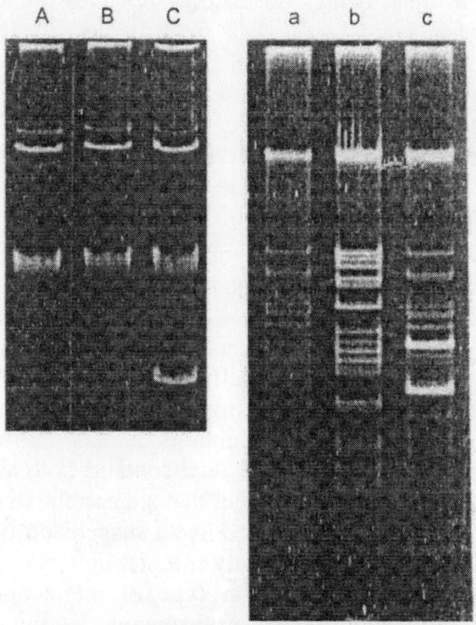


**Fig. 41** Plasmid finger printing: Agarose gel electrophoresis of plasmid DNA

As the migration rate of plasmid DNA in agarose gel is inversely proportional to the molecular weight, plasmids of different sizes appear as distinct bands in the stained gel. The molecular weight of each plasmid species can then be determined from a plot of the distance migrated by each species versus the log of the molecular weight of plasmid marker of known size that have been simultaneously put to agarose gel electrophoresis.

## 5.8 BLOOD GROUPS

Blood groups of animals are due to the presence of immunogenic molecules on the outer surface of their erythrocytes. An immunogen is a molecule that stimulates formation of antibodies that are specific to itself. Immunogen interacts with these antibodies when introduced into the circulatory system of an animal. Human erythrocytes manifest a variety of immunogenic molecules on their surface that give rise to a variety of different blood groups such as ABO, Rh, MN etc., of which, ABO and Rh blood groups are important in the human context.

### 5.8.1 ABO Blood Groups

It is Landsteiner (1900) who discovered human ABO blood groups for the first time. Human beings are categorized into four groups on the basis of A and B antigens on their erythrocytes. Erythrocytes of an individual belonging to group A have antigen A on their outer surface, whereas erythrocytes of an individual belonging to group B have antigen B on their outer surface. Erythrocytes of an individual belonging to group AB contain both the antigen A and B on their outer surface. An individual belonging to group O does not carry either antigen A or antigen B.

Blood serum is the liquid phase of blood in which different types of blood cells remain suspended in blood serum, which is the liquid phase of blood. The blood serum is isolated as supernatant at the top after the blood cells are separated by centrifugation. The blood serum of group A individual contains antibody B (briefly written as anti-B), while blood serum of group B individual contains antibody A (anti-A). The blood serum of group AB individual contains neither anti-A nor anti-B, whereas the serum of group O individual contains both anti-A and anti-B. Anti-A and anti-B are the antibodies that are specific to antigens A and B. When anti-A antibody is introduced into a suspension of erythrocytes having antigen, antigen and antibody interactions takes place causing agglutination of the RBCs. Similar type of interaction takes place resulting in agglutination when erythrocytes having antigen-B is introduced into erythrocytes suspension containing antigen-B. Agglutination is observed as clot formation due to irreversible aggregation of RBC in groups of variable sizes.

The blood from a person of blood group A (donor) can not be transfused into a person of blood group B (recipient) because anti-A present in the serum of recipient will interact with the antigen A on the RBCs of the donor's blood and it will get agglutinated. Similarly, blood from group B donor cannot be transfused into a recipient of group A (Table 12). However, a recipient having the blood group AB can be given blood from donor of either A or B blood groups as the serum of such recipients has neither anti-A nor anti-B. Therefore, the individuals with AB blood group are called universal recipients as they can accept blood from donor having either A, B, AB or O blood groups. Similarly, the RBCs of O blood group persons lack active A and B antigens. Hence, their blood can be safely transfused to recipients belonging to any of the blood group of the A, B, O system and they are called as universal donors.

**ABO antigens**

Several different types of antigens are present on the outer surface of human erythrocytes or RBCs, which give rise to the different human blood group systems. Similarly, other types of human cells, RBCs and other cells of other animals have a variety of antigen molecules on their surfaces. Thus, it is clear that every protein molecule has antigenic properties, which are expressed when introduced into an animal body that does not contain this protein molecule. It is of paramount significance to note that an animal, normally, does not have antibodies against those antigens, which are naturally present in its own body. An animal produces antibodies against only those antigens, which are foreign to its system. A group of antigens present on human erythrocytes is known as antigen A and antigen B, which are the basis for ABO blood groups. Antigens A and B are produced from antigen O or H, which are produced in all the individuals. Antigen O is added to proteins and the antigenetically-active portion of antigen A and B are carbohydrates. In the case of antigen A, it is galactosamine, while in antigen B, it is galactose.

**Inheritance of the ABO system**

It is reported that a single gene, normally, designated as I control the ABO blood groups system. This gene (i) has three alleles designated as $I^A$ $I^B$ and I°. Allele $I^A$ controls the production of the antigen A, allele $I^B$ controls antigen B, whereas allele I° does not produce an active antigen. These alleles are co-dominant; hence, the individuals having the genotype $I^AI^B$ have both the antigens A and B on their RBCs, and the blood group is called AB blood group. Individuals with the genotype $I^AI^A$ or $I^A$ i produce the antigen A and the blood group is known as A blood group, those with $I^BI^B$ or B (i) produce the antigen B and belongs to the blood group B. The persons with the genotype $I^AI^B$ have both the antigens A and B, and belong to the blood group AB while those having the genotype (ii) produce neither the antigen A nor B, and the blood group is called as O (Table 12).

**Table 12:** Different genotypes, their phenotypes and the compatible blood groups for transfusion in human ABO blood group system

| Blood group | Genotype | Antigen present on the surface of RBC | Antibody present in blood serum | Compatible transfusion from blood groups | Frequency population (%) |
|---|---|---|---|---|---|
| A | A/A<br>A i | A | Anti-B | A and O | 28 |
| B | B/B<br>BI | B | Anti-A | B and O | 27 |
| AB | A/B | A and B | Nil | A, B, AB and O (Universal recipient) | 05 |
| O | Ii | O | Anti-A and anti-B | O (Universal donar) | 40 |

In fact, gene i encodes a galactosyl transferase, which adds a further sugar group to the antigen O (or H). Antigen O is produced in all the individuals. Antigen O or H consists of a particular group of carbohydrate that is added to proteins. Allele IA produces N-acetyl-galactose transferase, which converts antigen O to antigen A. Similarly, allele IB encodes the enzyme galactose transferase, which generates antigen B from antigen O. Two enzymes N-acetyl galactose transferase and galactose transferase differ for four amino acids. Allele I, on the other hand, carries a small deletion produces an active enzyme and is unable to modify antigen O.

### 5.8.2 Rh Blood Groups

Human beings can he classified into Rh-positive and Rh-negative blood groups. The classification is based on the agglutination of their RBCs by an antibody produced in response to an antigen present on the RBCs of a species of monkeys called Macaca rhesus (earlier name was Rhesus macaca). The RBCs of Rh-positive individual are agglutinated by this antibody indicating that they posses the concerned antigen. In contrast, the Rh-negative individual does not agglutinate with this antibody suggesting that they lack active form of the concerned antigen. As this classification is based on the presence or absence of an antigen originally discovered in rhesus monkeys, this blood group system in human beings is referred as rhesus blood group or Rh blood group.

The antigen involved in this blood group is called as Rh antigen and is designated as Rho. Rh-positive persons produce an active Rho, which is present on their RBCs. However, the RBCs of Rh-negative persons do not have Rho in an active state. The Rh blood group differs from the ABO system in one important part. The blood serum of Rh-negative persons does not contain anti-Rho unless they have been previously exposed to this antigen. The Rho is highly immunogenic in human beings. Therefore, Rh blood group is also considered in the determination of blood transfusion compatibilities. The Rh antigen is analogous to *E. coli*. Anti-B protein functions as ammonium transporter and is a transmembrane protein. Further, Rh proteins function as ammonium transporter when they are expressed in appropriate mutant strains of yeast.

**Incidence of erythoblastosis foetalis**

The Rh blood group is responsible for the haemolytic anaemia of newborn called erythroblastosis foetalisis; when Rh-negative woman marries Rh-positive man. All the babies from such a couple or at least a few may be expected to be Rh-positive. Normally, the RBCs from the developing foetus do not move across the placenta into the blood stream of the mother. However, at the time of child birth, RBCs from the embryo usually migrate into the blood stream of the mother. If the embryo was Rh-positive and mother was Rh-negative, the Rho antigen present on these RBCs will induce a strong immunological response i.e., production of anti-Rho in the mother. In the subsequent pregnancies of such a mother, the RBCs of Rh-positive embryos will be exposed to the anti-Rho molecule present in the blood stream of mother, and they will freely migrate across the placenta into the blood stream of embryo. These anti-Rho molecules would attach to the RBCs of Rh-positive embryo and cause haemolysis producing anaemia or erythroblastosis foetails in such embryos. The intensity of haemocytic anaemia varies considerably and in cases of severe anaemia, babies may even fail to survive. Most of the newborns suffering from haemolytic anaemia can be saved through timely and appropriate medical care including blood transfusion.

The problem of haemolytic anaemia in Rh-positive babies is due to Rh incompatible marriages; however, wife Rh-negative and husband Rh-positive can be successfully overcome by the use of defective anti-Rho. The defective anti-Rho molecules specifically bind to Rho molecules present on the RBCs without causing agglutination. When Rh-negative mother gives birth to a Rh-positive baby, the mother is immediately injected with the defective anti-Rho molecules that bind to the Rho molecules present on the RBCs, which may have entered from the Rh-positive embryo into the blood stream of the mother. As a result, the Rho molecules are unable to induce production of anti-Rho in the Rho-negative mother since their antigenetically-active sites have already been occupied by the defective anti-Rh molecules. Therefore, such an Rh mother, who is given the injection of defective anti-Rho, immediately after the birth of an Rh-positive baby, would not develop anti-Rho and subsequently produce Rh-positive baby without the risk of haemolytic anaemia to the baby.

## 5.9 BIOCHEMICAL METHODS

Pathogenic microorganisms specially, bacteria can be identified by various biochemical methods along with other diagnostic methods. Various biochemical methods commonly employed for the identification of pathogenic bacteria are summarized in Table 13.

**Table 13:** Biochemical tests useful in the identification of pathogens

| Test | Principle | Procedure | Most common use |
|---|---|---|---|
| Carbohydrate fermentation | Acid and/or gas produced during fermentative growth with sugars or sugar alcohols | Broth with carbo-hydrate and phenol red as pH indicator; inverted tube for gas | Enteric bacteria differentiation |
| Catalase | Enzyme decomposes $H_2O_2$ | Add drop of $H_2O_2$ to dense culture and look for bubbles ($O_2$) | *Bacillus* (+) from *Clostridium* (-); *Streptococcus* (-) from *Micrococcus-Staphylococcus* (+) |
| Citrate utilization | Utilization of citrate as sole carbon source results in alkalinization of medium | Citrate medium with bromthymol blue as pH indicator. Look for intense blue colour (alkaline pH) | *Klebsiella-Enterobacter* (+) from *Escherichai* (-), *Edwardsiella* (-) from *Salmonella* (+) |
| Coagulase | Enzyme causes clotting of blood plasma | Mix dense liquid suspension of bacteria with plasma, incubate, and look for fibrin clot | *Staphylococcus aureus* (+) *S. epidermidis* (-) |
| Decarboxylases | Decarboxylation of amino acid releases $CO_2$ and amine | Medium enriched with amino acids, Bromocresol purple pH indicator becomes purple (alkaline pH) if there is enzyme action | Aid in determining bacterial group among the enteric bacteria |

| Test | Principle | Procedure | Most common use |
|---|---|---|---|
| β-Galactosidase Orthonitrophenyl-β-galactoside (ONPG) test | ONPG is an artificial substrate for the enzyme. When hydrolyzed, nitrophenol (yellow) is formed | Incubate heavy suspension of lysed culture with ONPG. Look for yellow colour | *Citrobacter* and *Arizona* (+) *Salmonella* (-). Identifying some *Shigella* and *Pseudomonas* species |
| Gelatin liquefaction | Many proteases hydrolyze gelatin and destroy the gel | Incubate in broth with 12% gelatin. Cool to check for gel formation. If gelatin is hydrolyzed, broth remains liquid on cooling. | To aid in identification of *Serratia, Pseudomonas, Flavobacterium, Clostridium* |
| Hydrogen Sulphide ($H_2S$) production | $H_2S$ produced by breakdown of sulphur amino acids or reduction of thiosulphate | $H_2S$ is detected in iron-rich medium by the formation of black ferrous sulphide (many variants: Kligler's iron agar and triple sugar iron agar and triple sugar iron agar also detect carbohydrate fermentation) | In enteric bacteria, to aid in identifying *Salmonella, Arizona, Edwardsiella* and *Proteus* |
| Indole test | Tryptophan from proteins converted to indole | Detect indole in culture medium with dimethyl-aminobenzaldehyde (red colour) | To distinguish *Escherichia* (+) from *Klebsiella* (-) and *Enterobacter* (-); *Edwardsiella* (+) from *Salmonella* (-) |
| Methyl red test | Mixed-acid fermenters produce sufficient acid to lower the pH below 4.3 | Glucose-broth medium. Add methyl red indicator to a sample after incubation | To differentiate *Escherichia* (+ culture red) from *Enterobacter* and *Klebsiella* (usually -, culture yellow) |
| Nitrate reduction | Nitrate alternate electron acceptor, reduced to $NO^-_2$ or $N_2$ | Broth with nitrate. After incubation, detect nitrite with α-naphthylamine-sulphanilic acid (red colour). If negative, | To aid in identification of enteric bacteria (usually +) |

| Test | Principle | Procedure | Most common use |
|---|---|---|---|
| | | confirms that $NO^-_3$ is reduced to $NO^-_2$. If no colour after zinc, then $NO^-_3$ reduced $N_2$ | |
| Oxidase test | Cytochrome c oxidizes artificial electron acceptor: tetramethyl (or dimethyl)-p-phenylenediamine | Broth or agar. Oxidase-positive colonies on agar can be detected by flooding plate with reagent and looking for blue or brown colonies | To separate Neisseria and Moraxella (+) from Acinetobacter (-). To separate enteric bacteria (all -) from pseudomonads (+). To aid in identifi-cation of Aeromonas (+) |
| Oxidation-fermentation (O/F) test | Some organisms produce acid only when growing aerobically | Acid production in top part of sugar-containing culture tube; soft agar used to restrict mixing during incubation | To differentiate *Micrococcus* (acid produced aerobically only) from *Staphylococcus* (acid produced anaerobically). To characterize *Pseudomonas* (aerobic acid production) from enteric bacteria (acid produced anaerobically) |
| Phenylalanine deaminase test | Deamination produces phenylpyruvic acid, which is detected in a colorimetric test | Medium enriched in phenylalanine. After growth, add ferric chloride reagent and look for green colour. | To characterize the genus *Proteus* and the *Providencia* |
| Starch hydrolysis | Iodine-iodide gives blue colour with starch | Grow organism on plate containing starch. Flood plate with Gram's iodine and look for clear zones around the colonies. | To identify typical starch hydrolyzers such as *Bacillus* spp. |
| Urease test | Urea ($H_2N$—CO—NH2) split to 2 $NH_3 + CO_2$ | Medium with 2% urea and phenol red indicator. Ammonia release raises pH, intense pink-red colour. | To distinguish *Klebsiella* (+) from *Escherichia* (-). To distinguish *Proteus* (+) from *Providencia* (-). |

| Test | Principle | Procedure | Most common use |
|---|---|---|---|
| Voges-Proskauer test | Acetoin produced from sugar fermentation | Chemical test for acetoin using α-naphthol | To separate *Klebsiella* and *Enterobacter* (+) from *Escherichia* (-). To characterize members of the genus *Bacillus* |

## REVIEW QUESTIONS

### I. Answer the following questions

1. What general guidelines should be followed in collecting and handling clinical specimens?
2. Describe the role of microscopy in the initial diagnosis of infectious disease.
3. How can nucleic-acid based detection methods be used in the diagnosis of a disease?
4. Name general classes of cell cultures.
5. Describe the methods of cultivation of virus.
6. When is Widal test used?
7. How haemagglutination can be used in the clinical laboratory?
8. What are the two methods of ELISA and how do they work?
9. Compare and contrast between direct agglutination and passive agglutination.
10. Discuss about fluorescent antibody assays.
11. Describe the RIA techniques.
12. Describe different methods that are used to collect the samples and how the collected samples are transported to the clinical laboratory.
13. Describe different methods involved in isolation of the microbial pathogens from clinical samples that are transported to the laboratory.
14. Give concise account of the most important biotechnical tests employed to identify the microbial pathogens in laboratory.
15. Describe different molecular techniques employed in the identification of microbial pathogens in the laboratory.
16. List three examples of direct fluorescent stain microscopic examination. Give one example in which each method would be used.

**II. Write short notes on the following**

1. Intubation
2. Needle aspiration
3. Swab
4. Catheter
5. Blood culture
6. Faecal culture
7. Culturing of wound abscesses
8. Culturing of anaerobic organisms
9. Neutralization
10. Serology
11. Bacteriophage typing
12. Bacteraemia and septicaemia
13. Immunologic rapid test kits
14. Plasmid finger printing

## 5.10 FURTHER READING

Baron EJ, Peterson LR, and Finegold SM 1994 *Bailey and Scott's Diagnostic Microbiology,* St.Loius: Mosby

Chapin K Clinical microscopy, in *Manual of Clinical Microbiology* (eds) Murray et al, Washington, DC: American Society for Microbiology

De Las Rivas, Marcobal BA, and Munoz R 2006 Development of a multilocus sequence typing method for analysis of *Lactobacillus plantarum* strains, *Microbiology* **152** 85–93

Dipersio JR et al 1996 Spread of serious disease producing M3 clones of Group A Streptococcus among family members and health care workers, *Clinical Infectious Diseases* **22** 490–495

Murray PR 1996 ASM Pocket guide to Clinical Microbiology, Washington, DC: ASM

Tumpey TM, Basler CF, Aguilar PV, Zeng H, Solorzano A, Swayne DE, Cox NJ, Katz JM, Taubenberger JK, Palese P, and Garcia-Sastre 2005 Characterization of the reconstruction of 1918 spanish influenza pandemic virus, *Science* **310** 77–80.

# 6

# Chemotherapy and Antimicrobials

## 6.1 INTRODUCTION

The process of treating human and animal diseases caused by different types of microorganisms by using a chemical is called as chemotherapy. The modern era of chemotherapy began with the work of the German physician Paul Ehrlich. He stated that a chemical with selective toxicity that would kill the pathogen and not human cells might be effective in treating disease.

Different aspects of chemotherapy are discussed under the following leads:

**General Characters of Antimicrobial Drugs:** A chemotherapeutic drug must have selective toxicity to be successful, that is, it must kill or inhibit the pathogen, but must not harm the host. The degree of selective toxicity may be expressed in relation to:

- The therapeutic dose: The concentration of drug required for clinical treatment of a particular disease
- The toxic dose: The concentration of the drug at which the drug becomes more toxic for the host
- The therapeutic index: The ratio of toxic dose to the therapeutic dose. The larger the therapeutic index, the better will be the chemotherapeutic agent.

A drug that inhibits the function of a pathogen, which is absent in eukaryotic organism's cell has a greater selective toxicity and a higher therapeutic index. For example, penicillin inhibits bacterial cell wall peptidoglycan synthesis, but it has a little effect on the host cells because they lack peptidoglycan in their cell wall. Therefore, penicillin therapeutic index is very high. Similarly, a drug may have low therapeutic index because it inhibits the same processes both in the pathogen and host. These undesirable effects on the host are called side effects. Due to severe side effects of some drugs, they should be administered with great care.

Based on the effectiveness of the drugs, they are grouped into two categories:

1. **Narrow-spectrum Drugs:** They are effective against a limited variety of pathogens e.g., Gentamycin effective against gram-negative bacteria only.
2. **Broad-spectrum Drugs:** They are effective against many different pathogens e.g., Ampicillin, effective against both gram-positive and gram-negative bacteria (Table 14 and 15).

**Table 14:** Broad and narrow spectrum of antibiotics

| Drug | Primary effect | Spectrum | Side effects[a] |
|---|---|---|---|
| Ampicillin | Cidal | Broad (gram-positive, some negative) | Allergic response (diarrhoea, anaemia) |
| Bacitracin | Cidal | Narrow (gram-positive) | Renal injury if injected |
| Carbenicillin | Cidal | Broad (gram-positive, many negative) | Allergic responses (nausea, amenia) |
| Cephalosporins | Cidal | Broad (gram-positive, some negative) | (Allergic responses, thrombophlebitis, renal injury) |
| Chloramphenicol | Static | Broad (gram-positive, -; rickettsia and Chlamydia) | Depressed bone marrow function, allergic reactions |
| Ciprofloxacin | Cidal | Broad (gram-positive, negative) | Gastrointestinal upset, allergic responses |
| Clindamycin | Static | Narrow (gram-positive, anaerobes) | Diarrhoea |
| Dapsone | Static | Narrow (mycobacteria) | (Anaemia, allergic responses) |
| Erythromycin | Static | Broad (gram-positive, some negative) | (Gastrointestinal upset, hepatic injury) |
| Gentamicin | Cidal | Narrow (gram-negative) | (Allergic responses, nausea, loss of hearing, renal damage) |
| Isonaizid (INH) | Static or cidal | Narrow (mycobacteira) | (Allergic reactions, gastrointestinal upset, hepatic injury) |
| Methicillin | Cidal | Narrow (gram-positive) | Allergic responses (renal toxicity, anaemia) |
| Penicillin | Cidal | Narrow (gram-positive) | Allergic responses (nausea, anaemia) |
| Polymyxin B | Cidal | Narrow (gram-negative) | (Renal damage, neurotoxic reactions) |
| Rifampin | Static | Broad (gram-positive, mycobacteira) | (Hepatic injury, nausea, allergic responses) |

| Drug | Primary effect | Spectrum | Side effects[a] |
|---|---|---|---|
| Streptomycin | Cidal | Broad (gram-positive, negative; mycobacteria) | (Allergic responses, nausea, loss of hearing, renal damage) |
| Sulphonamides | Static | Broad (gram-positive, negative) | Allergic responses (renal and hepatic injury, anaemia) |
| Tetracyclines | Static | Broad (gram-positive, negative; rickettsia and Chlamydia) | Gastrointestinal upset, teeth discolouration (renal and hepatic injury) |
| Trimethoprim | Cidal | Broad (gram-positive, negative) | (Allergic responses, rash, nausea, leucopenia) |
| Vancomycin | Cidal | Narrow (gram-positive) | Hypotension, neutorpenia, kidney damage, allergic reactions |

Occasional side effects are in parentheses. Other side effects not listed may also arise.

**Table 15:** Microbial sources of some antibiotics

| Microorganism | Antibiotic |
|---|---|
| Bacteria | |
| *Streptomyces* spp. | Amphotericin B |
| | Chloramphanicol (also synthetic) |
| | Erythromycin |
| | Kanamycin |
| | Neomycin |
| | Nystatin |
| | Rifampin |
| | Streptomycin |
| | Tetracyclines |
| | Vancomycin |
| *Micromonospora* spp. | Gentamicin |
| *Bacillus* spp. | Bacitracin |
| | Polymyxins |
| Fungi | |
| *Penicillium* spp. | Griseofulvin |
| | Penicillin |
| *Cephalosporium* spp. | Cephalosporins |

Drugs are also classified based on their action on the type of pathogens:

- Antibacterial: active against bacteria

- Antiviral: active against viruses
- Antifungal: active against fungi
- Antiprotozoal: active against protozoa

However, some of them are effective against more than one group of pathogens e.g., sulphonamides, which are active against certain bacteria and protozoa.

Chemotherapeutic drugs may be synthesized by microorganisms or manufactured artificially through a chemical synthesis. The former are called as natural drugs and the later as synthetic drugs. Several important drugs are natural (Table 16). Many antiviral and antiprotozoal drugs are synthetic.

**Table 16:** Some important natural antibiotics

| Antibiotic classification | Sub classification | Example |
|---|---|---|
| I. Carbohydrate containing compounds | Pure sugars | Nojirimycin |
| | Aminoglycosides | Streptomycin |
| | Orthosomycins | Everninomicin |
| | N-Glycosides | Streptothricin |
| | C-Glycosides | Vancomycin |
| | Glycolipids | Moenomycin |
| II. Macrocyclic lactones | Macrolide antibiotics | Erythromycin |
| | Polyene antibiotics | Candicidin |
| | Ansamycins | Rifampin |
| | Macrotetrolides | Tetranactin |
| III. Quinones and related compounds | Tetracyclines | Tetracycline |
| | Anthracyclines | Adriamycin |
| | Naphthoquinones | Actinorhodin |
| | Benzoquinones | Mitomycin |
| IV. Amino acid and peptide analogs | Amino acid derivatives | Cycloserine |
| | β-Lactam antibiotics | Penicillin, ceftriaxone |
| | Peptide antibiotics | Bacitracin |
| | Chromopeptides | Actinomycin |
| | Depsipeptides | Valinomycin |
| | Chelate-forming peptides | Bleomycin |
| V. Heterocyclic compounds containing nitrogen | Nucleoside antibiotics | Polyoxins |
| VI. Heterocyclic compounds containing oxygen | Polyether antibiotics | Monensin |
| VII. Alicyclic derivatives | Cycloalkane derivatives | Cycloheximide |
| | Steroid antibiotics | Fusidic acid |

| Antibiotic classification | Sub classification | Example |
|---|---|---|
| VIII. Aromatic compounds | Benzene derivatives | Chloramphenicol |
| | Condensed aromatics | Griseofulvin |
| | Aromatic ether | Novobiocin |
| IX. Aliphatic compounds | Compounds containing phosphorus | Fosfomycin |
| X. Quinolone compounds | 4-Quinolone | Nalidixic acid |
| | Fluoro-4-quinolones | Ciprofloxacin |
| XI. Oxazolidinone | Cyclic lactone | 2-Oxazolidinone |

Chemotherapeutic drugs may also be semi-synthetic, which are generally natural drugs, but are chemically modified by the addition of extra chemical group to make them less susceptible to inactivation by pathogens e.g., Ampicillin, carbenicillin and methicillin.

Chemotherapeutic drugs may be cidal or static. A cidal agent kills the pathogen. Its activity depends upon the concentration and may be static at low concentration while a static agent reversibly inhibits the growth i.e, if the agent is removed, the pathogen will recover and grow again. However, the effectiveness of the agent varies with the target agent; an agent may be cidal for one species and static for another. As static agents do not kill the pathogen, elimination of the infection depends upon the host's own disease-resistance mechanism. A static agent may not be effective if the host's resistance is too low.

Effectiveness of a chemotherapeutic agent can also be described with respect to its concentration as Minimal Inhibitory Concentration (MIC) and Minimal Lethal Concentration (MLC). The MIC is the lowest concentration of a drug that inhibits the growth of a pathogen while MLC is the lowest concentration of a drug that can kill a pathogen. A cidal drug kills a pathogen at a concentration two to four times greater than MIC; whereas a static agent kills a pathogen at much higher concentration.

**1. Determining the level of antimicrobial activity:** Assessment of antimicrobial effectiveness against a specific pathogen is essential for proper treatment. Therefore, testing is essential to know, which drugs are more effective against a pathogen and at what concentration. For this, the following tests are generally conducted in the laboratory:

- Dilution susceptibility tests
- Disk diffusion tests
- The Etest
- Measurement of drug concentrations in the blood

**2. Mechanism of action of antimicrobials:** Knowledge about the mechanism of drug action is necessary because it helps not only to explain the nature and degree of selective toxicity of individual drug, but also in the designing of new chemotherapeutic drugs (Fig. 42).

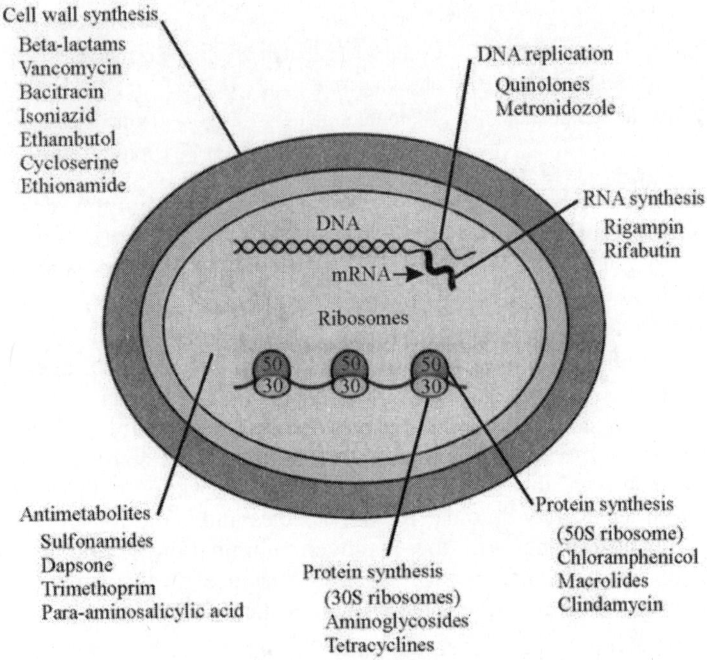


**Fig. 42.** Different mechanisms of action of antimicrobial agents

Chemotherapeutic drug can damage or kill the pathogen by interfering or inhibiting several life processes of pathogens, such as:

a. Inhibition of cell wall synthesis
b. Inhibition of protein synthesis
c. Inhibition of nucleic-acid synthesis
d. Disruption of cell membrane
e. Metabolic antagonism (Table 17)

**Table 17:** Mechanisms of action of antibiotics

| Drug | Mechanism of action |
|---|---|
| Cell wall synthesis inhibition<br>Penicillin<br>Ampicillin<br>Carbenicillin | Inhibits transpeptidation enzymes involved in the cross-linking of the polysaccharide chains of the bacterial cell wall peptidoglycan. Activate cell wall lytic enzymes. |

| Drug | Mechanism of action |
|---|---|
| Methicillin<br>Cephalosporins<br>Vancomycin<br>Bacitracin | Binds directly to the D-Ala-D-Ala terminus and inhibits the transpeptidation step.<br>Inhibits cell wall synthesis by interfering with action of the lipid carrier that transports wall precursors across the plasma membrane. |
| Protein synthesis inhibition | |
| Streptomycin<br>Gentamicin<br>Chloramphenicol<br>Tetracyclines<br>Erythromycin and clindamycin<br>Fusidic acid | Binds with the 30S subunit of the bacterial ribosome to inhibit protein synthesis and causes misreading of messenger RNA (mRNA).<br>Binds to the 50S ribosomal subunit and blocks peptide bond formation through inhibition of peptidyl transferase.<br>Binds to the 30S ribosomal subunit and interferes with aminoacyl-transfer RNA binding.<br>Binds to the 50S ribosomal subunit and inhibits peptide chain elongation.<br>Binds to Elongation Ffactor G (EF-G) and blocks translocation. |
| Nucleic-acid synthesis inhibition | |
| Ciprofloxacin and other quinolones<br>Rifampin | Inhibit bacterial DNA gyrase and thus, interfere with DNA replication, transcription and other activities involving DNA.<br>Blocks RNA synthesis by binding to and inhibiting the DNA-dependent RNA polymerase. |
| Cell membrane disruption | |
| Polymyxin B | Binds to the plasma membrane and disrupts its structure and permeability properties. |
| Metabolic antagonism | |
| Sulphonamides<br>Trimethoprim<br>Dapsone<br>INH | Inhibit folic acid synthesis by competition with p-aminobenzoic acid.<br>Blocks tetrahydrofolate synthesis through inhibition of the enzyme dihydrofolate reductase.<br>Interferes with folic acid synthesis.<br>May disrupt pyridoxal or NAD metabolism and functioning. Inhibits the synthesis of the mycolic acid 'cord factor'. |

a. **Inhibition of cell wall synthesis:** The cell walls of most of the bacteria are made up of peptidoglycan. Some of the antimicrobial agents, such as penicillins, cephalosporins, vancomycin and bacitracins, inhibit the synthesis of peptidoglycon molecules and thereby, interfere with the formation of bacterial cell wall, which lead to stoppage of bacterial multiplication, formation of new cells and bacterial growth. Such drugs have a high therapeutic index against bacteria because of their unique structure of bacterial cell wall, which is absent in eukaryotic cells.

b. **Inhibition of protein synthesis:** Some of the antimicrobial agents inhibit protein synthesis by binding to the prokaryotic ribosomes. Some drugs bind with 30S subunit, while other with 50S subunit of a ribosome. Several different steps in the protein synthesis can be affected such as aminoacyl tRNA binding, peptide bond formation, mRNA reading and translocation. For example, fusidic acid binds EF-G and blocks translocation, where as mucopyrocin inhibits isoleucyl tRNA synthesis. The therapeutic index of these drugs is high, but less than cell wall inhibitors. Streptomycin, gentomycin etc. (Table 17) belong to this category.

c. **Inhibition of nucleic acid synthesis and disruption of cell membranes:** The therapeutic drugs that inhibit synthesis of nucleic acids or damage cell membranes are not as selectively toxic as other antimicrobial agents. This is because the synthetic mechanism of nucleic acid or cell membrane structure, of prokaryotes and eukaryotes do not differ greatly. Quinolones and polymyxins are the good examples of this category of drugs. Quinolone inhibits DNA gyrase and thereby, interferes with DNA replication, repair and transcription. Polymyxins act as detergent or surfactants and disrupt the bacterial plasma membrane.

d. **Metabolic antagonism:** Several therapeutic drugs act as antimetabolites. They block the metabolic pathways by competitively inhibiting key enzyme. Sulphonamides (e.g., sulphamethoxazole, sulphamilamide and sulphacetamide) and many other drugs inhibit folic acid metabolism. They have a high therapeutic index because folic acid is not synthesized in human body and hence, it should be supplemented to the diet. However, several bacterial pathogens can synthesize their own folic acid and are, therefore, susceptible to folic acid inhibitors. Antimicrobial agents can also inhibit other synthetic pathways e.g., INH interferes with either pyridoxal or NAD metabolism.

**3. Factors influencing the effectiveness of antimicrobial drugs:** Drugs may be given in different ways to allow them to spread rapidly throughout the host body or kill the invading pathogen immediately. Normal routes of administration are called parenteral routes. A complex group of factors influence the effectiveness of the drug, some of them have been discussed below:

- The drug must be able to reach the site of infection. Here the mode of administration plays an important role. A drug may be given in two different routes:
  a. Oral administration
  b. Intramuscular or intravenous administration

Some drugs are not suitable for oral administration, such as penicillin G, because it is degraded in stomach by the stomach acids. Some other drugs such as gentamicin and other glycosides are not absorbed well through intestinal wall. Therefore, they must be given either intramuscularly or intravenously. Even though a drug is administered properly, it may fail to reach the site of infection. For example, blood clots and necrotic tissue, which prevent the spread of the drug and thereby, protect the pathogen present inside.

- The pathogen must be susceptible to the drug Penicillin. Other drugs are effective only when the pathogen is actively growing and they remain ineffective if the pathogen is in dormant stage as in abscesses. A pathogen, even though actively growing, may not be susceptible to the drug. For example, cephalosp-osins and penicillins, which inhibit the cell wall synthesis, do not harm Mycoplasma, because they lack cell wall.
- The drug must exceed the pathogen's MIC value if it has to be effective. The MIC depends upon the amount of drug given, the route of administration, speed of uptake and the rate at which the drug is cleaved or eliminated from the body. A drug generally remains longer in the body at a higher concentration when it is absorbed over an extended period and excreted slowly.
- The process of chemotherapy is made less effective due to development of resistance against the drugs by microorganisms.

## 6.2 CHEMOTHERAPEUTIC AGENTS

A brief account of the properties and mechanism of action of some of the important drugs is given below:

### 6.2.1 Penicillins

**Structure:** Structurally, penicillins are derivatives of 6-amino-penicillanic acid. Penicillin G or benzyl penicillin is the first antibiotic, which is used widely in medicine. The most important aspect of the penicillin molecule is the presence of β-lactam ring, which is essential for antimicrobial activity of penicillins. Penicillins differ from one another with respect to the side chain attached to the amino group (Fig.

43). Most of the penicillins loose their antimicrobial activity if the β-lactam ring is hydrolyzed. This is what precisely the enzyme penicillinase secreted by penicillin resistant bacteria does and makes the bacteria resistant to the action of penicillins. Functionally, they differ from each other in several ways:

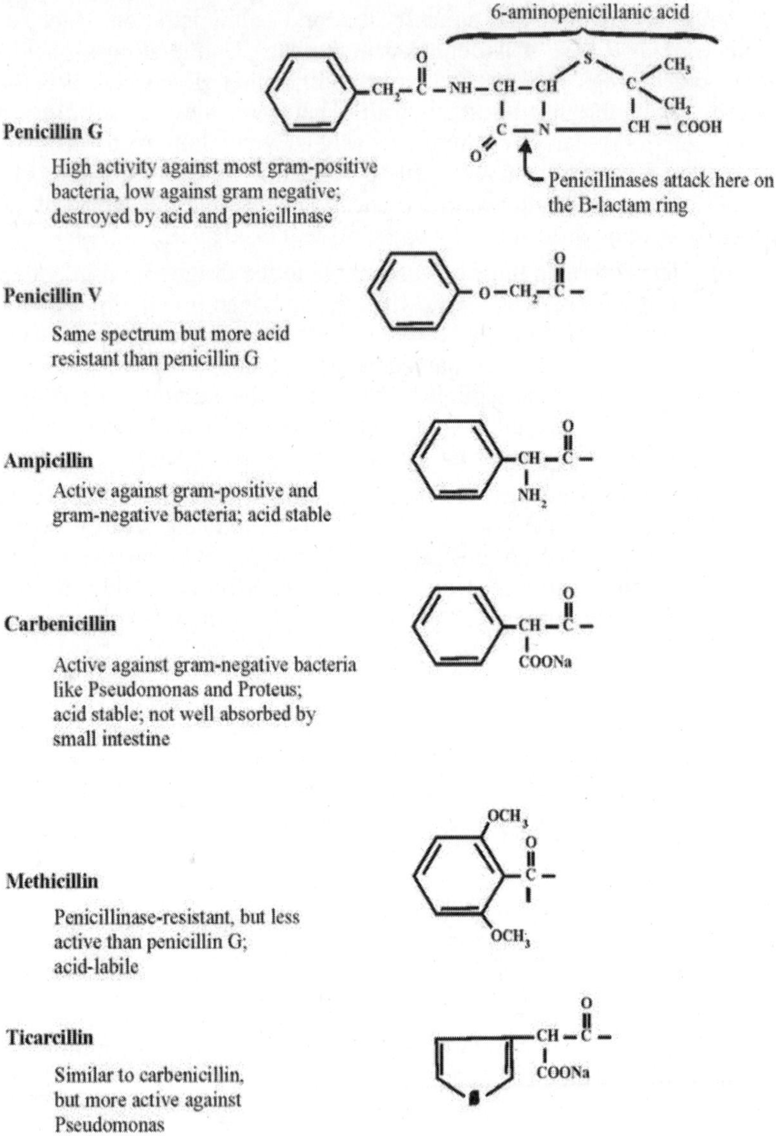


**Fig. 43** Structure and characterstics of different penicillins

- Penicillin-G is effective against gonococci, meningococci and several gram-positive pathogens such as streptococci and staphylococci. In these cases, it should be given intravenously or intramuscularly because penicillin is destroyed by gastric acids, if given orally.
- Penicillin V is acid-tolerant and can be given orally.
- Ampicillin, a semi-synthetic penicillin, is a broad-spectrum drug and acts well if given orally. It is effective against gram-positive bacteria such as *Haemophilus*, causing the disease haemophilia; *Salmonella*, causing typhoid; and *Shigella,* which causes dysentery.
- Carbesicillin and ticarcillin are also semi-synthetic and broad-spectrum antibiotic effective against *Pseudomonos* and *proteius.*
- Penicillinase-resistant penicillins such as methicillin, nafcillin and oxacillin are effective against penicillin-resistant pathogens.
- About 1–5% adults develop allergy to pencillins, which may sometimes lead to death.

**Mechanism of Action:** Mechanism of action of penicillins is more complex and still is not completely known. It is reported that the penicillins probably act on the cell wall in the following ways:

- Penicillinase inhibits the enzyme catalysing the transpeptidation, which would block the synthesis of a complete, fully cross-linked peptidoglycan and lead to osmotic lysis. This mechanism is in agreement with the observation that penicillins act only on growing cells of bacteria that are synthesizing new peptidoglycan.
- Penicillins bind to several proteins and may destroy bacteria by activating their own autolytic enzymes.
- Penicillins kill bacteria even in the absence of autolysins or murein hydrolases. Lysis occurs only after bacteria loose their viability.
- Penicillins may stimulate special proteins called bacterial holins that form holes or lesions in the plasma membrane, which results in membrane leakage and death.
- When holes are formed in the plasma membrane, murein hydrolases could move through the holes, disrupt transpeptidation and lyse the cell.

### 6.2.2 Cephalosporins

Cephalosporins are broad-spectrum antibiotics frequently given to patients who are allergic to penicillin. These are group of drugs isolated from the fungus C*ephalosporium* in 1948. They are similar to penicillin as they contain β-lactam structure, but they differ in having seven carbon ring. (Fig. 44) They are similar in action to penicillin and inhibit the cell wall synthesis.

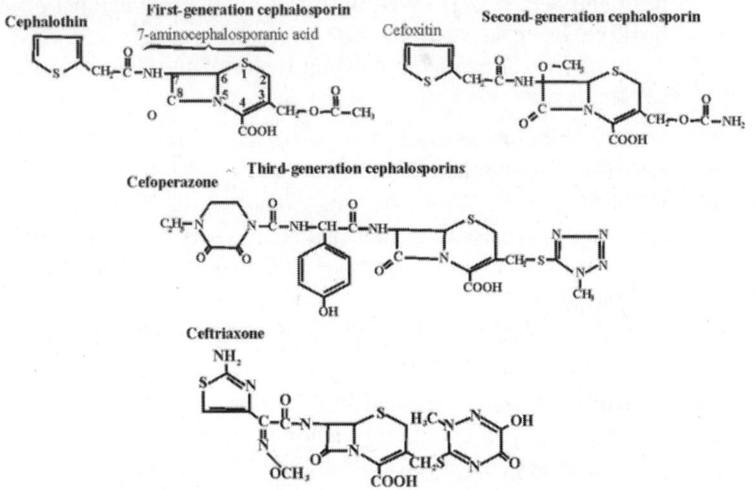


**Fig. 44** Different cephalosporin

There are three groups or generations of cephalosporins that differ in their mechanism of action. First group or generation of cephalosporins is more effective against gram-positive than gram-negative pathogens while the second group or generation drugs act against both gram-positive and gram-negative pathogens. On the other hand, third group or generation of cephalosporins are active against gram-negative pathogens only and also capable of reaching the CNS. They are administrated either by oral route or intramuscularly or intravenously. Important cephalosporins are cephalothin, cetfoxitin, ceftriaxone and cefoperazone. Cefoperazone is resistant to the action of B-lactamase and acts against many gram-negative bacteria including *Pseudomonos aeruginosa.*

### 6.2.3 Aminoglycosides

There are many important aminoglycoside antimicrobial agents; most of them are synthesized by actinomycetes. For example, streptomycin, kanamycin, neomycin and tobramycin are synthesized by the species of *Streptomyces* while gentmicin by *Micromonorpora purpuurea.*

Although many of them are structurally different, all contain a cyclohexane ring and amino sugar (Fig. 45). Aminoglycosides bind to the small subunit of ribosome and inhibit the protein synthesis either by directly inhibiting the protein synthesis or indirectly by misleading the genetic message carried by mRNA.

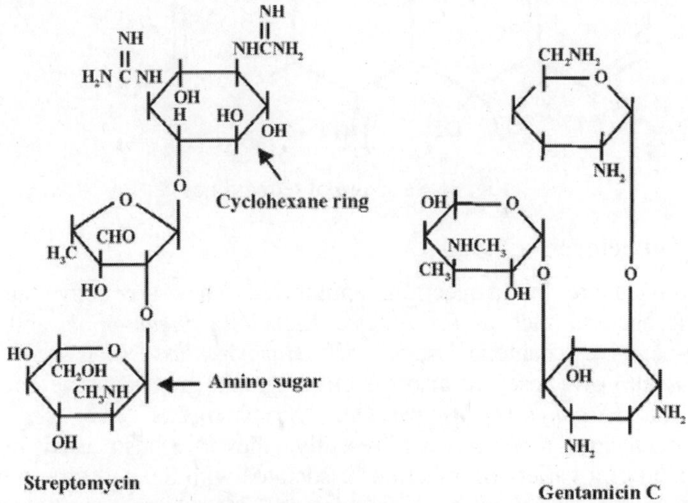


**Fig. 45** Structure of streptomycin and gentamicin

They are bactericidal and are more active against gram-negative pathogens. Though the action of streptomycin decreases greatly due to drug resistance, it is still used in the treatment of tuberculosis and plague. Infections caused by *Serratia, Eschereschia, Klebsiella* and *Proteus* are treated by gentamycin.

Aminoglycosides are relatively toxic and may cause deafness, loss of balance, nausea, renal damage and allergic manifestations.

### 6.2.4 Tetracyclines

Tetracyclines are broad-spectrum antibiotics active against both gram-negative, gram-positive bacteria, ricketsias, Chlamydia and Mycoplasma. These are a group of antibiotics consisting of four ring structure to which different side chains remain attached (Fig. 46). Oxytetracycline and chlortetracycline are naturally synthesized by the species of *Streptomyces* while others are semi-synthetic antibiotics. These drugs inhibit protein synthesis by directly binding the small subunit (30S) of the bacterial ribosome or by inhibiting the binding of aminoacyl tRNA to the ribosomal A site. As their action is only bacteriostatic, their effectiveness as therapeutic agents depends primarily on the host resistance against the pathogens. However, high doses may result into yellowing of teeth in children, diarrhoea, nausea, and damage to the liver and kidneys.

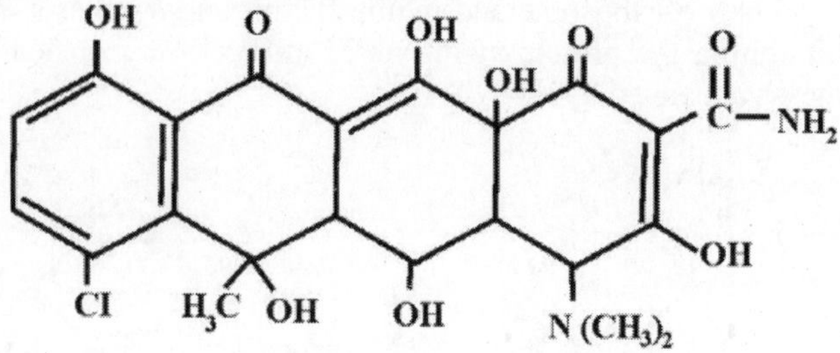


**Fig. 46** Structure of tetracylines

### 6.2.5 Quinolones

Quinolones are broad-spectrum antibiotics. They are active against enteric bacteria such as *E. coli and Klebsiella pneumoniae,* and also gram-negative bacteria such as *Harmophilus, Neisseria* and *Pseudomonos*. These are also effective against gram-positive bacteria such as *Staphylococcus aureus, Streptococcus pyogenes* and *Mycobacterium tuberculosis*. Presently, they are also used in the treatment of a variety of infections associated with Sexually-transmitted Diseases (STDs) caused by *Neisseria* and Chlamydias, gastrointestinal infections, respiratory tract infections, osteomyelitis and skin infections.

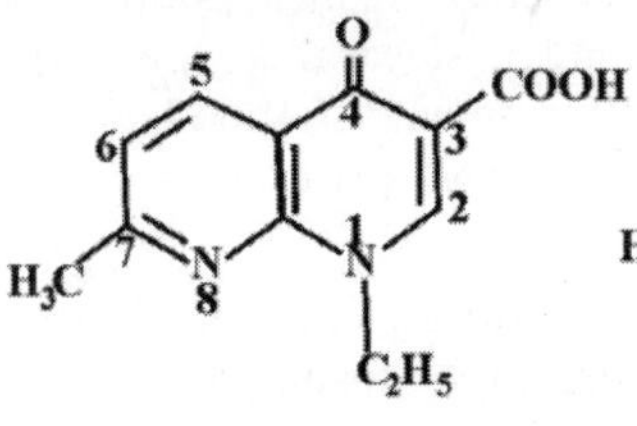


**Nalidixic acid**

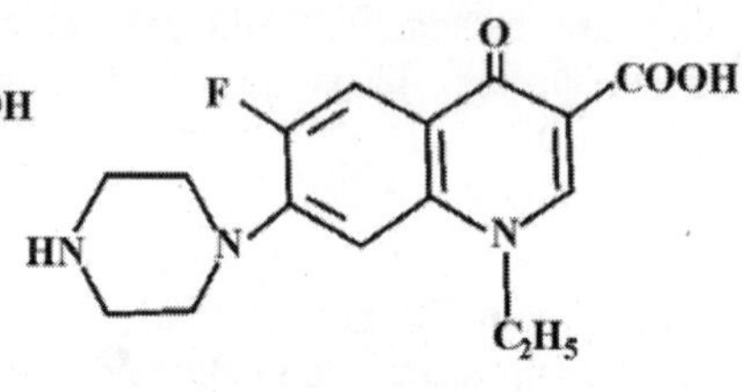


**Norfloxacin**

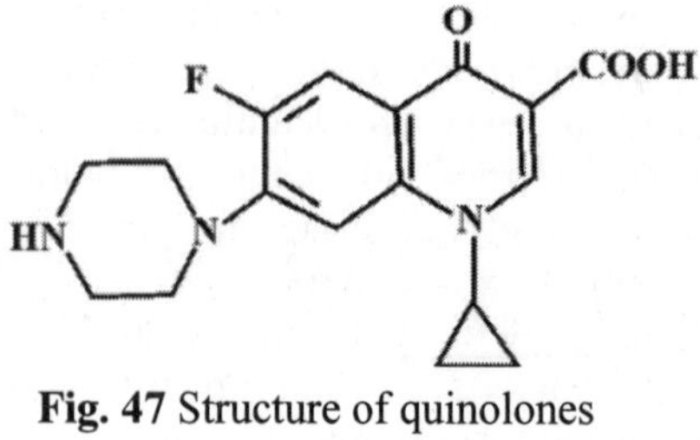


**Fig. 47** Structure of quinolones

Quinolones are synthetic drugs with the 4-quinolonring. The first quinolone, nalidixic acid (Fig. 47), was synthesized in 1962 followed by the production of fluoroquinolones. Ciprofloxacin, norfloxacin and ofloxacin, the three types of quinolones are currently used for the treatment of a variety of infections as mentioned above. They function well when given orally and sometimes cause adverse side effects in gastrointestinal system.

These inhibit the activity of DNA gyrase or topoisomerase-II by binding to DNA gyrase complex, the enzyme helps DNA in the separation of its strand by introducing negative twist (Fig. 48). An inhibition of DNA gyrase results in the disruption of DNA replication and repair, transcription, chromosomal separation during cell division, and other cell processes in which DNA is involved. Fluoroquinolones also inhibit topoisomerase-IV, which is required in DNA replication. They are basically bactericidal in action.

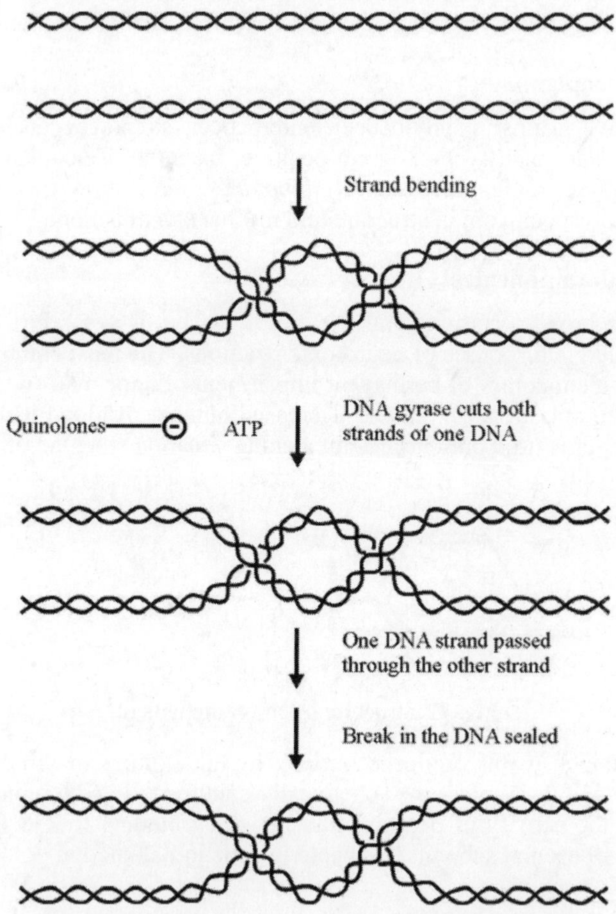


**Fig. 48** DNA gyrase action and quinolone inhibition

### 6.2.6 Vancomycin

This antibiotic is bactericidal and active against *Staphylococcus* and some species of *Clostridium, Bacillus, Streptococcus* and *Enterococcus*. It is given either orally or intravenously, and useful particularly in the treatment of antibiotic-resistant *Staphylococcus* and *Enterococcus* infections. However, vancomycin-resistant strains of *Enterococcus* are widespread and recently few cases of resistant strains of *Staphylococcus aureus* have also been reported.

Vancomycin is a glycopeptide, natural antibiotic produced by *Streptococcus orientalis*. It is a cup-shaped molecule consisting of a peptide linked to disaccharide. This drug inhibits peptidoglycan synthesis by disrupting the transpeptidation step that cross links the adjacent peptidoglycan strands. This results in the formation of a mechanically-weak peptidoglycan due to which cells lyse osmotically.

### 6.2.7 Teicoplanin

It is active against staphylococci, enterococci, streptococci, clostridia, *Listeria* and many other gram-positive bacteria. Teicoplanin is a glycopeptide antibiotic from *Actinoplanes teichomyceticus* that is similar to vancomycin in structure and mechanism of action.

### 6.2.8 Chloramphenicol

This is a broad-spectrum synthetic antibiotic, but is severely toxic and cause allergic responses or neurotoxic reactions. The most common side effect is a temporary or permanent impairment of bone marrow function leading to aplastic anaemia and decreased number of blood leukocytes. However, it is used only in life-threatening situation when no other drug is effective (Fig. 49).

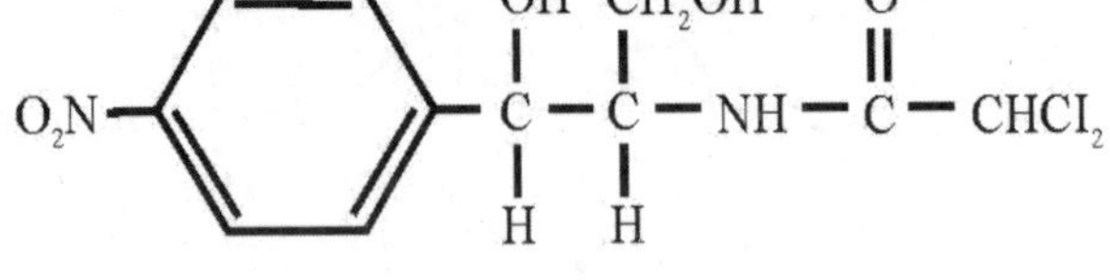


**Fig. 49** Structure of chloramphenicol

Although it was produced initially by the cultures of *Streptomyces venezuelae,* it is presently chemically-synthesized. Chloremphenicol inhibits the activity of peptidyl transferase by binding to 23S RNA on the 50S ribosomal subunit. It is bacteriostatic in its function.

### 6.2.9 Erythromycin

Erythromycin is a broad-spectrum antibiotic effective against Mycoplasma, many gram-positive bacteria and a few gram-negative bacteria. It is used for the patients allergic to penicillin, and in the treatment of diphtheria, whooping cough, and diarrhoea caused by *Campylobacter,* and pneumonia from *Laginella* or Mycoplasma infections.

Erythromycin is a macrolide antibiotic naturally synthesized by *Streptomyces erythraeus* containing a lactone ring with 12–22 carbon atoms linked to one or more sugar molecules. It inhibits peptide chain elongation during protein synthesis by binding to 23S rRNA of the 50 ribosomal subunit. It is bacteriostatic in function (Fig. 50).

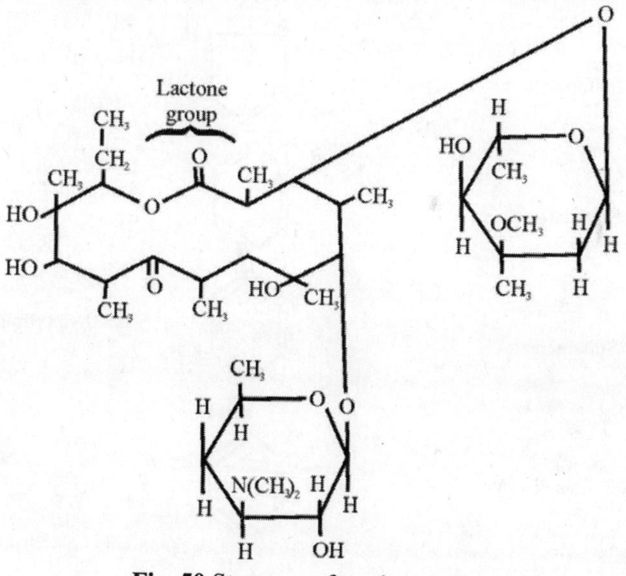


**Fig. 50** Structure of erythromycin

## 6.3 SULPHA DRUGS

It is also called sulphonamide and was the first successfully-used chemotherapeutic drug.

### 6.3.1 Structure

It was chemically synthesized by Gelmo (1908), Ersenberg (1913), and they studied the bactericidal properties of azo dyes with sulphonamide grouping. These drugs are effective against a large variety of pathogenic organisms. There are several types of sulphonamides, which differ structurally with different functional group (Fig. 51).

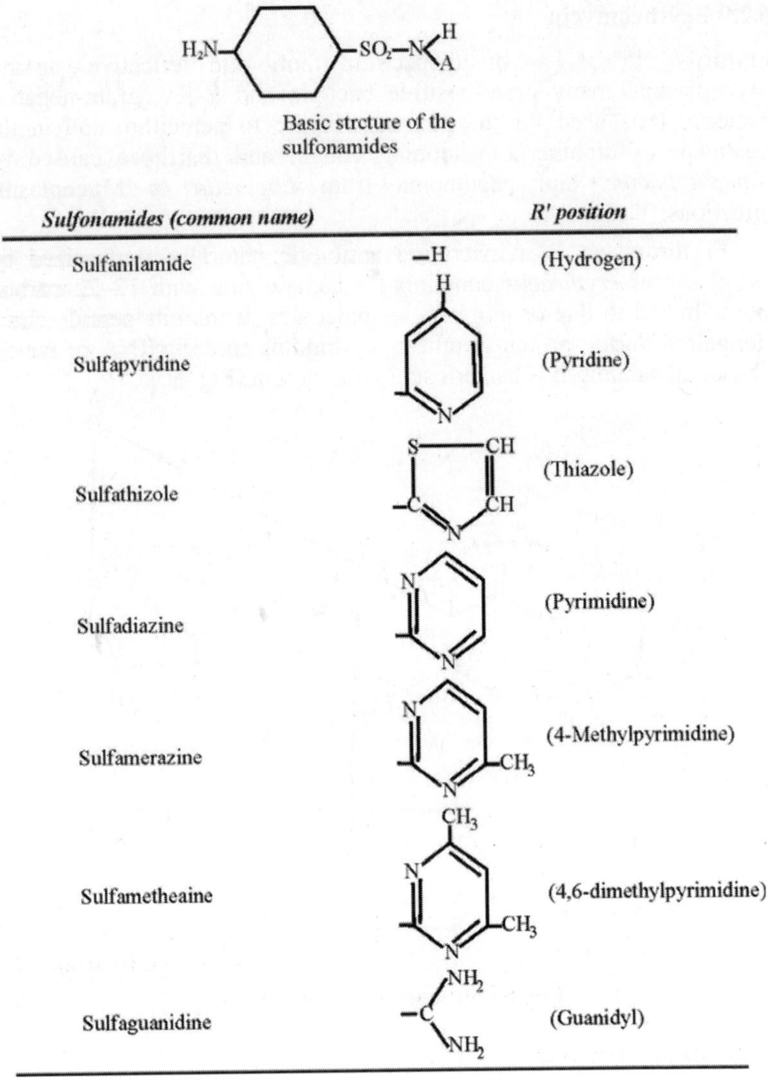


**Fig. 51** Structure of different sulphonamides

Sulphonamides are structural analogues and compete with metabolites formed in metabolic processes because of their similarity. However, they are slightly different so that they cannot participate in the cellular metabolism. They are structurally similar to P-aminobenzoic Acid (PABA), which is used in the synthesis of folic acid (Fig. 52 and Fig. 53).

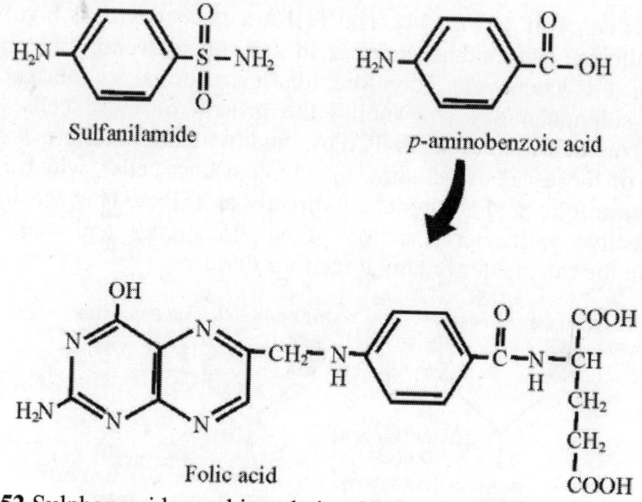


**Fig. 52** Sulphonamides and its relationship to the structure of folic acid

Sulfamethoxazole

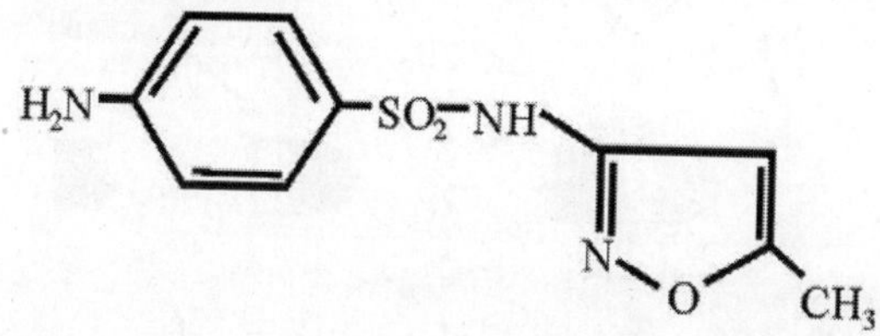

Sulfisoxazole

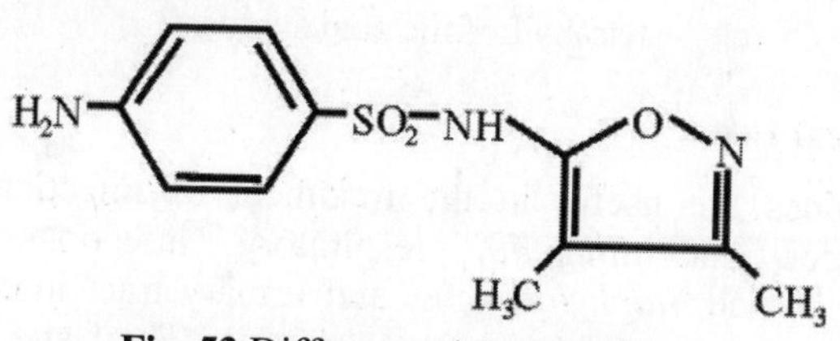

**Fig. 53** Different sulphonamides

### 6.3.2 Mechanism of Action

Many bacteria require PABA as a precursor for the synthesis of essential coenzyme Tetrahydrofolic Acid (THFA). The PABA is a structural part of THFA molecule. As sulphonamide and PABAare structurally quite similar, hence, instead of PABA if sulphonamide enters the metabolic process, there will be blockage of an essential cellular component's synthesis i.e., THFA (Fig. 54). The THFA as a coenzyme is involved in the synthesis of thymidine amino acid and other essential metabolites. Lack of this coenzyme, therefore, disrupts normal cellular activity. Hence, sulphonamides will inhibit the growth of those cells, which synthesize their THFA from PABA, and will not interfere with the growth of those cells (including mammalian host cells), which require the vitamin folic acid and reduce it directly to THFA. This accounts for the selective antibacterial action of sulphonamides and make them useful in the treatment of many infectious diseases.

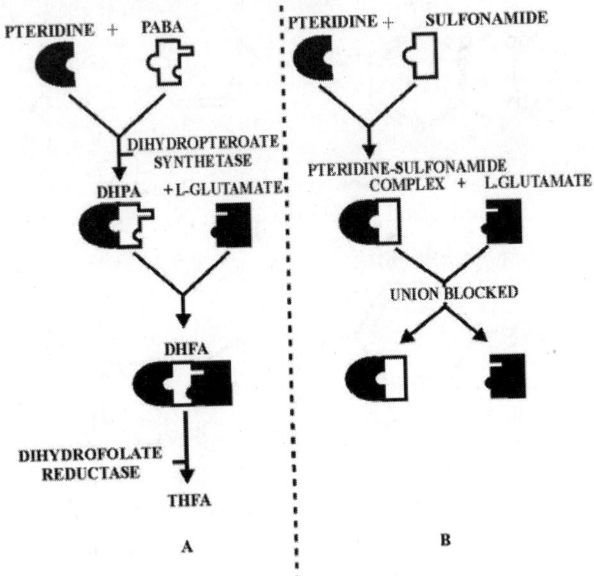


**Fig. 54** The mode of action of sulphonamides on inhibition of tetrahydrofolic acid synthesis

### 6.3.3 Clinical Use

Sulphonamides are useful in the treatment of infections caused by meningococci and *Shigella,* respiratory infections caused by *Streptococcus* and *Staphylococcus,* and urinary tract infections caused by many pathogenic gram-negative bacteria. They are useful in the prevention of rheumatic fever, bacterial endocarditis, wound infections and urinary tract infections following surgery or catheterization.

The effectiveness of sulphonamides is, however, limited because of sulphonamide resistance development in many pathogenic bacteria. Moreover, about 5% patients treated with sulphonamides experience adverse side effects in the form of allergic responses such as fever, hives and rashes.

## 6.4 ANTIVIRAL AGENTS

Antibiotics are not generally effective against viruses because viruses are intracellular and, therefore, to be effective, the chemotherapeutic agents must enter the host cells. They must not be toxic to the host cells, while exerting an inhibiting action on the viruses. This requires a high level of selective toxicity. In case of infections caused by bacteria, fungi or protozoa, the pathogen is acted upon by the drugs outside the host cells, and there are many more metabolic processes that can be interrupted in these microorganisms.

Viruses are intracellular obligate parasites that utilize host cell enzymes and biosynthetic processes for replication. Therefore, it is more difficult to inhibit viral replication without causing harm to the host cells. Early antiviral drugs were selective poisons that targeted the cells with rapid DNA and RNA synthesis. Newer antiviral drugs are targeted towards viral-encoded enzymes or structure of the virus that is important for replication. Many of these compounds are important biochemical inhibitors of viral-encoded enzymes.

Unlike the activity of antibacterial drugs, the action of antiviral drugs is limited to specific families of viruses. Newer antiviral drugs are directed in blocking the expression of the viral genetic material and immune resolution of the viral infection. Antiviral drugs are being developed against those viruses, which cause severe morbidity and deaths in humans. However, increased resistance against antiviral drugs, as occurred with antibacterial drugs, is becoming a severe problem because of the increased long-term treatment to immune compromised people such as people suffering with AIDS.

### 6.4.1 Interferon (IFN)

These are the most important chemotherapeutic agents generally employed for treating viral diseases. The IFNs are small GP substances. There are two types of IFNs, leukocyte IFNs and fibroblast IFNs. Cells exposed to these agents develop antiviral properties. They are reported to act by inhibiting the protein synthesis. Natural IFNs are not only expensive, but also are in short supply. Recent advances in recombinant DNA technology have increased the availability of IFN for both chemotherapeutic and experimental use. The IFN alpha (IFN-a) is one of the antiviral drugs and is produced by leukocytes in response to viral infection.

### 6.4.2 Amantadine

It is a low molecular weight compound, which is very active against influenza A virus. It is not effective against influenza B. The incidence of influenza A infection is greatly reduced by the use of this drug. The amantadine interferes with the uncoating of virus particles and the subsequent release of their nucleic acids (Fig. 55).

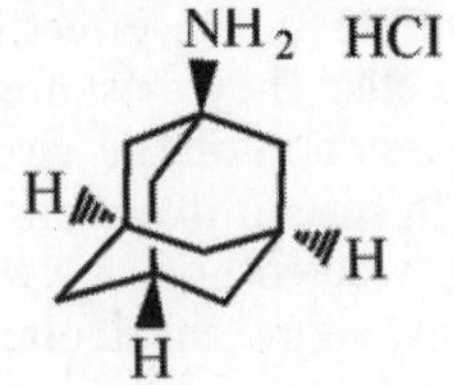


**Fig. 55** Structure of amantidine

### 6.4.3 Base Analogues

Base analogues are the chemical compounds, structurally similar to nitrogen bases adenine, guanine, cytosine and thymine, and their nucleosides. They preferentially inhibit the action of viral polymerases. These drugs are generally activated by phosphorylation by cellular or viral kinases. Selective inhibition of viral replication results from:

- enhanced nucleoside phosphorylation.
- preferential binding of the drug to viral rather than cellular DNA polymerase.
- more rapid synthesis of DNA in infected cells.

### 6.4.4 Acyclovir (ACV)

It is a base analogue of guanosine nucleoside. It differs from guanosine nucleoside by having an acyclic hydroxyethoxy methyl as side chain instead of ribose or deoxyribose sugar (Fig. 56). The ACV is used in the treatment of diseases caused by HSV such as encephalitis and disseminated herpes etc and VZV. As ACV is not toxic to uninfected host cells, its use as prophylactic treatment to prevent recurrent outbreaks especially in immune suppressed people is permitted. A recurrent episode may be prevented if it is used within 48 h of initiation of the disease or before the onset of inflammatory responses. It can also be used for the treatment of VZV disease, but higher doses are required as it is less sensitive to ACV, and also it is less-efficiently phosphorylated by the VZV thymidine kinase.

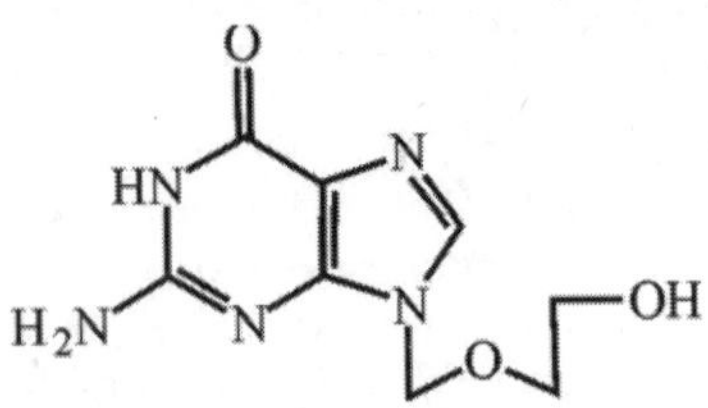


**Fig. 56** Structure of acyclovir

**Mode of action:** The viral thymidine kinase activates the drug by phosphorylation and the host cell enzymes complete the process with the formation of triphosphate (Fig. 57). As there is no initial phosphorylation in the uninfected cells, there is no active drug to inhibit cellular DNA synthesis or to cause toxicity. The ACV competes with the viral guanosine triphosphate causing termination of the growing viral DNA chain because there is no 3-hydroxyl group on the ACV molecule to allow the chain elongation, and inactivates DNA polymerase. The selectivity and minimal toxicity of ACV is also a reason for its greater use by the viral DNA polymerase than by cellular DNA polymerase. However, mutation in either the thymidine kinase or the polymerase can produce ACV-resistant strains. These strains are less virulent than the parent strain, but can still cause disease in immunocompromised patients, especially those with AIDS.

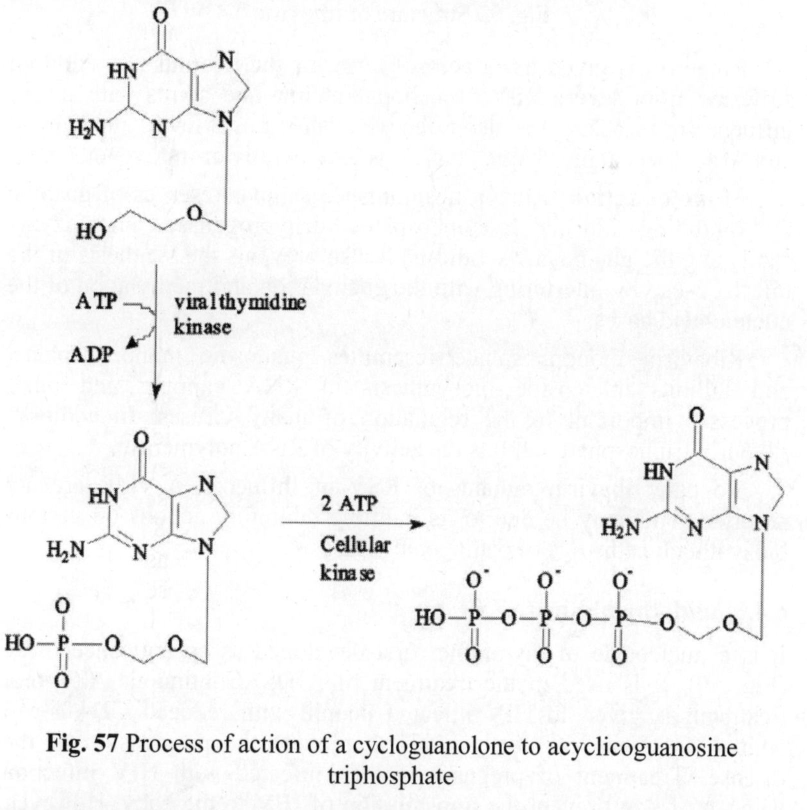


**Fig. 57** Process of action of a cycloguanolone to acyclicoguanosine triphosphate

### 6.4.5 Ribavirin

It is an analogue of guanosine nucleoside, but differs from it in having open and incomplete base ring. Similar to other nucleoside analogues, it is activated by phosphorylation. The drug is active against a broad range of viruses (Fig. 58).

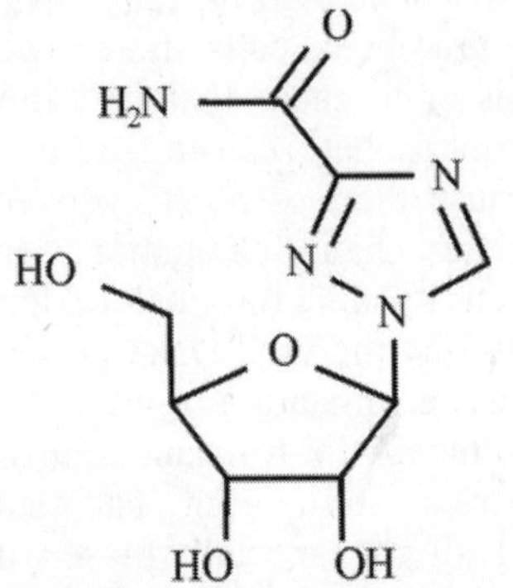

**Fig. 58** Structure of ribovirin

Ribavirin is given in an aerosol form for the treatment of children suffering from severe RSV bronchopneumonia and adults with severe influenza or measles. It is also effective against Lassa fever, and Korean and Argentina haemorrhagic fever. It is given orally or intravenously.

**Mode of action:** Ribavirin diminishes cellular reserves of guanine by inhibiting inosine monophosphate dehydrogenase, an enzyme catalysing the guainosine synthesis. It also prevents the synthesis of the mRNA 5-cap by interfering with the guanylation and methylation of the nucleic-acid bases.

Ribavirin monophosphate resembles guanosine monophosphate, and inhibits nucleoside biosynthesis in RNA capping and other processes important to the regulation of many viruses. In addition, ribavirin triphosphate inhibits the activity of RNA polymerase.

To date ribavirin mutants of RSV or influenza A virus are not reported. This may be due to its multiple inhibitory actions on various biosynthetic pathways of pathogenic viruses.

### 6.4.6 Azidothymidine (AZT)

It is a nucleoside of thymidine, first developed as an anticancer drug (Fig. 59). It is used in the treatment of AIDS. Continuous AZT oral treatment is given to HIV infected people with reduced CD4 T-cell counts (less than 500 cells / ml) to prevent the progression of the disease. Treatment of pregnant women infected with HIV infection using AZT can prevent the transmission of HIV to the baby. However, sometimes it becomes toxic and develops nausea or life-threatening bone marrow toxicity.

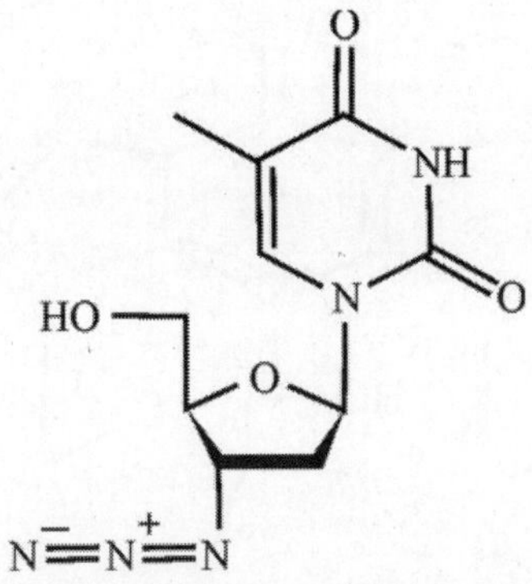

**Fig. 59** Structure of azidothymidine

High error rate of HIV polymerase creates extensive mutations and promotes the development of resistant strains. However, this problem can be overcome to some extent by the administration of multiple drug therapy.

**Mode of action:** AZT was first used in the HIV infection treatment. It acts by inhibiting the activity of the HIV enzyme, reverse transcriptase. Its action starts with its phosphorylation by the host cell enzymes. The DNA synthesis of HIV is inhibited by this drug because it lacks 3-hydroxyl group necessary for DNA chain elongation. It does not harm the host cells because the host cell DNA polymerase exhibit much lower sensitivity to the drug when compared to the sensitivity of HIV reverse transcriptase enzyme.

### 6.4.7 Ganciclovir (GCV)

It is a base analogue of guanosine. It differs from ACV by the presence of single hydroxylmethyl group in the acyclic side chain, which confers considerable activity against CMV (Fig. 60).

Though GCV is used in the treatment of retinitis, oesophagitis, colitis and pneumonia caused by CMV, its use is primarily restricted to the treatment of CMV infections in AIDS patients because the drug can cause bone marrow toxicity.

**Mode of action:** Presence of hydroxylmethyl group in the acyclic side chain confers inhibitory activity against CMV. However, CMV does not encode thymidine kinase required for initial phosphorylation of GCV, but a viral-encoded protein kinase phosphorylates GCV. Once activated by phosphorylation, GCV inhibits all herpes virus polymerases. The viral DNA polymerase has greater affinity for the drug than the cellular DNA polymerase.

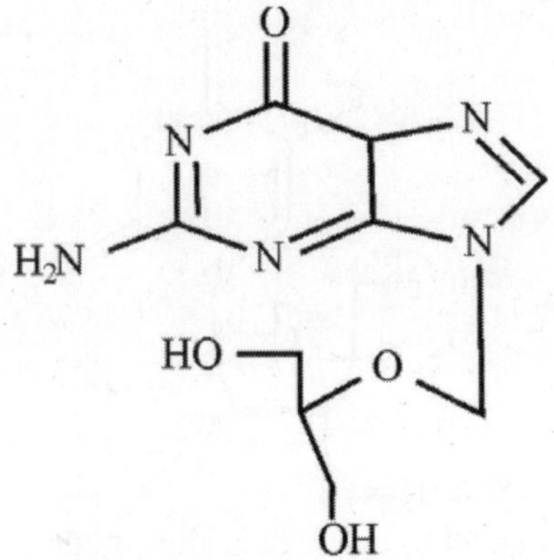


**Fig. 60** Structure of ganciclovir

### 6.4.8 Adenine Arabinoside (AraA)

It is also called as vidarabine. It is a purine nucleoside analogue identical in structure to adenosine except that arabinose is substituted for ribose as the sugar residue (Fig. 61).

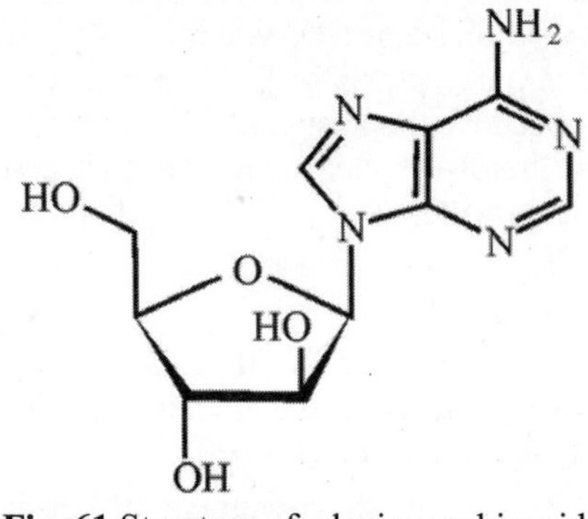


**Fig. 61** Structure of adenine arabinoside

The AraA was the principal antiviral drug used in the treatment of herpes virus infections until ACV was introduced. It is also active against VZV, but not against CMV. However, this drug is rapidly deactivated by the host cell enzymes to an inactive metabolite, which is insoluble in most of the physiological solutions. This property makes it difficult to give an effective inhibitory dose. Resistance can develop as a result of the mutation of the viral DNA polymerase.

**Mode of action:** It is phosphorylated by host cellular enzymes, specially adenosine kinase, even in uninfected cells, and thus, has a greater potential for causing toxicity than ACV. Although both cellular and viral DNA polymerases are inhibited by AraA triphosphate, the viral enzyme is 6–12 times more sensitive. It is incorporated into viral and cellular DNA.

### 6.4.9 Lamivudine

Chemically, it is called as 2-deoxy 3-thiacytidine. Structurally, it has modified sugar attached to the cytidine nucleotide. It is a base analogue of cytidine nucleotide and prevents DNA chain elongation (Fig. 62).

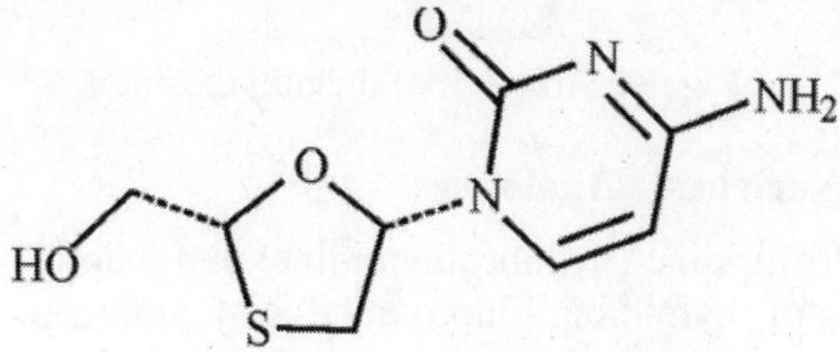


**Fig. 62** Structure of lamivudine

### 6.4.10 Dideoxycytidine

It is a base analogue of cytidine nucleoside, which lacks 3-hydroxyl group in the sugar residue. It inhibits HIV reverse transcriptase by preventing DNA chain elongation and also inhibits HIV replication (Fig. 63).

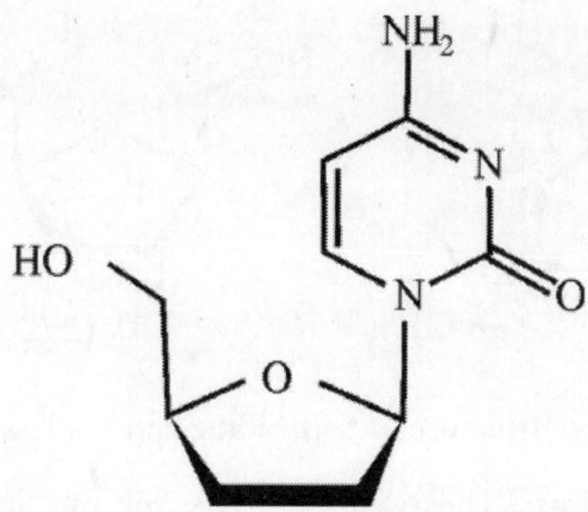


**Fig. 63** Structure of dideoxycytidine

### 6.4.11 Dideoxyinosine

It is also called as Diadenosine. It is a nucleoside analogue of adenosine that is converted to dideoxy adenosine triphosphate, and inhibits HIV reverse transcriptase (Fig. 64). All these three drugs, lamivudine, dideoxycytidine and dideoxyinosine, are used for the treatment of AIDS patient when AZT fails to control the disease.

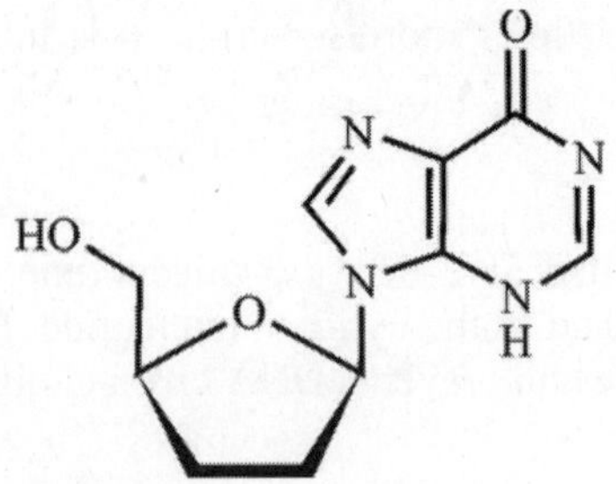

Fig. 64 Structure of dideoxyinosine

### 6.4.12 Other Nucleoside Analogues

Fluorouracil, trifluridine (trifluorothymidine) and iododeoxy uridine are base analogues of thymidine. Fluorouracil is an antineoplastic drug that kills rapidly-growing cells, but it has also been used externally for the treatment of warts caused by human papillomavirus. Iododeoxy uridine was the first drug used for humans against HSV. It was initially used externally for the treatment of herpes keratitis. However, this disease is presently treated with trifluridine (Fig. 65).

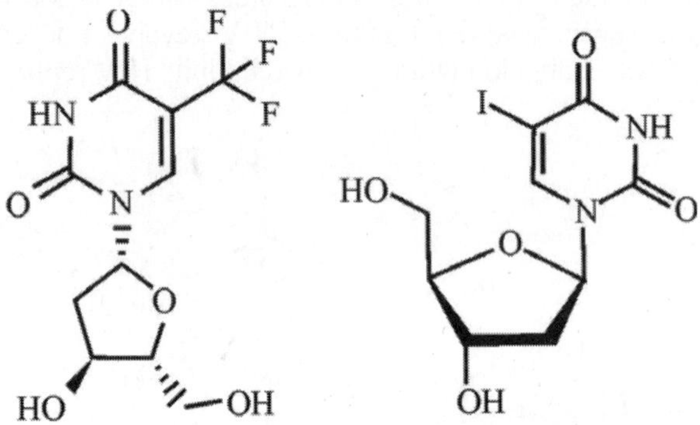

**Fig. 65 a.** Structure of trifluridine and b. Fluorouracil

**Mode of action:** These three drugs inhibit the biosynthesis of thymidine, which is an essential nucleotide for the synthesis of new DNA or replaces thymidine and becomes incorporated into the viral DNA. This action inhibits further synthesis of the virus or causes extensive misleading of the genome leading to mutation and inactivation of the virus. These drugs also target the cells in which extensive DNA replications is occurring, as in cells infected with HSV. Further investigations are still going on to discover those nucleoside base analogues having antiviral properties; as a result they can be clinically used against the herpes viruses, HBV and HIV.

## 6.5 ANTIFUNGAL ANTIBIOTICS

Fungi are eukaryotic and more similar to human cells than to the bacterial cells. Therefore, treatments of fungal infections are generally less successful than that of the bacterial infections. Many drugs that inhibit or kill fungi are; therefore, quite toxic for humans. Moreover, many fungi have a detoxification mechanism that modifies many antibiotics by hydroxylation. As a result, the added antibiotic prevents fungal growth as long as repeated application maintains high levels of unmodified antibiotics. Despite having relatively-low therapeutic index, a few drugs are useful in the treatment of some of the major fungal diseases. Effective fungal agents act either by extracting steroids or prevent their synthesis.

Fungal infections are generally divided into two types, superficial mycosis and systemic mycosis. Treatment of these two types of diseases is quite different. The following drugs are usually employed for the treatment of superficial mycosis:

- **The Drugs Containing Imidazole:** Miconazole, Ketaconazole and clotrimazole contain imidazole, and are broad-spectrum drugs used for the treatment of dermatophytic infections such as athlete's foot, oral and vaginal Candidiasis. They are available in the form of creams or solutions. They act by disrupting fungal membrane permeability and inhibit sterol synthesis.
- **Clotrimazole:** These are used in the treatment of oral and vaginal candidiasis, cryptococcal meningitis, athlete's foot and coccidioidal meningitis. There are three derivatives such as miconazole (Fig. 66) ketoconazole (Fig. 67), and clotrimazole, which are broad-spectrum agents available as creams and solutions for the treatment of dermatophytic infections. Adverse effects are uncommon.

**Miconazole** **5**

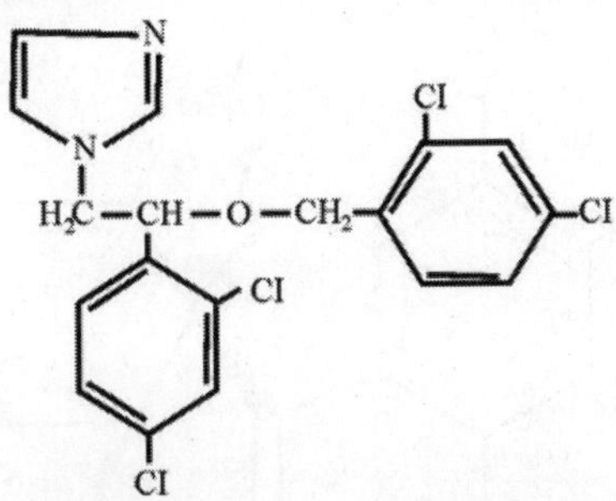


**Fig. 66** Structuire of miconazole

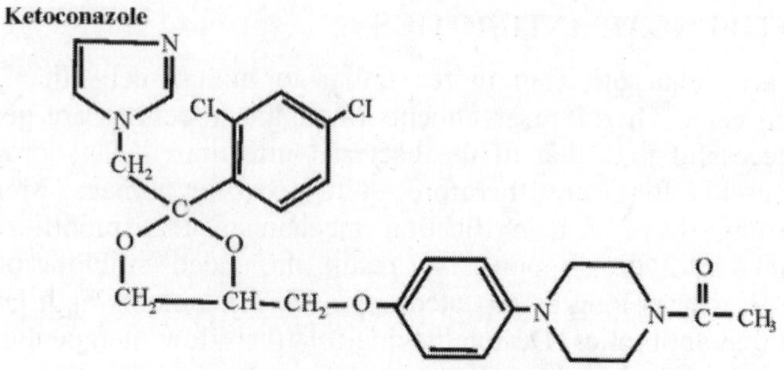


**Fig. 67** Structuire of ketoconazole

- **Nystatin:** It is polyene antibiotic produced by strain of *Streptyomyces noursei* through fermentation. It is used to control *Candida* infections of the skin, vagina, or alimentary tract. It binds sterols and damages the membrane leading to membrane leakage (Fig. 68).

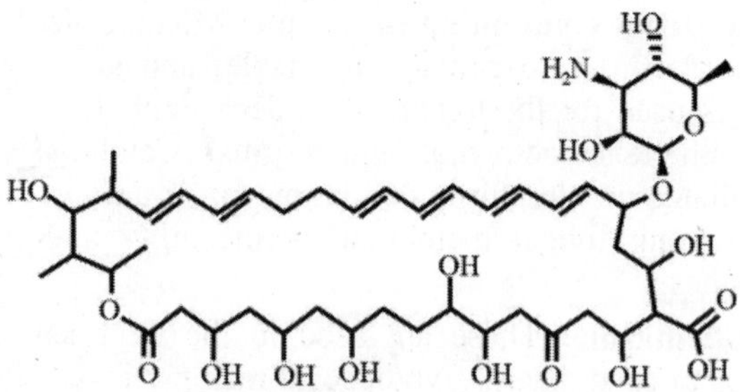


**Fig. 68** Structure of nystatin

- **Griseofulvin:** An antifungal antibiotic produced by Penicillium griseofulvum. It is given orally to treat chronic dermatophytic infections. It inhibits cell division by disrupting the mitotic spindle and, also inhibits protein and nucleic-acid synthesis (Fig. 69).

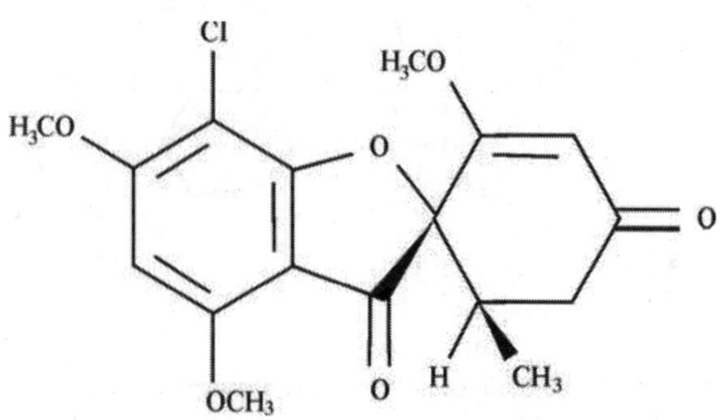


**Fig. 69** Structure of griseofulvin

Systematic fungal infections are very difficult to control and can sometimes end in deaths. The following three drugs are commonly used for treating systematic mycosis:

- **5-Flucytosine**: It is a synthetic drug generally given through oral route and is effective against many systematic diseases although resistance often develops quickly. The fungi generally convert it into 5-fluorouracil and this is incorporated in place of RNA, resulting in the disruption of RNA functioning (Fig. 70). It causes certain side effects such as liver damage, nausea, diarrhoea, skin rash etc.

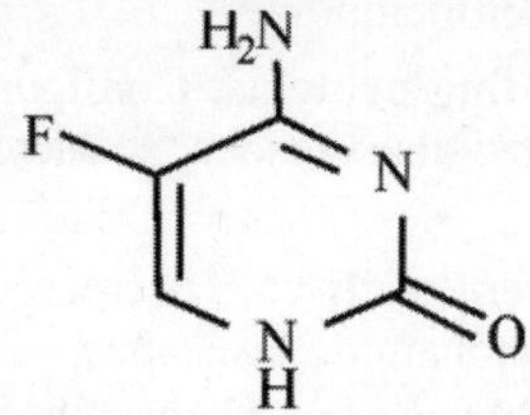


**Fig. 70** Structure of 5-flucytosine

- **Amphotericin B:** It is produced by *Streptomyces* sp. It is quite toxic and therefore, used only for serious life-threatening infections. It acts by binding the sterols in fungal membrane due to which membrane permeability is disrupted, which causes leakage of cellular contents (Fig. 71).
- **Fluconazole:** It is used in the treatment of candidiasis, cryptococcal meningitis and coccidioidal meningitis. Adverse effects are uncommon.

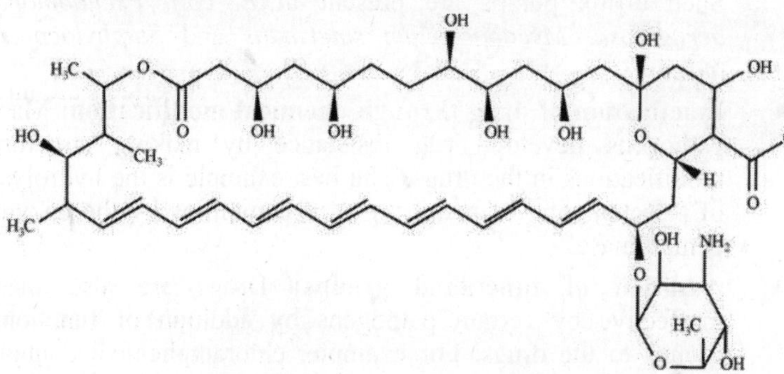


**Fig. 71** Structure of amphotericin B

## 6.6 DRUG RESISTANCE

Drug resistance is one of nature's never ending processes. The organisms develop resistance to tolerate new environmental conditions such as administration of drugs. Bacteria develop drug resistance in several different ways. It is to be noted here that a particular mechanism of drug resistance is not confined to a single group of drugs. Two different bacteria may use different mechanisms of drug resistance to the same drug or the same group of drug. Resistant mutants arise spontaneously and are then selected.

Some of the methods employed by microorganisms to develop drug resistance include:

- **Preventing the entry of drug:** Many gram-negative bacteria are unaffected by penicillin-G because it cannot penetrate the envelope of outer membrane.
- **Change in binding proteins:** Configural changes in penicillin binding proteins also render a bacterial cell resistance to penicillin.
- **Decrease in permeability:** A decreased permeability of a drug can lead to sulphonamide resistance.
- **Presence of high content of mycolic acid:** Mycobacteria develop drug resistance dueto the presence of high amount of mycolic acid in a complex lipid layer present outside the peptidoglycan layer, which is also impermeable to many drugs.
- **Pumping the drug out of the cell:** Some pathogens possess plasma-membrane-bound translocases, called efflux pumps that expel drugs out of the cell. They can pump out many different drugs because they are non-specific therefore, are called multidrug resistance pumps. They work as drug / proton anti porsters, that is, proton enters the cell as the drug leaves. Such efflux pumps are present in *E. coli, Pseudomonas aeruginosa, Mycobacterium smeymatis* and *Staphylococcus aureus.*
- **Inactivation of drug through chemical modification:** Many pathogens develop drug resistance by making structural modifications in the drugs. The best example is the hydrolysis of β-lactam ring of many types of penicillins by the enzyme penicillinase.
- **Addition of functional groups:** Drugs are also made ineffective by certain pathogens by addition of functional groups to the drugs. For example, chloramphenicol contains two hydroxyl groups (Fig 49), which can be acetylated (addition of acetyl groups) in a reaction catalysed by the enzyme chloramphenicol transacetylase with acetyl CoA.
- Similarly, aminoglycocides can be inactivated by structural modification with the acetylation of amino group, addition of

hydroxyl groups in the place of phosphates catalysed by phosphate transferases or adenyl groups catalysed by adenyl transferases.

- **Modification of target enzyme or cell organelle:** As each antimicrobial agent acts on a specific target/cell organelle, drug resistance arises when the target enzyme or organelle is modified so that it is no longer susceptible to the drug. For example, the affinity of ribosomes for erythromycin and chloramphenicol can be decreased by a change in the 23S rRNA to which they bind. Structural changes in the enzyme that uses paraaminobenzoic acid (PABA) during folic acid synthesis, bacteria become resistant to sulphonamides. Similarly, *Mycobacterium tuberculosis* becomes resistant to the drug rifampin due to mutation that alters the β-subunit of its RNA polymerase.
- **Adoption of alternate pathways:** Pathogenic bacteria may develop drug resistance by adopting an alternate pathway to bypass the existing pathway inhibited by the drug. For example, some pathogenic bacteria are resistant to sulphonamides because they use preformed folic acid instead of synthesizing it from PABA.
- **Increased production of the target metabolitic:** Some pathogenic bacteria resist sulphonamides by increased production of the folic acid.

### 6.6.1 Origin of Drug Resistance and its Transmission

Origin of drug resistance in pathogenic bacteria is related to change in genes, which are present on both bacterial chromosome and plasmids – extra chromosomal DNA molecules are generally present outside the bacterial chromosome. Pathogenic bacteria develop drug resistance because of spontaneous mutation in the bacterial chromosome, which generally, do not occur frequently. Such mutations usually result in a structural change in streptomycin receptor protein present on bacterial ribosome. However, many mutant pathogens are destroyed by natural host resistance mechanisms.

A pathogenic bacterium can also develop drug resistance due to the presence of a plasmid in it containing one or more resistant genes. Such plasmids are called resistant plasmids or R-plasmids. Plasmid resistance often codes for enzymes, which destroy or modify drugs e.g., hydrolysis of penicillin, or acetylation of chloramphenicol and aminoglycoside drugs. Such plasmid-associated resistance genes have been reported to confer resistance to a wide range of antibiotics such as cephalosporins, chloramphenicol, penicillins, aminoglycosides, erythromycin, tetracycline, sulphonamides etc. Such R-plasmids may be transferred to other bacterial cells quite rapidly in a bacterial population through normal gene-exchange processes such as conjugation, transformation and transduction. As a single plasmid may carry many resistance genes for several drugs, a pathogen population can become resistant to several antibiotics simultaneously.

The development of drug resistance can be discouraged by employing certain strategies such as:

- Chemotherapeutic drugs, especially broad-spectrum drugs should be used when definitely needed.
- Administration of two drugs simultaneously so that each drug will prevent the development of resistance to the other.
- Administration of a drug in higher concentration to destroy susceptible bacteria and most spontaneously formed mutants that may develop during treatment.
- Search for entirely new antibiotics.

## REVIEW QUESTIONS

### I. Answer the following questions

1. Briefly discuss about the mechanism of action of antimicrobial drug.
2. How do chemotherapeutic agents kill or damage the bacterial pathogens?
3. Define structural analogues, – 6 – amino penicillinic acid, penicillanase, aminoglycoside and nucleotide.
4. Discuss about the general chemical composition, structure, mechanism of action, behavioural properties and significant problems of cephalosporins.
5. Write an essay on antifungal drugs.
6. What are sulpha drugs? Give their chemical structure and mode of action against pathogens.
7. Discuss about the chemical structure and mode of action of penicillin.
8. How do different penicillins differ from one another?
9. Describe the structure and mode of action of any three antiviral drugs.

### II. Write short notes on the following

1. Bacteriostatic

2. Bactericidal
3. Aminoglycosides
4. Tetracycline
5. Vancomycin
6. Teicoplanin
7. Erythromycin
8. Base analogues
9. Ganciclovir
10. Adenine arabinoside
11. Lamivudine
12. Dideoxyinosine

## 6.7 FURTHER READING

De Clercg E 2005 Antivirals and Antiviral Strategies, Nature Reviews Microbiology **2** 704–720

Kucers A and Bennet NM 1989 The use of antibiotics: a comprehensive review with clinical symptoms, Philadelphia: Lippincott

Mandell GL, Bennet JE, and Dolin R 1995 Principles and Practice of infectious diseases, New York: Churchill Livingstone

Murray P et al 1995 Manual of Clinical Microbiology, Washington, DC: American Society for Micrbiology.

# 7

# Susceptibility Tests

## 7.1 INTRODUCTION

Bacterial pathogens isolated from clinical specimens are usually tested against selected antibiotics to assess the range of susceptibility of the pathogen and thus, to guide the appropriate antibiotic treatment of bacterial infections. Susceptibility tests are, generally, done from single pure colonies formed within 24 h on solid agar media. However, they are subjected to a further incubation period ranging from 18–24 h, though culture results may be available within 24 h after receipt of the bacterial specimen. A further period of 18–24 h incubation time is required for getting suitable sensitivity results.

In certain cases, direct sensitivity tests are done either in liquid broth with bacterial growth or specimen itself, which are setup at the same time along with the subculturing of specimen on agar plate. Although it may be possible to get quick results by direct sensitivity tests, there are certain disadvantages such as:

- The antibiotics tests may become inappropriate in controlling the pathogenic bacterium.
- It is difficult to inoculate the correct number of bacteria onto the agar surface as the inoculum.
- There is a possibility of presence of more than one type of bacteria in the inoculums (mixed inoculums), which makes it difficult to interpret the results accurately.

There are different methods to determine the susceptibility of bacteria to various antibiotics, which include:

1. **Break-point Sensitivity Tests:** A known and defined concentration of antibiotic is called as break point. This break-point concentration of antibiotic is fed into an agar plate. Bacteria are then inoculated at a particular point on the agar surface and incubated for 18–24 h. Inoculation is usually done with multipoint inoculator that permits many different strains of bacteria to be tested on the same plate. Bacteria that are resistant to break-point concentration of the antibiotic will

grow and form colonies while sensitive ones will not grow. A control agar plate with number of added antibiotics is also set up for checking growth of viable bacteria (Fig. 72).

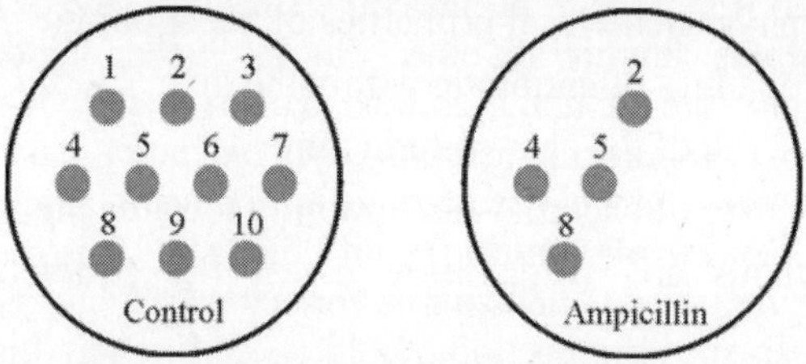


**Fig. 72** Break-point sensitivity tests: Control sensitivity agar plate with no added antibiotic; Agar plate incorporating ampicillin; Strains 1, 3, 6, 7, 9, and 10 are sensitive to ampicillin; Strains 2, 4, 5, and 8 are resistant

With the help of multi point inoculator it is possible to test more than 30 different strains of bacteria against a wide range of antibiotics in one batch. Therefore, this process is rapid and less time consuming when compared to other methods, especially disk tests. Sometimes, the same antibiotic may be used at two different concentrations in separate agar plates e.g., 1 and 4 mg/litre concentrations. This facilitates for recognizing bacteria into sensitive or moderately sensitive or resistant types.

2. **Disk sensitivity tests:** A small and spherical disk of filter paper, with appropriate size, pre-impregnated with a known concentration of antibiotic is used in this process. The disk is placed on the surface of an agar plate, which has already been inoculated with a suspension of sample bacteria. The plates are then incubated for a period of 18–24 h. During this period, the antibiotic present in the disk diffuses out of the disk into the agar, along a concentration gradient. If the inoculated bacterial strain is sensitive to the antibiotic, a zone of inhibition is formed around the disk (Fig. 73).

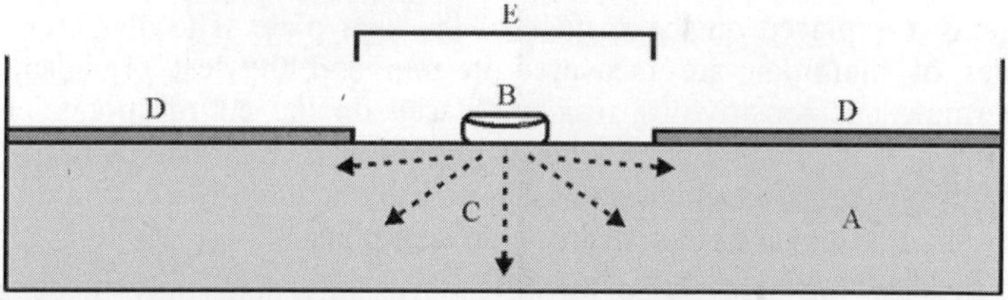


**Fig. 73** Disk sensitivity test: A- Agar; B- Antibiotic disc; C- Antibiotic diffused into agar along concentration gradient; D- Bacterial growth on surface of agar after 18 h of incubation; E- Zone (diameter) of inhibition

The diameter of the inhibitory zone depends upon a number of factors such as:

- The concentration of bacteria in the inoculum
- The physicochemical properties of the antibiotic
- The quantity of antibiotic within the disk
- The depth of the agar plate in mm
- The range of bacterial susceptibility towards the antibiotic

Two methods are available to assess the antibiotic sensitivity pattern:

**1. The strokey's method or comparative disk test:** Control and test organisms are used in this method. Both of them are inoculated onto the surface of an agar plate. The control organism of known sensitivity to the antibiotic being tested is used. With the help of the diameter of inhibitory zone around test and control organisms, it is possible to compare the sensitivity of these two organisms against the antibiotic (Fig. 74).

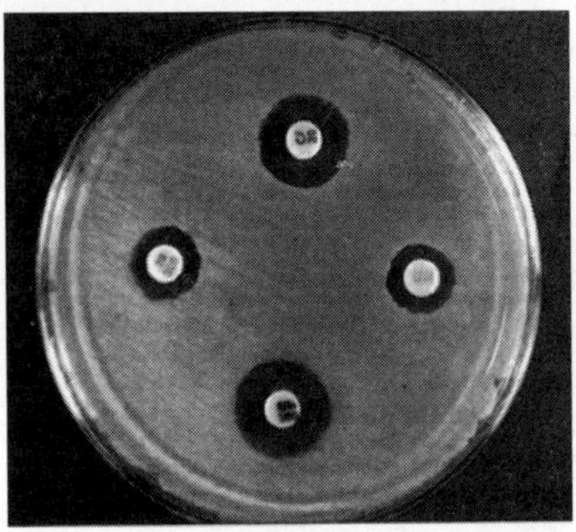

**Fig. 74** Comparative disk test (strokey's method)

**2. Standardized disk testing:** In this process, carefully standardized agar plates and the standardized inoculum of the test organism is placed on the surface of the agar plate. The diameters of zones of inhibition are measured in mm and the test organism is determined as sensitive or resistant based on the cut-off points. For example, if the diameter of zone of inhibition is less than 18 mm, the organism is declared as resistant. It is possible to employ a maximum of only six different antibiotic disks on an agar plate.

**3. Test for MIC:** Test for MIC is usually done only in certain instances when a very precise determination of the *in vitro* susceptibility of a bacterial strain is required e.g., in the treatment of streptococcal endocarditis or pneumococcal meningitis. This test is also performed to assess the overall range of susceptibility of an antibiotic against different strains of the same bacterial species, particularly when evaluating or developing new antibiotic agent. The relative action of an antibiotic against a single or group of organism is generally expressed with the terms MIC50 or MIC100. They signify the lowest concentration of an antibiotic, which is capable of inhibiting 50% or 90% of the bacterial cells tested respectively.

The MIC is determined conventionally, by employing a series of doubling dilutions of the antibiotic in a liquid culture medium to produce a range of concentrations in test tubes called macrodilutions or in a microtiter tray, called microdilutions. Bacterial test strains are inoculated into all these test tubes with series of antibiotic concentrations and incubated for 18–24 h. Bacterial growth is assured by visible turbidity after incubation period. The MIC is the lowest concentration of the antibiotic where there is no bacterial growth (Fig. 75).

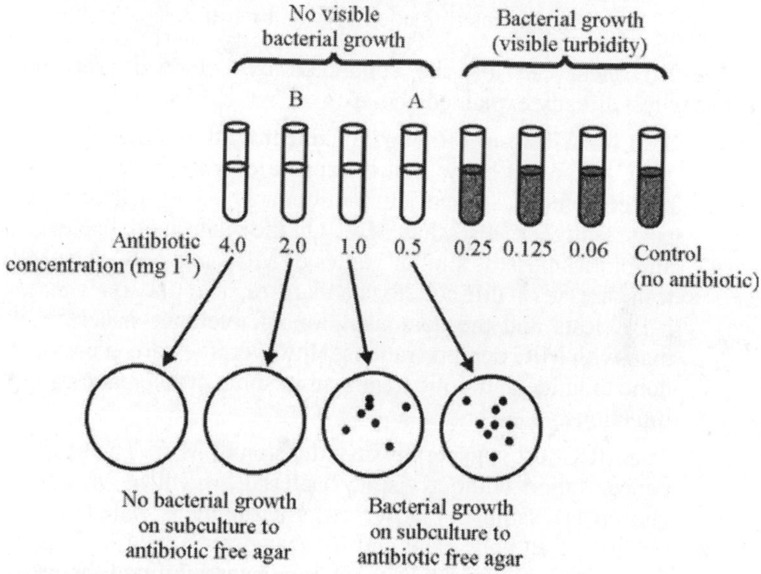


**Fig. 75** MIC and MBC testing: The minimum inhibitory concenbtration (MIC) of the antibiotic is 0.5mg/l (Tube A); the minimum bactericidal concentration (MBC) is 2.0 mg/l (Tube B)

An alternative method for determining MIC is by the use of commercially-available E-test strips. These are the specialized antibiotic-impregnated strips. Concentration gradient is marked within the strip and numerical gradations are marked along the margins of the strip to indicate MIC. These E-test strips are placed on the surface of the agar plates, which are subsequently incubated for 18–24 h. During incubation, antibiotic diffuses into the agar forming a zone of inhibition.

The MIC is determined by measuring the point at which the edge of the zone of inhibition crosses the E-test strip (Fig. 76).

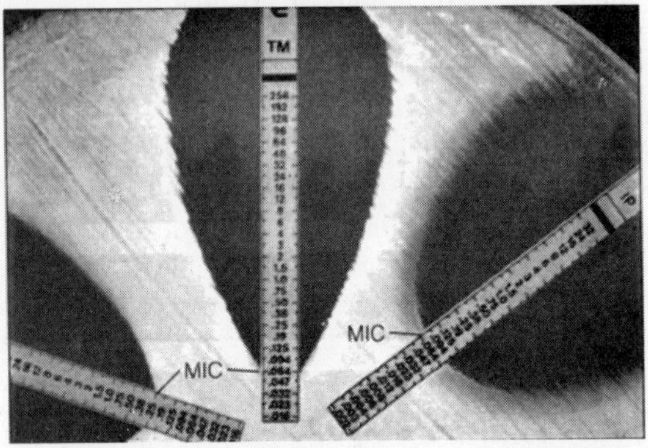


**Fig. 76** Determination of MIC by E-test

The MIC test can also be conducted by extended break-point sensitivity test already explained above.

a. **Test for Minimum Bacterial Concentration (MBC):** It is the MBC of an antibiotic that is capable of killing 99% of tested bacterial cells. Some antibiotics are bacteriostatic, which possess higher MBC than MIC. On the other hand, bactericidal antibiotics possess similar values of MBC and MIC. The MBC tests are very difficult to standardize, and the relevance of MBC tests and the demonstration of tolerance is less useful than with MIC demonstrations. However, they are occasionally done to guide antibiotic treatment in some problematic cases of infections.

   The MBC test is generally an extension of MIC. The antibiotic concentration without visible bacterial growth in MIC test is chosen as a sample of MBC test. The sample is plated on to the surface of an agar plate and incubated for 18–24 h. This will determine whether the bacteria have been inhibited the growth even though they are viable or killed.

b. **Tests for Detection of Bacterial Resistance Mechanisms:** These tests help in guiding suitable antibiotic treatment. These tests can be used to predict the results of conventional sensitivity tests especially, when a specific resistance mechanism is detected. These tests generally do not require overnight incubation and therefore, the results may be available at an earlier stage to guide antibiotic therapy. Some common examples of detection of bacterial resistance are given in the Table 18 and Table 19.

**Table 18:** Detection of bacterial resistance mechanisms aganist drugs

| **Resistance mechanism** | **Antibiotic** | **Genetic basis of resistance** | **Mechanism** |
|---|---|---|---|
| Reduced permeability | Penicillins | Chromosomal | *Pseudomonas aeruginose*<br>Enteric Bacteria |
| Inactivation of antibiotics (e.g., penicillinase; modifying enzymes methylases, acetylases and phosphroylases; and others) | Penicillins<br><br>Chloramphenicol<br><br>Aminoglycosides | Plasmid and chromosomal<br><br>Plasmid and chromosomal<br><br>Plasmid | *Staphylococcus aureus*<br>Enteric Bacteria<br>*Neisseria gonorrhea*<br>*Staphylococcus aureus*<br>Enteric Bacteria<br>*Staphylococcus aureus*<br>*Staphylococcus aureus* |
| Alteration of target (e.g., RNA polymerase, rifamycin; ribosome, erythromycin and streptomycin; DNA gyrase, quinolones) | Erythromycin<br>Rifampiycin<br>Streptomycin<br>Norfloxacin | Chromosomal | Enteric Bacteria<br>Enteric Bacteria<br>Enteric Bacteria |
| Development of resistant biochemical pathway | Sulphonamides | Chromosomal | *Staphylococcus aureus*<br>Enteric Bacteria |
| Efflux (pumping out of cell) | Tetracyclines<br>Chloramphenicol | Plasmid<br>Chromosomal | *Staphylococcus aureus*<br>Enteric Bacteria |

**Table 19:** Detection of bacterial resistance mechanisms

| **Resistance mechanisms** | **Organisms** | **Method of detection** | **Comment** |
|---|---|---|---|
| β-Lactamase production | *Haemophilus influenzae* | Rapid 'stick' test (Hydrolysis of nitrocefin) | Predicts resistance to ampicillin and amoxicillin |
| | *Neisseria gonorrhoeae* | Rapid 'stick' test (Hydrolysis of nitrocefin) | Predicts high level resistance to ampicillin |
| Methicillin resistance (altered PBP2') | *Staphylococcus aureus* | Latex agglutination for PBP2'<br>Detection of mec A gene by PCR | Predicts resistance to β-lactam antibiotics (MRSA) |
| Rifampicicn resistance | *Mycobacterium tuberculosis* | Detection of rpo B gene mutations by PCR | Detects 95% of rifampicicn-resistant *M. tuberculosis* strains |

PBP: Penicillin-binding Protein; MRSA: Metahcillin-resistant *Staph. Aureus*

**4. Clinical relevance of *in vitro* antibiotic sensitivity test:** It should be emphasized that *in vitro* antibiotic sensitivity tests are to be taken only as a guide to decide the suitable antibiotic treatment. If a laboratory report indicates that organism X is sensitive to antibiotic Y, it does not mean that Y will not work on X and *vice versa.* Though *in vitro* tests are designed to reflect the *in vivo* situation, they can never be taken account of the human and bacterial biological variables.

While interpreting laboratory reports of sensitivity test results, a number of factors are to be considered:

- An antibiotic with clear *in vitro* activity may not work clinically, because there are many pharmacokinetic and other factors to consider in choosing most appropriate antibiotic treatment.
- The organism identified in the laboratory tests may not be the primary organism causing the infection.
- Many infections subside spontaneously. If a patient has already responded positively and clinically to an antibiotic treatment then it is not necessary to change the antibiotic even though the laboratory test indicates that the organism is resistant one.

## REVIEW QUESTIONS

**I. Answer the following questions**

1. Discuss different methods of disk sensitivity tests.
2. What is minimum inhibitory concentration? How does it differ from minimum bacterial concentration? Explain.
3. What factors are to be considered while interpreting sensitivity results of laboratory?
4. Describe structure and mode of action of any three antiviral agents.
5. Give an account of disk diffusion test for antimicrobial drug susceptability.

**II. Write short notes on the following:**

1. Break-point sensitivity test
2. Detection of bacterial resistance mechanism
3. Base analogues
4. Ganciclovir
5. Adenine arabinoside
6. Lamivudine
7. Dideoxyinosine

8. E-test
9. Importance of antimicrobial susceptibility test
10. Agar diffusion technique
11. Describe the Kirby-Bauer method of testing the susceptibility of antimicrobial drug.
12. Discuss the importance of antimicrobial drug susceptibility.
13. Describe broth dilution technique.

## 7.2 FURTHER READING

De Clercg E 2005 Antivirals and antiviral strategies, *Nature Reviews Microbiology* 2 704–720

Harbarth S and Samore MH 2005 Antimicrobial resistance determinants and future control, *Emerging Infectious diseases* 11 794–801

Walsh FM and Amyes SGB 2004 Microbiology and drug resistance mechanism of fully resistant pathogens, *Current Opinion in Microbiology* 7 439–44.

# 8

# Preventive Control of Diseases

## 8.1 INTRODUCTION

Humans are constantly subjected to different types of diseases caused by different types of microorganisms such as bacteria, viruses, fungi and protozoa. Most of the bacterial diseases of humans can be successfully treated and controlled by chemotherapeutic drugs called antibiotics. Similar chemotherapeutic agents are also available for the treatment and control of most parasitic and fungal diseases. However, such chemotherapeutic agents are not available for the treatment of many viral diseases and also certain bacterial diseases. But they can be prevented from manifestation and spread among human population by following certain preventive measures called as immunization, which is carried out by the use of vaccines. Therefore, 'prevention is better than cure' holds good as far as viral diseases are concerned. Such vaccines are presently available for the prevention of some important and life-threatening viral diseases such as rabies, hepatitis-B, polio, smallpox, Japanese encephalitis, influenza, diphtheria, tetanus, pertusis etc.

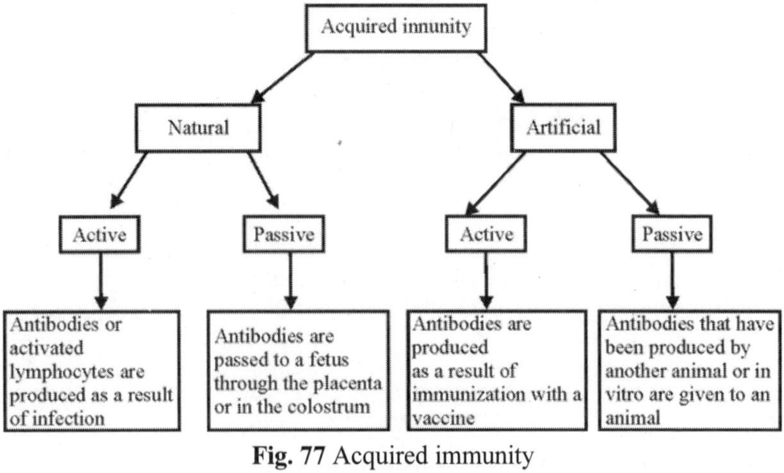


**Fig. 77** Acquired immunity
(Active and passive immunity)

## 8.2 ACTIVE AND PASSIVE IMMUNIZATION

Any process that helps to develop resistance in a host (immunity) to a specific disease is called immunization. There are two types of immunization – active and passive (Fig. 77).

### 8.2.1 Passive Immunization

A process in which rapid temporary protection or treatment of a person is carried out by injecting purified antibody or antibody containing serum is called as passive immunization. Passive immunization is used:

- To prevent a disease after a known exposure e.g., needle stick injury with HBV-contaminated blood.
- To ameliorate the symptoms of an ongoing disease.
- To protect immunosuppressed patients.
- To block the action of bacterial toxins and prevent the diseases they cause.

Immune serum globulin preparations derived from seropositive humans or animals (horses) are available as prophylaxis for several bacterial and viral diseases (Table 20). Human serum globulin is prepared from pooled plasma and contains normal stock of antibodies for an adult. Human Ig is preferred to animal Ig because of the attenuated reduced risk of a hypersensitivity reaction. Such Ig preparations are available for HBV (HBIG), VZV (VZIG), rabies (RIG) and tetanus (TIG) (Table 21).

**Table 20:** Immune globulins and antitoxins used for immunization against specific organisms and diseases

| Organism/Disease | Agent |
|---|---|
| Black widow spider bite | Horse antivenom |
| Snake bite | Horse antivenom |
| Acute respiratory failure (RSV) | Monoclonal antibodies |
| Botulism | Botulinum antitoxin (equine) |
| Diphtheria | Diphtheria antitoxin (equine) |
| Hepatitis A (Hep a) and Hep B | Pooled human immune gamma globulin |
| Measles | Pooled human immune gamma globulin |
| Rabies | RIG administered around the wound in addition to being injected intramuscularly |
| Tetanus | TIG |
| Eczema | Vaccinia immune globulin |
| Immunocompromised individuals | VZIG |

* Immuneglobulins and antitoxins are administered intramuscularly unless otherwise noted.

**Table 21:** Immune globulins available for postexposure prophylaxis*

| Disease | Source |
|---|---|
| Hep A | Human |
| Hep B | Human† |
| Measles | Human |
| Rabies | Human† |
| Chickenpox, zoster | Human† |
| Tetanus | Human, † equine |
| Botulism | Equine |
| Diphtheria | Equine |

*Immune globulins to other agents may also be available.

† Specific, high-titre antibody is available and is the preferred therapy.

### 8.2.2 Active Immunization

This is a process in which an immune response and immunological memory are stimulated in response to an immunogen or an antigen. Active immunization can be achieved with the use of either inactivated or live vaccines.

**Immunization with inactivated vaccines:** Inactivated vaccines are used to confer protection against most bacteria and the virus that may be highly virulent or cause recurrent infection or develop cancer. They can be produced through chemical treatment (formation) or through heat inactivation of bacteria, bacterial toxins or viruses or through the purification of components or subunits of the disease-causing agents.

These vaccines are generally given along with an adjuvant (A substance when injected together with an antigen increases antibody production) such as alum, which stimulates their immunogenicity. The adjuvants also influence the type of immune response produced by the vaccine [T-helper 1 (TH1) or T-helper 2 (TH2)]. Attenuated forms of Cholera Toxin (CT) and *E. coli* lymphotoxin are being developed as possible adjuvants that promote sIgA production after intranasal or oral immunization.

Immunity is not, generally, life long because it is humoral and not cell-mediated, and does not give a local IgA response. This immunity requires booster doses and a large dose of inactivated vaccine. Inactivated vaccines are generally safe, except in people who possess allergic reactions to vaccine components. For example, many vaccines that are produced in eggs are allergic. Therefore, these vaccines cannot be given to those people who are allergic to eggs (Table 22).

**Table 22:** A comparison of inactivated (killed) and attenuated (live) vaccines

| Major characteristics | Inactivated vaccine | Attenuated vaccine |
|---|---|---|
| Booster shots | Multiple booster required | Only a single booster |
| Production | Virulent microorganisms inactivated by chemicals or irradiation | Virulent microorganisms grown under adverse condition or passed through different hosts until avirulent |
| Reversion tendency | None | May revert to a virulent form |
| Stability | Very stable, even where refrigeration is unavailable | Less stable |
| Type of immunity induced | Humoral | Humoral and cell-mediated |

**Table 23:** Properties of live and inactivated vaccines

| Property | Live | Inactivated |
|---|---|---|
| Route of administration | Natural or injection | Injection |
| Dose of virus cost | Low | High |
| Number of doses | Single | Multiple |
| Need for adjuvant | No | Yes |
| Duration of immunity | Long term | Short term |
| Antibody response | IgG, IgA | IgG |
| Cell-mediated immune response | Good | Poor |
| Heat lability in tropics | Yes | No |
| Interference | Occasional | None |
| Side effects | Occasional mild symptoms | Occasional |
| Reversion to virulence | Rarely | None |

There are three major types of inactivated bacterial vaccines (Fig. 23):

1. Toxoid
2. Inactivated bacteria
3. Protein subunits of the bacterial capsule (Table 24)

Similarly, inactivated vaccines are also available for polio, Hep A, rabies, influenza and other viruses.

**Table 24:** List of bacterial vaccines

| Bacterial name (disease) | Vaccine components | Indications |
|---|---|---|
| *Corynebacterium diphtheriae* (diphtheria) | Toxoid | Children and adults |
| *Clostridium tetani* (tetanus) | Toxoid | Children and adults |
| *Bordetella pertussis* | Killed cell or cellular components | Children |
| | | Children |

| Bacterial name (disesase) | Vaccine components | Indications |
|---|---|---|
| (pertussis) *H.influenzae* B (Hib; Haemophilus disease) | Capsule polysaccharide; capsule polysaccharide-protein conjugate | |
| *N. meningitidis* A and C (meningococcal disease) | Capsule polysaccharide | People at high risk (e.g., those with asplenia), travellers to epidemic areas (e.g., military personnel), children |
| *S.pneumoniae* (pneumococcal disease) | Capsule polysaccharide | People at high risk (e.g., those with asplenia) |
| *Vibrio cholerae* (cholera) | Killed cell | Travellers at risk to exposure |
| *S. typhi* (typhoid) | Killed cell, polysaccharide | Travellers at risk to exposure, household contacts, sewage workers |
| *Bacillus anthracis* (anthrax) | Killed cell | Handlers of imported fur |
| *Yersinia pestis* (plague) | Killed cell | Veterinarians, animal handlers |
| *Francisella tularensis* (tularemia) | Live attenuated | Animal handlers in endemic areas |
| *Coxiella burnetii* (Q fever) | Inactivated | Sheep handlers, laboratory personnel working with *C.burnetii* |
| *Mycobacterium tuberculosis* | Live attenuated Bacilli Calmette-Guerin (BCG) (*M.bovis*) | Not recommended in US |

A subunit vaccine can be developed after identification of the bacterial or viral components that express protective immune response. In the process of subunit vaccine preparation, either the immunogenic component is isolated from the bacterium-infected cells or virus-infected cells by biochemical methods or the vaccine is prepared through genetic engineering involving the expression of cloned viral gene in bacteria or eukaryotic cell. For example, the HBV subunit vaccine was initially prepared from surface antigen obtained from human sera of chronic carriers of the virus. The antigen is purified, chemically treated and absorbed on to alum, and then used as vaccine.

Subunit vaccines against *Haemophilus influenzae B, Neisseria meningitidis, Salmonella typhi* and *Streptococcus pneumoniae* are prepared from capsular polysaccharides.

**Immunization with live vaccines:** Live vaccines are prepared using organisms with limited ability to cause disease (avirulent or attenuated). These vaccines are used to protect people against viral diseases, especially those caused by enveloped viruses, which require T-cell immune responses for the infection to develop. Immunization with a live vaccine resembles the natural infection, and produces both humoral and cellular immunity. These vaccines usually generate long-lived (generally lifelong) immunity against a viral disease because they initiate TH1 and TH2 immune responses. This again depends upon the type and administration. However, there are certain problems with live vaccines, such as:

- The vaccine virus is still being dangerous and causes disease in immunosuppressed people or pregnant woman – who do not have immunological resources even for a weakened virus infection.
- The vaccine may revert to a virulent viral form. However, this rarely happens.

Vaccine for typhoid (with attenuated *S.typhi* strain Tv21a), BCG vaccine (with attenuated strain of *M.bovis*) and tularemia vaccine belong to bacterial live vaccines. Smallpox vaccine (with attenuated vaccinia virus), Sabina vaccine for polio (consists of three types of attenuated polioviruses) and the Measles, Mumps and Rubella (MMR) vaccine (vaccine with attenuated viruses of the three diseases) belong to viral live vaccines (Table 28) Recommended course and indications for vaccination of children are given in the Table 26. One has to take the following precautions for immunization (Table 25).

**Table 25:** Precautions to be taken in vaccine use

| **Problems with Vaccine Use** |
|---|
| • Live vaccine can occasionally revert to virulent forms.<br>• Interference by other organisms may prevent the infection produced by a live virus vaccine e.g., rubella prevents poliovirus replication.<br>• Vaccination of an immunocompromised person with a live vaccine can be life threatening.<br>• Side effects to vaccination can occur; these include hypersensitivity or allergic reaction to the antigen, to non-microbial material in the vaccine, or to contaminants (e.g., eggs).<br>• Vaccine development and liability insurance for the manufacturer are very expensive.<br>• Organisms with many serotypes are difficult to control by vaccination. |

**Table 26:** Vaccination course schedule to children

| Age vaccine | Birth | 1 mo | 2 mo | 4 mo | 6 mo | 12 mo | 15 mo | 18 mo | 24 mo | 4–6 y | 11–12 y | 14–16 Y |
|---|---|---|---|---|---|---|---|---|---|---|---|---|
| H B | | Hep B | Hep B | | Hep B | | | | | | Hep B | |
| Diphtheria, Tetanus, Pertussis | | | DTaP | DTaP | DTaP | | DTaP | | | DTaP | Td | |
| *Haemophilus influenzae* type b (Hib) | | | Hib | Hib | Hib | | Hib | | | | | |
| Polio | | | Inactivated Polio Vaccine (IPV) | IPV | | | IPV | | | IPV | | |
| MMR | | | | | | | MMR | | | MMR | MMR | |
| Chickenpox (Varicella) | | | | | | | varicella | | | | varicella | |
| Hep A | | | | | | | | | | | Hep A in selected areas | |

mo-Months; Y-Years; DTap- diphtheria, tetanus;Td- tetanus-diphtheria

Live virus vaccines consist of type virus or virus mutants (attenuated) or virus from other species that share antigenic determinants or genetically-engineered viruses lacking virulence properties.

Wild type viruses are usually attenuated by growth in embryonated eggs or tissue culture cells at non-physiological temperature (32–34ºC) and away not from the selective pressures of the host immune response. These conditions select or allow the growth of viral strains (mutants) (Table 27):

1. That are less virulent because they grow poorly at 37ºC e.g., measles vaccine
2. Do not replicate in any human cell (host range mutants)
3. Can replicate at benign site, but do not spread in the body
4. Bind or replicate in the target tissue characteristically affected by the disease e.g., polio vaccine
5. Replicates in the gastrointestinal tract, but does not reach or infect the brain.

**Table 27:** Purified macromolecule vaccines for human use

| Type of purified macromolecule (disease or microorganism) | Form of vaccine |
|---|---|
| Capsular polysaccharide | |
| Hib | Polysaccharide-protein conjugate (HbCV) or bacterial polysaccharide (HbPV) |
| *Neisseria meningitidis* | Polysaccharides of serotypes A/C/Y/W-135 |
| *Streptococcus pneumoniae* | 23 distinct capsular polysaccharides |
| Surface antigen | |
| Hep B | Recombinant surface antigen (HbsAg) |
| Toxoids | |
| Diphtheria | Inactivated exotoxin |
| Tetanus | Inactivated exotoxin |

List of currently-used live viral vaccines are given in Table 28.

**Table 28:** List of currently-used viral live vaccines

| Virus | Vaccine components | Indications |
|---|---|---|
| Polio | Inactivated (IPV, Salk vaccine) | Children |
| | Attenuated (Oral Polio Vaccine (OPV), Sabin vaccine) | Children |
| Measles | Attenuated | Children |
| Mumps | Attenuated | Children |
| Rubella | Attenuated | Children |
| Varicella zoster | Attenuated | Children |
| Influenza | Inactivated | Adults, especially medical personnel and the elderly |
| Hep B | Subunit | Newborns, healthcare workers, high-risk groups (e.g., promiscuous people, intravenous drug users) |
| Hep A | Inactivated | Children, childcare workers, travellers to endemic areas, Native Americans and Alaskans |
| Adenovirus | Attenuated | Military personnel |
| Yellow fever | Attenuated | Travellers at risk to exposure, military personnel |
| Rabies | Inactivated | Post exposure<br>Pre exposure: veterinarians, animal handlers |
| Smallpox | Vaccinia | No longer necessary |

| Virus | Vaccine components | Indications |
|---|---|---|
| Japanese encephalitis | Inactivated | Travellers at risk to exposure |
| Eastern, Western, Russian spring-summer encephalitis viruses | Inactivated | Military personnel |

## 8.3 VACCINES – NATURAL AND RECOMBINANT

Vaccines are suspensions of killed or modified pathogenic micro-organisms or specific fractions isolated from the microorganisms that when injected into an animal, produce immunity to a particular disease. Vaccines are classified into two types based on the nature of the suspension of microorganisms (Table 22) and genetic manipulations induced in them – natural vaccines and recombinent vaccines.

**Table 29:** Characteristic features of a good vaccine

- Organism causes significant illness.
- Organism exists as only one serotype.
- Antibody blocks infection or systemic spread.
- Organism does not have oncogenic potential.
- Vaccine is heat stable; hence, it can be transported to endemic areas.

### 8.3.1 Natural Vaccines

Preparations used for inducing passive and active immunization belong to natural vaccines. Refer to **Chapter 11** for details about immunization.

### 8.3.2 Recombinant Vaccines

Genetic engineering can be applied in many different ways for the production of vaccines. Such vaccines are called recombinant vaccines. A pathogenic genome can easily be modified by recombinant DNA technology because *in vitro* genetic techniques are more precise and more powerful than traditional *in vivo* techniques. For example, in some cases, one can simply delete genes involved in virulence, but leave those genes whose products express an immune response giving a recombinant live attenuated vaccine. By using recombinant DNA technology, it is possible to add genes to a virus that will specifically confer immunity to a host against a viral disease.

Vaccinia virus can be used as a vector for the preparation of recombinant vaccine that can be used against small pox disease. Cloning in vaccinia virus is done using an *E.coli* plasmid containing a fragment of the vaccinia virus – thymidine kinase genes. An appropriate foreign

DNA is inserted into the plasmid, and the recombinant plasmid is transformed into a host cell whose own thymidine kinase gene is inactive, but that has previously been infected with wild-type vaccinia virus. If homologous recombination occurs between plasmid DNA and vaccinia virus DNA, recombinant virus can be obtained containing an inactivated thymidine kinase gene. An active thymidine kinase leads to growth inhibition by the compound 5-bromo deoxy uridine. Therefore, recombinant vaccinia virus can be selected by allowing viral replication to occur in the presence of this inhibitor (Fig. 78). Although such recombinant viruses do not express thymidine kinase, they can still infect human cells and they do express the foreign genes cloned into them. It is reported that some recombinant vaccinia viruses can carry genes from four different viruses.

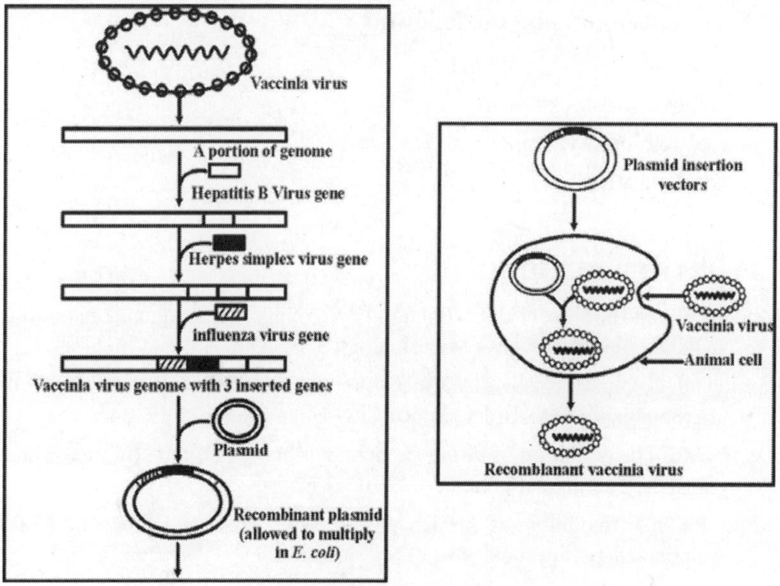


**Fig. 78** Recombinant vaccines

**DNA Vaccines:** DNA vaccines are also a kind of recombinant vaccines. They would be safe and inexpensive unlike viral vaccines. They would also escape surveillance by the host immune system.

Though vaccines of various types have been successful against a wide variety of infectious diseases, there are other important diseases for which production of vaccine is difficult such as malaria and AIDS. Though genetic engineering techniques might yield effective vaccines against these diseases, trials are being made to develop DNA vaccines to control these diseases. DNA vaccines use the genetic material of the pathogen itself to immunize. This genetic material may be in the form of defined fragments of the pathogen's genome or genes from the pathogen cloned into plasmid or viral vector. In the later case, DNA or even RNA is used in the vaccine. A number of such DNA vaccines are undergoing clinical trials and many more are being developed.

## REVIEW QUESTIONS

### I. Answer the following questions

1. Differentiate between active and passive immunization.
2. Discuss the process of immunization with inactivated vaccines.
3. What are live vaccines? Describe immunization process with live vaccines.

### II. Write short notes on the following

1. Vaccine
2. Natural vaccines
3. Recombinant vaccines
4. DNA vaccines

## 8.4 FURTHER READING

Aktories K and Barbieri JT 2005 Bacterial cytotoxins: Targeting and eukaryotic switches, Nature Reviews Microbiology **3** 397–340

Casadevall A and Pirofski LA 2003 The damage response framework of microbial pathogenesis, Nature Reviews Microbiology **1** 17–24

Day T 2002 The evolution of virulence in vector borne and directly transmitted parasites, Theoretical Population Biology **62** 199–213

Ewald P 2004 Evolution of virulence, Infectious Disease Clinics of North America **18** 1–15

Stevens JM, Galyor EE, and Stevens MP 2006 Actin dependent movement of bacterial pathogens, Nature Reviews Microbiology **4** 91–101

Waldvogel FA 2004 Infectious diseases in the 20$^{th}$ century old challenges and new opportunities, International Journal of Infectious Diseases **8** 5–12.

# 9

# Diseases of Man

## 9.1 AIR-BORNE DISEASE

Air is a dynamic medium for transmission of the pathogens especially, disease-causing microorganisms. Most of the air-borne pathogenic microorganisms cause respiratory tract diseases. The infections caused by these organisms tend to occur in epidemic form, appearing explosively and attacking large number of people within a short time. The disease incidence generally increases during rainy and winter seasons when people usually occupy crowded quarters. The causative organisms occur in secretions of the nose and throat of infected individuals, and can be transmitted directly to healthy individuals by aerosols. Aerosols are fine droplets that remain suspended in air for a time, which are generated by sneezes, cough or even talking. Microorganisms that cause respiratory tract infections can also be transmitted indirectly through fomites such as drinking glasses, eating utensils and handkerchief that have recently been used by an infected person.

Aerosols generated from nasal and throat discharges of a patient can settle out rapidly from air on various surfaces such as clothes, bed and bed sheets used by a patient, and become dry. Disturbance of these dried aerosols by bed making, handling patient's clothes, handkerchiefs, and even sweeping a floor in the patient's room can generate dust particles, which add microorganisms to the circulating air. The survival of microorganisms for relatively longer periods in dust creates a significant health hazard, particularly in hospital areas, and can contribute to nosocomial diseases (disease acquired in hospitals). Tubercle bacilli have been isolated from the dust of sanitoria, and diphtheria bacilli and haemolytic streptococci have been found in floor dust near the patients or carriers harbouring these microorganisms. The common air-borne diseases are infections of nose, throat, and lungs etc, the organs associated with respiratory system. The air-borne pathogenic bacteria are the causative agents of the disease pneumonia. Effective diagnosis, prevention and treatment are important for these air-borne diseases for good health.

### 9.1.1 Tuberculosis

Tuberculosis is a respiratory tract infection and mostly endemic in nature. It is still a leading cause of human deaths in most of the countries. Evidence of tuberculosis existence in many forms has been found in Egyptian mummies 3,000 years old, proving that this disease has caused untold sufferings to mankind since antiquity. Robert Koch (1882) isolated the causative organism of human tuberculosis i.e., *Mycobacterium tuberculosis* and proved that pure cultures of it would produce the infection in experimental animals from which the organism could subsequently be recovered.

There are several types of tubercle bacilli, the human, bovine and avian species, and those found in cold-blooded animals and some are saprophytic. Of particular importance to human are the human and bovine varieties, as both of these can cause infection in man. The avian variety is not pathogenic for humans. The human strain, almost exclusively a parasite of man, is responsible for over 90% of all the cases of tuberculosis, whereas the bovine-type produces tuberculosis in man through ingestion of infected beef or milk from an infected cow. This later type is generally manifested as tuberculosis of the bone or lymph system. Based on the site of infection in the human body, tuberculosis is recognized as pulmonary tuberculosis, bone tuberculosis, brain tuberculosis, kidney tuberculosis etc.

**Causal organism:** It is a chronic and long-lasting human disease. It is caused by species of *Mycobacterium* (Table 30). Two distinct species of this genus cause tuberculosis in humans.

**Table 30:** Classification of *Mycobacteria* pathogenic organisms for humans

| Organism | Pathogenicity | Frequency in US |
|---|---|---|
| **Tuberculosis complex** | | |
| *M. tuberculosis* | Strictly pathogenic | Common |
| *M. leprae* | Strictly pathogenic | Uncommon |
| *M. africanum* | Strictly pathogenic | Rare |
| *M. bovis* | Strictly pathogenic | Rare |
| *M. ulcerans* | Strictly pathogenic | Rare |
| **Runyon Group I** | | |
| *M. kansasii* | Usually pathogenic | Common |
| *M. marinum* | Usually pathogenic | Uncommon |
| *M. simiae* | Usually pathogenic | Uncommon |
| *M. asiaticum* | Usually pathogenic | Rare |
| *M. intermedium* | Sometimes pathogenic | Rare |

| Organism | Pathogenicity | Frequency in US |
|---|---|---|
| **Runyon Group II** | | |
| *M. szulgai* | Usually pathogenic | Uncommon |
| *M. scrafulaceum* | Sometimes pathogenic | Uncommon |
| *M. xenopi* | Sometimes pathogenic | Uncommon |
| *M. lentiflavum* | | Rare |
| *M. interjectum* | Unknown | Rare |
| **Runyon Group III** | | |
| *M. avium* | Strictly pathogenic | Common |
| *M. intracellulare* | Strictly pathogenic | Common |
| *M. genavense* | Strictly pathogenic | Uncommon |
| *M. haemophilum* | Usually pathogenic | Uncommon |
| *M. malmoense* | Usually pathogenic | Rare |
| *M. shimoidei* | Sometimes pathogenic | Rare |
| *M. celatum* | Sometimes pathogenic | Rare |
| *M. branderi* | Sometimes pathogenic | Rare |
| **Runyon Group IV** | | |
| *M. fortuitum* | Sometimes pathogenic | Common |
| *M. chelonae* | Sometimes pathogenic | Common |
| *M. abscessus* | Sometimes pathogenic | Uncommon |
| *M. mucogenicum* | Sometimes pathogenic | Uncommon |
| *M. peregrinum* | Sometimes pathogenic | Rare |

1. *Mycobacterium tuberculosis* var-*hominus* causes pulmonary tuberculosis in 90% of the humans.
2. *M. tuberculosis* var-*bovis* causes bone, joints and lymph nodes tuberculosis in humans.

Apart from these bacteria, paratuberculosis causes diarrhoea in cattle and sheep while *M. tuberculosis* var-*avian* causes tuberculosis in poultry and certain other birds.

Species of *Mycobacterium* are rod shaped, non-motile, non-spore forming, aerobic and gram-negative with size 0.2–0.6 x 1.00–10.00 µm. They may grow as an individual cell or in colonies or remain in filamentous form. Generally, individual cells remain encapsulated in a capsule, which contain cord factor that give virulence to the bacterium.

The cell wall is rich in lipids and hence, hydrophobic. It is resistant to many disinfectants, detergents, common antibiotics and laboratory stains (Fig. 79). It is slow-growing and requires 3–8 wk of incubation to form colonies on Lowenstein-Jenson agar medium, either colourless or buft coloured (Fig. 80 and Fig. 81). Once stained, the bacilli cannot be decolourized with acid solutions, hence called as acid-fast bacilli, because of the presence of chemically-complex cell wall. The bacterium is fastidious and the disease results in high rate of mortality.

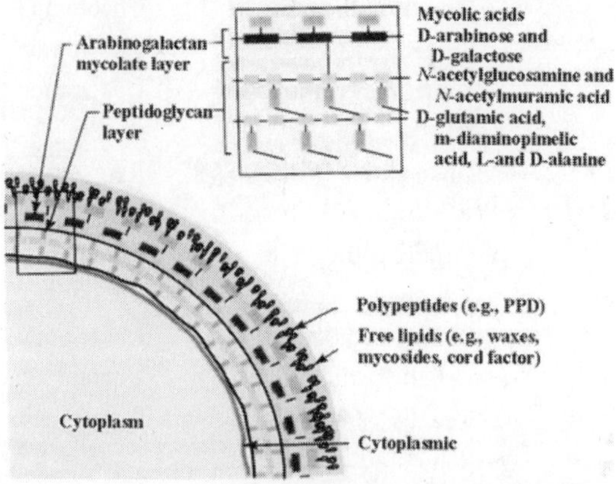


**Fig. 79** Structure of cell wall of *Mycobacterium tuberculosis*

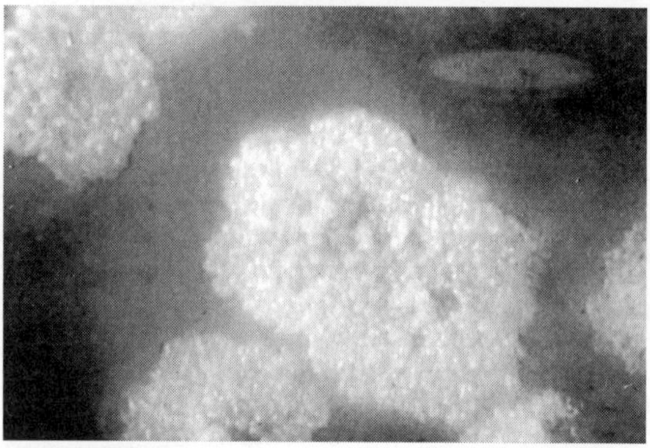

**Fig. 80** *M. tuberculosis* colonies on Lowenstein-Jensen Agar

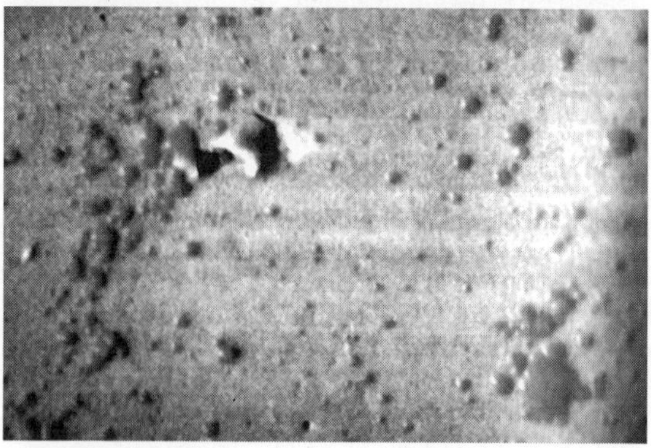

**Fig. 81** *M. kansaii* colonies on Middlebrook Agar

**Pathogenesis:** The infection is acquired by the inhalation of infectious air-borne particles, which reach alveoli of lungs through trachea and bronchi. At this site, the bacteria penetrate into the macrophages and replicate freely.

Intracellular replication of the bacteria stimulates both helper (CD4+) T-cells and cytotoxic (CD8+) T-cells. Activated CD4+ cells produce antibodies, and release gamma IFN and other cytokines that activate macrophages, which can engulf and kill the mycobacteria. The cytotoxic T-cells can lyse inactivated macrophages with replicating bacteria and also kill them.

If the number of bacteria is small then the CD4+ and CD8+ TH-cells are active, and produce antigens and other toxins. After some time, the macrophages are stimulated and the pathogenic bacteria are destroyed with minimal tissue damage, and no disease develops. However, if the number of bacterial cells is more, the activity of TH-cells becomes ineffective leading to lowered immunity response resulting in inactivation and lyses of macrophages. After this, there are further cycles of penetration of inactivated macrophages by the bacteria, and their continuous replication and subsequent lysis of macrophages. Meanwhile, circulating macrophages, lymphocytes and cellular debris are attracted to the site of distruction of infected macrophages by the bacteria. These histological changes result in the formation of multinucleated giant cells of fused macrophages called langhans cells. Accumulation of these langhans cells results in the formation of opaque nodular bodies in the lungs called tubercles.

**Table 31:** Summary of *M. tuberculosis* infections

| **Bacterial factors** | **Geography/season** |
|---|---|
| Diseases: localized pulmonary infections or dissemination to other body sites such as bone marrow, spleen, kidneys, CNS.<br>Bacteria may multiply to produce symptomatic disease or become dormant after initial replication. Dormant bacilli can be reactivated, followed by progression to disease with declining immunity or late in life.<br>Virulence factors: Organism able to replicate in non-activated macrophages; can survive for years in dormant state.<br>Pathology depends on host's response to infection rather than produced by bacterial toxins or enzymes.<br>Lipid-rich cell wall makes mycobacteria | Worldwide, with highest incidence in underdeveloped countries.<br>No seasonal incidence.<br>Modes of control<br>Therapy prolonged with combination of effective antibiotics. (Resistance develops rapidly if single antibiotic is used.)<br>Prophylaxis with INH or another effective antibiotic for exposed people with a positive skin-test result.<br>Vaccination with BCG used as immunoprophylaxis in areas with high incidence of endemic disease.<br>Implementation of appropriate public health measures to ensure prompt |

resistant to disinfectants, detergents, common antibacterial antibiotics and traditional stains.

**Transmission**

Humans are the only natural reservoir.

Person-to-person exposure by means of infectious aerosols

**Who is at risk?**

Anyone exposed to person with active tuberculosis, particularly immunocompromised patients, the homeless, drug and alcohol abusers, and institutionalized people.

Risk of disease related to infectious dose and patient's immunological competence.

detection and treatment of newly-infected patients.

Many such tubercles are formed initially in an infected lung. If treatment is not given at this stage, the tubercles grow in size, and the macrophages and other cells present in them are destroyed with pus formation. Due to this, the tubercles are converted into amorphous cheese-like masses called caseation necrosis. If suitable treatment is given even at this stage, the disease can be prevented from further progress. However, in the absence of treatment, the infection progresses and several tubercles coalesce to form an area of dead tissue, which can be detected as an opaque spot in chest x-ray. As the area of the dead tissue expands, it may erode the wall of a bronchus; as a result the bacteria begin to appear in blood streaked sputum coughed by the patient. At this stage, the bacilli may enter into the blood vessels and get transported to various parts of the body establishing secondary infection in these parts (Table 31).

**Clinical symptoms of the disease:** As the disease progresses, the patient may develop:

- Loss of apetite
- Fatigue
- Weight loss
- Night sweats
- Persistent worsening cough
- Evening mild fever
- Blood streaked sputum

Death ultimately results when sufficient damage has occurred in the lungs or other vital organs.

**Epidemiology:** Humans are the only natural reservoirs of tuberculosis although it can be seen in primates and laboratory animals such as guinea pigs. The disease spreads by close person-to-person contact through the inhalation of infectious aerosols generated by coughing of the infected patient. These particles trapped on the mucosal surface are moved down by the ciliary movement of the respiratory organs. Increased infection is seen in homeless, drugs and alcohol abusers, prisoners and people infected with HIV. Sometimes, the tuberculosis bacilli remain dormant in these people. In such cases, they not only become carriers, but also potential source for the spread of the infection.

It is estimated that one-third of the world's population is infected with *M.tuberculosis*. Each year 8 million new cases and 3 million deaths are reported worldwide.

**Diagnosis:** Various laboratory tests are used for the diagnosis of tuberculosis (Table 32).

**Table 32:** Laboratory tests for the diagnosis of tuberculosis

| |
|---|
| 1. Detection |
| (i) Skin test |
| (ii) Microscopy |
| (a) Carbol fuchsin acid-fast stain test |
| (b) Fluorochrome acid-fast stain test |
| (iii) Direct nucleic-acid probes |
| (iv) Culture |
| (a) Solid agar or agar-based media |
| (b) Broth-based media |
| 2. Identification |
| (i) Morphological properties |
| (ii) Biochemical reactions |
| (iii) Analysis of cell wall lipids |
| (iv) Nucleic acid probes |
| (v) Nucleic acid sequencing |

**1. Skin test:** An extract of harmless protein antigen called Purified Protein Derivative (PPD) from *M.tuberculosis* is injected (0.1 μg equal to 5 tuberculin units) into the intradermal layer of the patient's skin. A positive test is indicated by an appearance of a red swollen zone at the site of the injection within 48 h.

**2. Microscopic studies:** The microscopic detection of acid-fast bacilli in clinical specimens is the most rapid way to confirm the disease. A clinical specimen is stained with carbolfuchsin (Ziehl-Nielsen or kinyoun methods) decolourized with an acid-alcohol solution and then counter stained. The specimens are observed with a light microscope. The bacilli appear in light pink colour.

**3. Fluorochrome acid-fast stain:** This method is more sensitive because the specimen can be scanned rapidly under low magnification for fluorescent areas and then acid fast bacilli are confirmed using higher magnification.

**4. Radiographic studies:** Chest x-ray studies are carried out in pulmonary tuberculosis. Exposed x-ray film shows the presence of opaque masses, which are the regions of calcification of lungs.

**Prevention and control**

**1. Treatment:** In general, patients are given multiple antibiotics for an extended period, a minimum of 6–9 mo. At present, the recommended treatment consists of INH, rifampin and pyrazinamide for 2 mo followed by an additional 4 mo of therapy with INH and rifampin alone.

**2.** If the patient is from a community with drug resistant to tuberculosis, ethambutol or streptomycin is added to the initial therapy. Treatment is more prolonged in HIV-infected patients.

**3. Prevention:** A live attenuated vaccine (BCG vaccine) from *M.bovis* that induces cell-mediated immunity against tuberculosis is used to control the disease, which is about 80% effective. However, this vaccination is not followed in those countries where the incidence of tuberculosis is low.

### 9.1.2 Influenza

Influenza is not only an air-borne disease, but also an acute infectious disease of the respiratory tract, which occurs in sporadic, epidemic and pandemic form. The name 'influenza' is said to have been given by Italians during the epidemic of 1358, which they thought to be the malevolent influence of the heavenly bodies or of inclement weather.

The modern history of the disease may be considered to date from the pandemic of 1889–1890, during which Pfeffer isolated *Haemophilus influenzae,* and claimed it to be the causative agent and bacillus form. However, later studies claimed that Pfeffer's bacillus was not the causative agent of influenza. Smith Andrewer and Laidlaw (1993) proved that the influenza was caused by a virus and named it as influenza virus.

Further studies revealed about the existence of three serotypes of influenza virus. They are designated as serotype-A, serotype-B and serotype-C. Serotype-A causes pandemics whereas serotype-B causes sporadic as well as an epidemic influenza while serotype-C is endemic throughout the world, and causes very mild and inapparent infection. Among all the infections, epidemic influenza is important in the temperate regions, where it strikes during winter months producing more deaths in the aged, and in those with heart and lung diseases. In the tropics, epidemic influenza does not exhibit the winter prevalence and occur frequently in the monsoon season.

Influenza is an important and challenging disease as far as humans are concerned since it was responsible for the production of devastating pandemics in the past. It is for this reason that worldwide surveillance is maintained on the disease by World Health Organization (WHO). Influenza pandemics have been recorded at regular intervals from 1173. Pandemic of modern times dates back to 1889 while the latest occurred in 1977. Influenza disease also occurs in animals and birds in nature.

**The causal organism:** The influenza disease commonly called flu is caused by influenza viruses. There are three types of influenza viruses called A, B and C. They are pleomorphic appearing spherical (diameter from 8–120 μm) or filamentous (up to several μm in length). The virions are enclosed in an envelope consisting of two GPs, HA NA. The HAs are spike-like and project from the envelope while NAs are mushroom-like and lie among HA. The genome of A and B viruses consist of eight different helical nucleocapsid segments, each of which contain a single-stranded, negative-sense RNA associated with the Nucleoprotein (NP) and the transcriptase (RNA polymerase components PB1, PB2 and PA). This ribonucleoprotein is wound up to form a rounded mass and is covered by lipid containing envelope. Influenza virus has only seven genomic segments (Fig. 82 and Table 33).

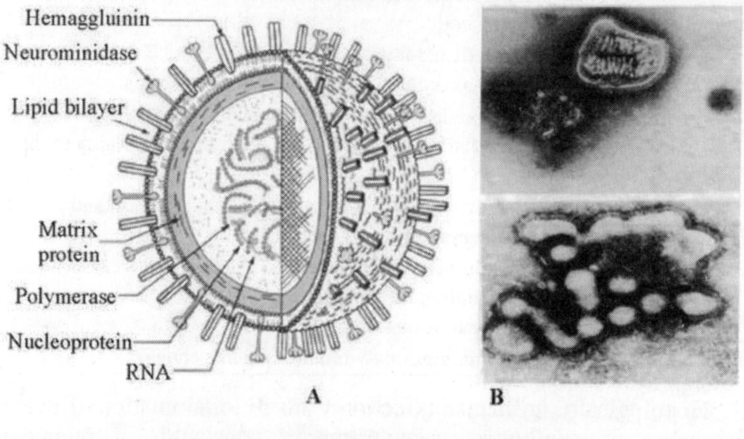


**Fig. 82** A. Model of influenza virus and B. Electron micrograph of influenza virus

**Table 33:** Characteristic features of influenza A and B viruses

- The HA GP is the viral attachment and fusion protein, and it exhibits neutralizing, protective antibody responses.
- Influenza transcribes and replicates its genome in the target cell nucleus, but assembles and buds from the plasma membranes.
- Enveloped virion has a genome of eight negative-sense RNA nucleocapsid segments.
- The segmented genome promotes genetic diversity caused by mutation and reassortment of segments on infection with two different strains.
- The antiviral drugs amantadine and rimantadine inhibit an uncoating step, and target the M2 membrane protein.

**Pathogeneses:** Influenza initially establishes a local upper respiratory tract infection. It targets and kills mucus-secreting, ciliated and other epithelial cells causing the loss of this primary defence system. The NAs facilitate the infection by cleaving mucus membranes, and thereby, providing access to virions in the tissue. Later the virus spreads to lower respiratory tract where it causes severe desquamation of bronchial or alveolar epithelium. In addition to impairing the natural defence mechanism of the respiratory tract, influenza infection may lead to development of secondary bacterial or viral pneumonia infection (Fig. 82 and Table 34).

**Table 34:** Disease mechanism of influenza A and B viruses

- Virus can establish infection of upper and lower respiratory tract.
- Systemic symptoms are due to the IFN and lymphokine response to the virus. Local symptoms are due to epithelial cell damage including ciliated and mucus-secreting cells.
- The IFN and cell-mediated immune response (NK and T-cell) are important for immune resolution and immunopathogenesis.
- Infected people are predisposed to bacterial superinfection because of the loss of these natural barriers and induction of bacterial adhesion to the epithelial cells.
- Antibody is important for future protection against infection and is specific for defined epitopes on HA and NA.
- The HA and NA of influenza A virus can undergo major (reassortment, shift) and minor (mutation, drift) antigenic changes to ensure the presence of immunologically-naive, susceptible people.
- Influenza B virus undergoes only minor antigenic changes.

Histologically, influenza infection leads to inflammation of mucosal membrane of respiratory tract primarily consisting of monocytes, lymphocytes and few neutrophils. This leads to development of alveolar emphysema and necrosis of the alveolar walls.

Protection against infection is associated with development of antibodies to HAs and NAs, which are specific to each strain of influenza, but the cell-mediated immune response is more general.

Recovery is associated with the production of IFN and the development of cell-mediated immune responses. T cell responses are important for effective recovery and immunopthogenesis.

**Epidemiology:** New influenza strains are generated through mutations and reassortment especially, in influenza A virus. The genetic diversity of influenza A is due to the presence of segmented genomic structure, and its ability to infect and replicate in humans and many animal species, including birds and pigs (Table 35).

**Table 35:** Epidemiology of Influenza A and B viruses

**Disease viral factors**

- Influenza is an enveloped virus and is inactivated by detergents.
- Segmented genome facilitates major genetic changes; especially, to targets of humoral immune response, HA and NA.
- Influenza infects many vertebrate species, including mammals and birds.
- Coinfection with animal and human strains of virus can generate very different virus strains by genetic reassortment.
- Transmission of virus often precedes symptoms.

**Transmission**

- Virus is spread by inhalation of small aerosol drops expelled during talking, breathing and coughing.
- Virus prefers cool, less-humid atmosphere e.g., winter hacking season.
- Virus is extensively spread by school children.

**Who is at risk?**

- Seronegative people
- Adults: classic 'flu syndrome'
- Children: asymptomatic to severe respiratory tract infections
- High-risk groups: elderly, immunocompromised people, and people with underlying cardiac or respiratory problems (including people with asthma and smokers)

**Geography/Season**

- There is worldwide occurrence: Epidemics are local, pandemics are worldwide.
- Disease is more common in winter.

**Modes of control**

- Amantadine and rimantadine have been approved for prophylaxis or early treatment.
- Killed vaccine contains predicted yearly strains of influenza A and B viruses.

Hybrid viruses are created by coinfection of a cell with different strains of influenza A virus, which facilitates the genomic segments to randomly associate resulting into new hybrid viruses. For example, a highly-pathogenic reassortant virus for seals was generated by simultaneous infection of seal with two avian influenza viruses. This type of reassortment is thought to be a source of pathogenic human strains (Fig. 83).

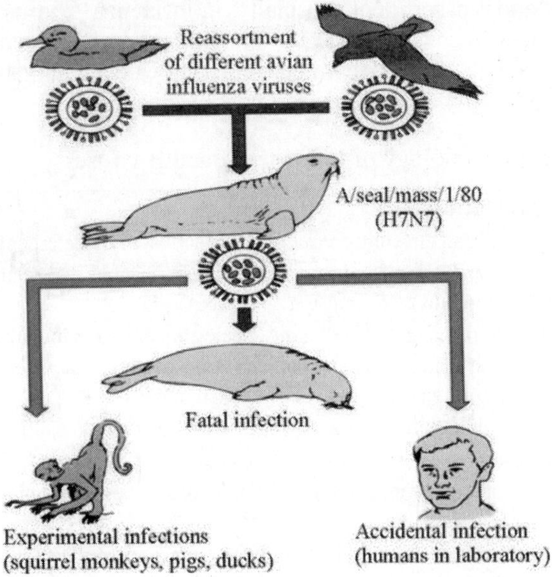


**Fig. 83** Reassortment of genome fragments of influenza virus

Minor antigenic changes resulting from mutation of HA and NA genes are called antigenic drifts, which occur for every 2–3 y causing local outbreaks of influenza A and B infection. Major antigenic changes resulting from the reassortment of genomes between different strains, including animal strains, are called as antigenic shift, which occur only with the influenza A virus. This antigenic shift results in the occurrence of pandemics, which occur on an average for every 10 y (Table 36).

**Table 36:** Influenza pandemics resulting from antigenic shift

| Year of pandemic | Influenza A subtype |
|---|---|
| 1918 | HSWN1 probable swine flue strains |
| 1947 | H1N1 |
| 1957 | H2N2 Asian flu strain |
| 1968 | H3N2 Hongkong flu strain |
| 1972 | H1N1 |

Influenza virus infection spread readily through air-borne droplets expelled during talking, breathing and coughing (Fig. 84). The most susceptible population are children, immunosuppressed people including pregnant women, and the elderly people. People with lung and heart diseases are at highest risk for influenza infection.

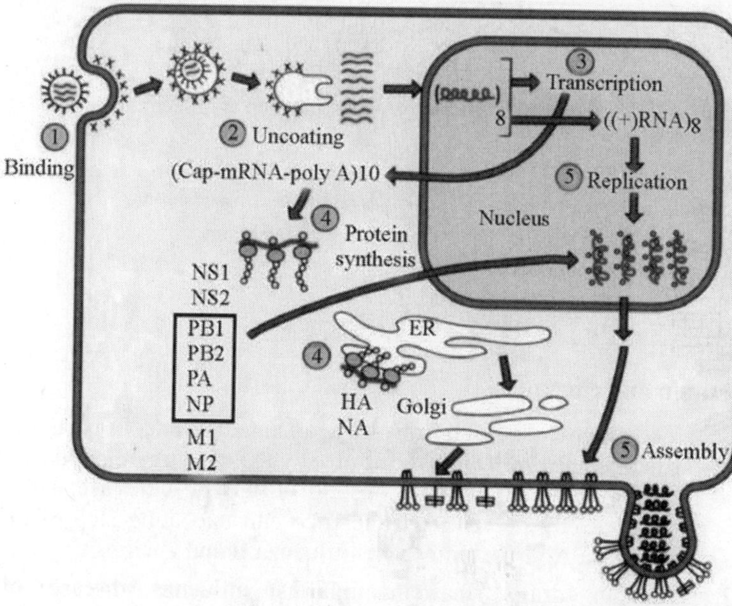


**Fig. 84** Replication of influenza virus

**Clinical symptoms:** Infection may range from asymptomatic to severe based on the degree of immunity to the infecting strain of virus. Clinical symptoms appear after an incubation period of 1–4 d. Infection starts with general weakness and headache followed by sudden onset of fever, severe muscle pains, nasal discharge, cough and sore throat.

As the influenza virus hampers the natural defences of respiratory tract, there is a tendency of development of secondary bacterial infection, mostly pneumonia.

**Diagnosis:** Influenza viruses are obtained from respiratory secretions. The virus is generally isolated from primary monkey kidney cell culture or Madin Darby canine kidney cell lines. These viral cultures are employed in different diagnostic tests (Table 37).

**Table 37:** Laboratory tests for the diagnosis of influenza virus infection

| Test | Detects |
|---|---|
| 1. Cell culture in primary monkey kidney or Madin Darby canine kidney cells | Presence of virus, limited Cytopathological Effects (CPEs) |
| 2. Haemadsorption to infected cells | Presence of HA on cell surface |
| 3. Haemagglutination | Presence of virus in secretions |
| 4. Haemagglutination inhibition | Type and strain of influenza virus or specificity antibody |
| 5. Antibody inhibition of haemadsorption | Identification of influenza type and strain |
| 6. Immunofluorescence ELISA | Influenza virus antigens in respiratory secretions or tissue culture |
| 7. Serology: haemagglutination inhibition, haemadsorption inhibition, ELISA, Immunofluorescence, complement fixation | Seroepidemiology |

## Prevention and control

1. **Treatment:** The antiviral drug, amantadine and its analogue rimantadine, are effective for prophylaxis and treatment during the first 24–48 h after the onset of influenza A disease. These drugs act by preventing adherence and uncoating step of the influenza A virus, but not the influenza B and C viruses.
2. Another drug, ribavirin, inhibits influenza disease by preventing the synthesis of guanosin monophosphate required for nucleic acid synthesis.
3. **Prevention:** The disease can be prevented by vaccination. Vaccination against influenza is accomplished by use of formalin-inactivated specific virus of the year. Immunity is based on the development of antibodies against the HA antigens, thereby, preventing viral attachment to host cells. As immunity is subtype specific, a mixture of several of the agglutinin subtypes, most likely causing infection, is used in the vaccine development.

## REVIEW QUESTIONS

### I. Answer the following questions

1. Discuss briefly about pathogenicity, clinical symptoms and diagnosis of tuberculosis.
2. Discuss epidemiology, clinical symptoms, diagnosis and treatment of influenza.

## II. Write short notes on the following

1. Air-borne disease
2. Structural characters of causal organism of tuberculosis
3. Structure of influenza virus
4. Peculiar characters of influenza A and B viruses
5. Different types of tuberculosis bacilli

## 9.2 FOOD AND WATER-BORNE DISEASES

Various types of food items also constitute an important transportation medium for the transmission of pathogenic microorganisms to humans and result in food-borne diseases. Most of the food-borne diseases are intestinal diseases.

Based on their epidemiology, food-borne diseases can be classified into two major categories:

1. **Food-borne intoxication**: The causative organisms produce an exotoxin in food. When a person consumes the food, the toxin is ingested and it gives rise to the disease e.g., botulism and staphylococcal food poisoning.
2. **Food-borne infection**: The causative organisms are swallowed, which subsequently grow within the body and cause disease.

Just like food items, water also constitutes an ideal transportation fluid for the transmission of pathogenic microorganisms to humans and result in water-borne diseases. Most of these diseases are intestinal infections such as typhoid fever, shigellosis and cholera.

Water-borne infections are generally acquired by the consumption of polluted water containing faecal matter from patients or healthy carriers. When human faeces pollute a municipal water supply or other common source of drinking water, the outbreak of intestinal disease tends to be of the epidemic type.

Sometimes, non-intestinal diseases also occur, not from drinking water, but also from the water contaminated with the urine from infected domestic or wild animals. This occurs when these animals swim in the farm ponds, lakes and such other water bodies e.g., Leptospirosis caused by the Leptospira. The Leptospira can penetrate the conjuctiva of the eye, cause abrasion in the skin or mucous membrane of the nose and mouth, kidney damage.

### 9.2.1 Cholera

Cholera is a gastrointestinal disease caused by *Vibrio cholerae*. The disease spreads by contaminated food and water. The name '*vibrio'* is derived from the characteristic vibratory motility exhibited by the organism. The organisms are present in sea and surface waters all over the world. *Vibrio cholerae* was first isolated by Robert Koch (1883) from cholera patients in Egypt.

Cholera was endemic in the Ganges and Brahmaputra Deltas in Bengal from very early times. Till $19^{th}$ century, cholera was confined to endemic areas. Cholera spread far and wide in the world between 1817 and 1923, and caused six pandemics. The most recent pandemic, the seventh, originated from Indonesia in 1966, spread in the whole of Asia.

Epidemic cholera has a seasonal distribution, the epidemic season being different in different areas, but remains constant in any one area. For example, in Dhaka, Bangladesh and Philippines, the cholera season follows monsoon and reaches peak during November to January while in Calcutta, the epidemic season is in the hot dry months of March to May and the disease disappears with the onset of rains in June. Cholera spreads like a wild fire in areas of crowded localities with insanitation and unhygienic conditions.

**Causal organism:** The genus *Vibrio* consists of more than 34 species of curve-shaped bacilli. About 11 species have been reported as causing human infections (Table 38), of which *V. cholerae, V. parahaemolyticus* and *V. vulnificus* are the most prominent. They are gram-negative, facultative anaerobic, motile and fermentative bacilli.

**Table 38:** Details of *V. cholerae* infections

**Bacterial factors**

- Disease: cholera
- Virulence factors: see table 2
- Two biotypes: *V. cholerae* O1 biotype cholera causes more severe disease than biotype el tor.
- 139 O antigens have been characterized; only strains with O1 or O139 can produce CT and cause pandemic disease.

**Transmission**

- Bacteria can grow in estuarine and marine environment, particularly one with chitinous shellfish.
- Asymptomatic human carriers serve as sources for the contamination of water or food.

- Disease is caused by the ingestion of large numbers of organisms in contaminated water or food.
- Infectious inoculum less in patients whose gastric acidity is reduced.

**Who is at risk?**

- People living in communities with poor sanitation
- People who consume raw shellfish, particularly oysters

**Geography/Season**

- Contaminated estuarine and marine waters
- *V. cholerae* gastroenteritis is not typically seasonal, but bacterial levels increase in contaminated waters during the warm months. The incidence of wound infections and septicaemia caused by other vibrios increases during warm months.

**Modes of control**

- Fluid and electrolyte replacement is crucial.
- Antibiotic therapy reduces the bacterial burden and exotoxin production; doxycycline (adults), trimethoprim-sulphamethoxazole (children), or furazolidone (pregnant women) is administered.
- Improved hygiene is critical for control.
- Killed cholera vaccine is of limited value for preventing O1 infections and of no value for preventing O139 infections.

*Vibrio* species can grow on a variety of simple media with a broad temperature range of 18–37°C. *Vibrio cholerae* is the etiological agent of cholera. Members of the species are subdivided on the basis of their somatic O antigens into 139 serovarieties, of which *V. cholerae* O1 and *V. cholerae* O139 cause classic cholera, which can occur in epidemics or pandemics. Virulance factors of *V. cholerae* O1 and O139 are precised in Table 39 and 40. *V. cholerae* O1 is further subdivided into two biotypes, biotype *el tar* and biotype *cholerae*. Other strains of *V. cholerae* can also cause human disease, but not in epidemics. *V. parahaemolyticus* can also be subdivided on the basis of differences in the somatic O antigens. Source of infection and clinical symptoms in human beings caused by different species of *Vibrio* are listed in Table 39 and 40.

**Table 39:** Virulence factors of *V. cholerae* O1 and O139

| Virulence factor | Biological effect |
|---|---|
| CT | Hypersecretion of electrolytes and water |
| Coregulated pilus | Adherence to mucosal cells |
| Accessory colonization factor | Adhesion |
| Heamagglutination-protease (mucinase) | Releases bacteria from mucosal cells |
| Zona occludens toxin | Exotoxin |
| Accessory cholera enterotoxin | Exotoxin |
| Flagellum | Motility |
| Sideropores | Iron sequestration |

**Table 40:** Different species of *Vibrio* associated with human diseases

| Species | Source of infection | Clinical disease |
|---|---|---|
| *V. cholerae* | Water, food | Gastroenteritis |
| *V. parahaemolyticus* | Shellfish, seawater | Gastroenteritis, wound infection, bacteraemia |
| *V. vulnificus* | Shellfish, seawater | Bacteraemia, wound infection, cellulitis |
| *V. alginclyticus* | Seawater | Wound infection, external otitis |
| *V. hollisae* | Shellfish | Gastroenteritis, wound infection, bacteraemia |
| *V. fluvialis* | Seawater | Wound infection |
| *V. damsel* | Seawater | Wound infection |
| *V. furnissii* | Unknown | Gastroenteritis |
| *V. metschnikovii* | Unknown | Bacteraemia |
| *V. cincinnatiensis* | Unknown | Bacteraemia, meningitis |
| *V. carchariae* | Seawater | Wound (shark bite) |

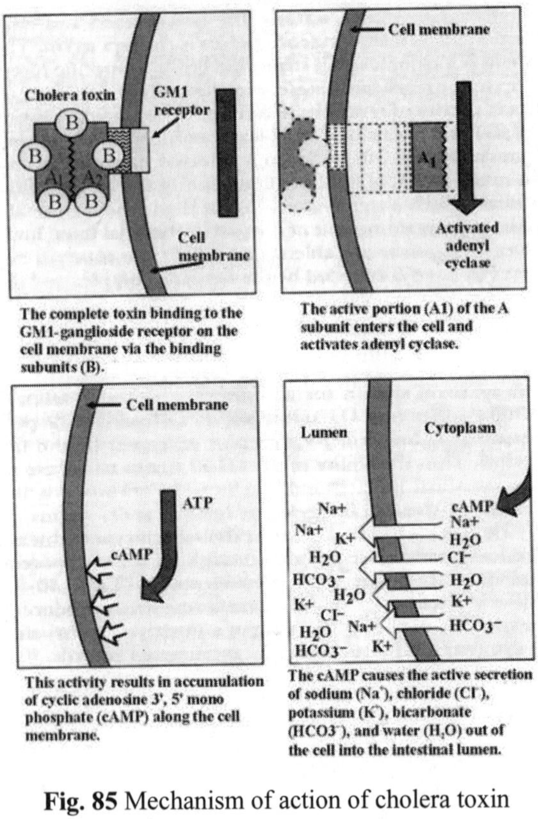


**Fig. 85** Mechanism of action of cholera toxin

**Pathogenesis:** About seven chromosomal genes are involved in the virulence of *V. cholerae* O1 serovariety. The best studied and most important of these virulence factors is the CT. The toxin consists of one A subunit and five B subunits, which binds to specific receptors in the small intestine, enters the mucosal cells and effects a series of reactions that result in the rapid secretion of sodium, potassium and bicarbonates into the intestinal lumen. Severely-infected patient can lose about one litre fluid per hour. When the disease is at its peak, *V. cholerae* O139 serovariety possesses the same virulence complex as that of *V. cholerae O*1 serovariety. Hence, it is as virulent as the later variety (Fig. 85).

Other pathogenic species include two marine species; *V. parahaemolyticus,* which causes an acute gastroenteritis contracted by consumption of raw or inadequately-cooked seafood and *V. vulnificus,* which causes wound infection as well as fatal septicaemias.

**Epidemiology:** *Vibrio* species including *V. cholerae* grow naturally in estuarine and marine environment all over the world. All *Vibrio* species survive and replicate in contaminated water with increased salinity and at temperatures of 10–30°C, pathogenic vibrios can also flourish in waters with chitinous shellfish.

Cholera is spread by contaminated water and food. It is more prevalent in communities with poor sanitation. Symptomatically, an infected human can also be an important reservoir for this organism in areas where cholera is endemic (Table 38). Cholera is a disease of antiquity, and has been the cause of untold suffering and death. It was endemic in parts of India for centuries, but began to invade the rest of the world in the early part of the $19^{th}$ century causing a series of pandemics between 1817 to 1923. A seventh pandemic caused by the *eltor* biovariety began in 1960 in Hong Kong, and by 1971 had spread to the Philippines, Indonesia, the Middle East, Russia, Africa, Spain and Portugal. The Pakistan–India war of 1971 with its massive population shifts resulted in thousands of deaths due to cholera.

Infections caused by *V.parahaemolyticus* and *V.vulnificus* result from the consumption of improperly-cooked sea food, particularly oysters or exposure to contaminated water, and gastroenteritis caused by these vibrios persists throughout the year.

**Clinical symptoms:** The clinical manifestation of cholera caused by *V.cholerae* O1 biovariety begin 2–3 d after ingestion of the vibrios, with abrupt onset of profuse watery diarrhoea (rice water stool) and vomiting. A severely-infected patient can loose on an average one litre of fluid per one hour, resulting in loss of severe fluid and electrolytes. This can lead to dehydration, metabolic acidosis (bicarbonate loss), and hypokalemia and hypovolemia (potassium loss) with cardiac and renal failure.

The mortality rate is 60% in untreated patients, but less than 1% in patients promptly treated to replace lost fluids and electrolytes.

Disease caused by *V. cholerae* O139 strain is worse than caused by *V. cholerae* O1 with higher incidence of severe dehydration and death.

The gastroenteritis caused by *V. parahaemolyticus* ranges from self-limiting diarrhoea to a mild cholera-like illness. Headache, abdominal cramps, nausea, vomiting and low-grade fever may persist for 72 h or more.

**Diagnosis:** Disease diagnosis is made by isolating *V. chlolerae* from the diarrhoeic stool by culturing it on selective media with pH of 8.5, which inhibits most other intestinal bacteria. Identification of the bacterial cells is carried out by biochemical and agglutination tests with *V. cholerae* antiserum.

**Prevention and control**

- The disease is treated with prompt supplement of fluids and electrolytes to replace the lost fluid due to diarrhoea.
- Antibiotic therapy, although of secondary value, can reduce exotoxin production and eliminate the vibrios. Generally, adults are treated with doxycycline or tetracycline, pregnant woman with furazolidone and children with trimethoprim-sulphamethoxazole.
- Effective control of the disease is possible by improvement in sanitation conditions, involving adequate sewage management, use of purification systems to eliminate the contamination of the drinking water supply and appropriate steps to prevent contamination of food.
- Immunization by a vaccine consisting of killed whole cells, will only confer 50% protection for about 3–6 mo.

### 9.2.2 Typhoid

Typhoid is a food and water-borne infection of man. It is a gastrointestinal disease caused by the bacterium *Salmonella typhosa*. The disease is transmitted by two methods: 1. Indirect transmission, and 2. Direct transmission. Indirect transmission occurs through contaminated water and food, which have been wasted from patients or carriers and thus, pollute the water and food. The common housefly is also responsible for the transfer of the typhoid bacterium from waste to water and food. Direct transmission occurs from the excreta of infected person to the healthy person by hands or fomites.

Typhoid is also called as typhoid fever. The typhoid bacillus was first observed by Ebereth in 1980 in the mesentric node and spleen of fatal cases of typhoid fever, and was isolated by Gaffky in 1984. Initially, this bacillus was called as *Eberhella typhi*. It was subsequently realized that typhoid bacillus resembles the genus *Salmonella* and therefore, it was redesignated as *Salmonella typhosa*, the genus *Eberhella* was then abolished.

Typhoid is endemic in all parts of India. It is reported that out of every 1, 00,000 people, 150–300 are affected with typhoid. The proportion of typhoid to paratyphoid A is about 10:1. Paratyphoid B and C are rare. The disease occurs in all ages, but most common in the 5–20 age groups.

**Causal organism:** Typhoid is caused by one of the species of the genus *Salmonella.* There are more than 2,000 species of *Salmonella.* They were originally classified as separate species, but it is now generally accepted that they represent serotypes or/and serovars of a single species *Salmonella enteric.* Various subspecies are also recognized, but most of the serotypes that infect mammals are found in subspecies, also designated as *enteric.* The full correct name is *S. enteric* subspecies *enteric,* serotype enteriditis.

All the serovars are characterized by different combinations of O antigens, which are heat-stable outer membrane poly saccharides and H antigens, which are heat-lable flagellar protein antigens. The O antigens are designated by the numbers 1–67. There are two major categories of H antigens called phase 1 and phase 2. Each phase 1 antigen occurs only in a few serovars, which are designated by letters a to z. Fewer kinds of phase 2 antigens occur, but they are widely distributed among serovars and are usually designated by numbers. In addition to O and H antigens, a Vi antigen, a capsular antigen, occurs in *S.typhi,* and few other serovars. Some of the *Salmonella* serovars were given names as if they are species while others are designated only by their antigenic formulas, which are written as 6,7: r :1,7, representing O antigens 6 and 7, phase 1 antigen r phase 2 antigens 1 and 7 (Table 41).

**Table 41:** Some *Salmonella* serovars commonly encountered in human infections

| | Kauffman-white | | |
|---|---|---|---|
| **Serovar name** | **O group** | **Formula** | **Human disease** |
| *S. paratyphi* A | A | 1, 2, 12:a:- | Enteric (paratyhoid) fever or gastroenteritis |
| *S. typhimurium* B | B | 1, 4, 5, 12 : 1 :1, 2 | Gastroenteritis, septicaemia or focal infection |
| *S. choleraesuis* | C | 6,7 : C : 1, 4 | Enteric fever or gastroenteritis |
| *S. typhi* | D | 9, 12 [vi] : d: - | Enteric (typhoid) fever |
| *S. enteritidis* | D | 1, 9, 12 : g, m :- | Gastroenteritis |

Salmonellas cause three kinds of infections – enteric fevers (typhoid and paratyphoid fever), gastroenteritis and septicaemia. Each of these has certain characteristic features.

The species that cause enteric fever or typhoid generally belong to *Salmonella,* which belongs to the bacterial family *Enterobacteriaceae.* Species of *Salmonella,* which cause typhoid, paratyphoid are facultative and anaerobic, gram-negative, motile bacillus-form able to grow on simple media and, is distinguished from other members of the family by biochemical and antigenic characters.

**Pathogenesis:** Relatively little is known about the role of specific virulence factors, clinical manifestation of the disease and epidemiology of the infection although much is known about the genetic relationship of *Salmonella* strains.

After ingestion and passage through the stomach, *Salmonella* cells are able to invade and replicate in the cells lining the lumen of terminal part of the small intestine. The adherence of the bacteria to the microvilli on the cell surface stimulates the rearrangement of actin, a receptor protein, with subsequent engulfment of the bacteria within endocytic vesicles.

The bacteria remain in the vesicles and replicate until the cell is lysed. Chromosomal invasin genes mediate the attachment and penetration of the bacteria into the cell. The bacterial cells are also protected from stomach acids and the acidic pH of the stomach cells by an acid-tolerance response gene.

Catalase and superoxide dismutase are other factors that protect the bacteria from intracellular killing. The inflammatory response establishes the infection to the gastrointestinal tract, mediates the release of prostaglandins, and stimulates cellular AMP and active fluid secretion.

From the intestine, the bacteria eventually reach the blood stream. Here, the lysis of some of them due to the combined action of antibodies and complement, results in the liberation of endotoxin, which causes generalized disease symptoms such as fever. Some of the bacteria are excreted in the urine, others are removed from the blood by macrophages of the liver, spleen, lymph nodes and bone marrow, in which they survive and multiply. Some bacteria move from the liver to gall bladder and bile ducts, and are transferred into the intestine, where they establish a secondary infection and may cause diarrhoea. Bacterial excretion in the faeces may occur for weeks, and some persons become chronic carriers of typhoid and paratyphoid diseases (Table 42).

**Table 42:** Summary of *Salmonella* infections

**Bacterial factors**

- Diseases: enteritis, septicaemia, enteric fever
- Virulence factors: ability to invade and replicate in enterocytes; protection from acid-mediated killing by Acid Tolerance Response (ATR) gene product; produce catalase and superoxide dismutase
- Animals are main reservoir for human disease, except for infection with bacteria responsible for typhoid and paratyphoid fevers (human reservoir).
- Asymptomatic carriage with *S. typhi* and less commonly with other species

**Transmission**

- Ingestion of contaminated food products (e.g., poultry, eggs, dairy products)
- Direct faecal-oral spread in children
- Person-to-person spread with contaminated food or water (typhoid or paratyphoid fevers)

**Who is at risk?**

- Anyone who consumes foods contaminated with large numbers of salmonellae is at risk, particularly children younger than 5 y and adults older than 60 y, patients with reduced gastric-acid levels, and immunocompromised patients (especially, patients with AIDS)
- Foreign travellers to areas where hygienic practices are sub-standard and people exposed to S.typhi carriers

**Geography/Season**

- Worldwide
- More common in summer and autumn months

**Modes of control**

- Symptomatic treatment rather than antibiotics to prevent prolongation of carriage
- Proper preparation and refrigeration of foods
- Improved hygiene

**Epidemiology:** Many strains of *Salmonella* have no host specificity, and cause disease in both humans and non-human hosts. However, serotypes such as *S. typhi* and *S. paratyphi* are highly specific to humans and do not cause disease in non-human hosts.

Typhoid infection occurs through ingestion of food or water contaminated with *S. typhi* or *S. paratyphi.* Food products and water generally get contaminated with faecal matter, and they serve as source for the disease. The incidence of disease is greater in children less than five years and adults more than sixty years. The disease is prevalent during summer and autumn months when contaminated foods are consumed at outdoor social gatherings. The most common source of human infections are poultry, eggs, dairy products and other contaminated food items. A large inoculum of about $10^6$–$10^8$ bacteria is required for the development of disease.

**Clinical symptoms:** Four forms of *Salmonella* infections are caused in humans – enteric fever, gastroenteritis, and septicaemia and asymptomatic colonization.

Gastroenteritis is most common form and symptoms appear within 6–48 h after the contaminated food is consumed. Nausea, vomiting and non-bloody diarrhoea, fever, abdominal cramps, myalgias and head ache are also common. Symptoms can persist from 2 d to 1 wk before spontaneous resolution.

All *Salmonella* species can cause septicaemia. The chances of septicaemia are more in children and aged persons, and in patients with AIDS

The species of *Salmonella* responsible for causing typhoid and paratyphoid fevers are maintained by human carriers for more than one year after symptomatic disease; the carriage will develop in 1 – 5% patients within the gall bladder as the reservoir.

The *S. typhi* causes typhoid disease while *Salmonella paratyphi* causes a form of typhoid called as paratyphoid A, B and C. Ten to fourteen days after the ingestion of the bacteria, the patients experience gradually-increasing fever, headache, muscular pains, malaise and anorexia. These symptoms persist for a week, which are followed by gastrointestinal symptoms such as intestinal inflammation and intestinal ulcers.

**Diagnosis:** Laboratory diagnosis of typhoid fever is based on isolation of the organism from blood samples during the first and second week or from stool samples during second and third week. Diagnosis is also done by biochemical and serological tests.

**Prevention and control:** Typhoid disease can be treated effectively with the antibiotics ampicillin and chloremphenicol.

Recovery from typhoid fever confers lifelong immunity, probably of the cell-mediated type, and a second attack in the same individual is rare.

Immunization can be achieved by vaccination with acetone-killed whole cells of *S. typhi* and can confer protection for 3 y. Live attenuated vaccines, which are given orally, can also be used.

### 9.2.3 Hepatitis-A

The term 'hepatitis' refers to a primary infection of the liver by hepatitis virus type-A or type-B and other hepatitis viruses presently designated as non-A and non-B. A liver infection occurring either sporadically in epidemics, affecting mainly children and young adults, and transmitted by the faecal-oral route is called as hepatitis-A or infective or infectious hepatitis.

In 1973, Feinstone and coworkers demonstrated this virus in faeces of experimentally-infected human volunteers using immunoelectron microscopy. Hepatitis-A is food and water-borne, and subacute disease of global distribution. In temperate regions, it is prevalent during autumn-winter season while in tropics, no such seasonal prevelence is evident. In India, the disease is associated with periods of heavy rainful.

**Causal organism:** Hepatitis-A is a viral disease caused by HAV. It is also called as acute epidemic hepatitis, infective hepatitis and short-incubation hepatitis.

The HAV belongs to the group of picorna viruses. It has a 27 μm naked icosahedral capsid surrounding a positive-sense single-stranded RNA genome consisting of approximately 7,470 nucleotides. The HAV genome has Viral genomic Protein (VPg) attached to the 5' end and polyadenosine attached to the even 3' end. The capsid is more stable to acid and other treatment than other picorna viruses. The HAV is resistant to detergents, acid pH and temperature upto 60°C, and can survive for many months in fresh water and salty water (Table 43). There is only one serotype of HAV (Fig. 86).

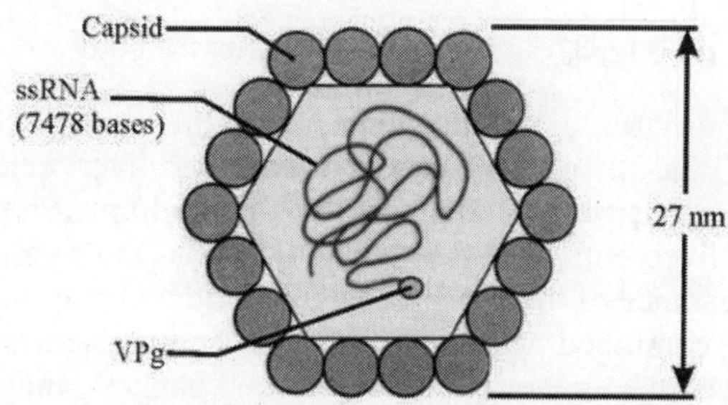


**Fig. 86** The picorna virus structure of HAV: The icosahedral capsid is made up of four viral polypeptides (VP1 to VP4). Inside the capsid is a single-stranded, positive-sense RNA (ssRNA) that has a genomic viral protein (VPg) on the 5' end

**Table 43:** Medically-important organisms of E*nterobacteriaceae*

| | |
|---|---|
| *Citrobacter* species | *Proteus* species |
| *Enterobacter* species | *Salmonella* species |
| *Escherichia* species | *Serratia* species |
| *Klebsiella* species | *Shigella* species |
| *Morganella* species | *Yersinia* species |

**Pathogenesis:** The HAV, after its ingestion, enters into the blood stream through oropharynx or the epithelial lining of the intestine and reaches the target – the parenchymal cells of the liver – where it replicates. The viruses produced in these cells are released into the bile from where they are released into the stool 10 d before the disease symptoms appear or antibody can be detected. The stool with HAV serves as a potential source for the spread of the disease (Fig. 87).

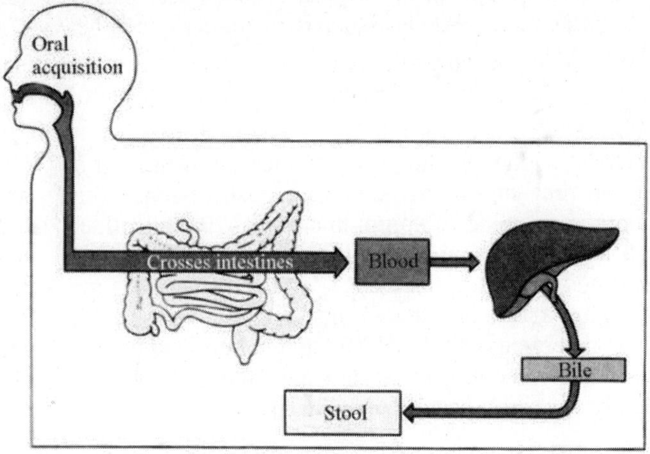


**Fig. 87** Spread of HAV within the body

When the virus is within the liver cell, it replicates without producing CPEs. In the absence of lysis of liver cells, the virus establishes a persistent infection. Although IFN limits viral replication, the infected liver cells are lysed by NK cells and cytotoxic T cells leading to clearance of the virus. Clearance of the viruses is also brought about by the combined action of antibody, complement and antibody-dependent cellular cytotoxicity. These changes induce immune pathology. Jaundice, resulting from damage to the liver, occurs when antibodies are produced, and the cell-mediated immune response to the virus occurs with the appearance of jaundice. The patient feels better, but the liver remains tender and palpable. Jaundice persists for 1–3 wk. An antibody protection against reinfection is life long. However, HAV neither initiates chronic infection nor hepatic cancer.

**Epidemiology:** Approximately 40% of acute hepatitis cases are caused by HAV, and the infection is more prevalent in children, about 90% in the recorded cases, but milder in intensity and asymptomatic. The HAV released into the stool by a patient is the important source for the spread of the disease. The transmission of the virus is similar to that of poliovirus, that is, the faecal-oral route (Table 44).

**Table 44:** Characteristics of HAV

| |
|---|
| **Stable to:** |
| Acid of pH 1 |
| Solvents: (ether, chloroform) |
| Detergents |
| Saltwater, groundwater (months) |
| Drying (stable) |
| Temperature |
| 4°C: weeks |
| 56°C for 30 min: stable |
| 61°C for 20 min: partial inactivation |
| **Inactivated by:** |
| Chlorine treatment of drinking water |
| Formalin (0.35%, 37°C, 72 h) |
| Peracetic acid (2%, 4 h) |
| B-propiolactone (0.25%, 1 h) |
| Ultraviolet radiation (2 $\mu W/cm^2/min$) |

Infections with HAV are endemic in nursery schools, mental institutions, and all establishments and societies where there is a high risk of faecal contamination. Raw or improperly-treated sewage can contaminate the water supply and shell fish, which are the important source of the HAV. A relatively high incidence of HAV infection is directly related to poor hygienic conditions and overcrowding.

**Symptoms:** Clinical symptoms develop from immune-mediated damage to the liver. As already noted, the disease in children is generally milder than that in adults and is usually asymptomatic.

The symptoms occur abruptly 15–50 d after exposure and intensity for 4–6 d before manifestation of jaundice.

Initial symptoms include fever, fatigue, nausea, loss of apetite and abdominal pain. Symptoms generally disappear during the jaundice period. Shedding of virus in the stool preceeds the onset of symptoms and stops before the disappearance of the symptoms (Table 45).

**Diagnosis:** The best way for the diagnosis of HAV is by the estimation of anti-HAV immunoglobulin M (IgM) by ELISA or RIA. Virus isolation is not preferred because efficient tissue culture systems for growing the virus are not available.

**Prevention and control:** No treatment is available for HAV. Prophylaxis with immune serum globulin given before or early in the incubation period (two weeks after exposure), is 80–90% effective in preventing the disease.

It can also be prevented to some extent by a killed HAV vaccine or attenuated live HAV vaccine. The best way to control the disease is by interrupting the faecal or oral spread of the virus by:

- Avoidance of potentially-contaminated water or food especially, uncooked shellfish.
- Proper hand washing of those engaged in day care centres, hotels, mental hospitals and other such centres, is very important.
- Chlorine treatment of drinking water, which is able to kill the virus.

**Table 45:** Epidemiology of HAV infections

| |
|---|
| **Disease/viral factors** |
| Capsid viruses are very resistant to inactivation. |
| Contagious period extends from before to after symptoms. |
| Virus may cause asymptomatic shedding. |
| **Transmission** |
| Virus can be transmitted via faecal-oral route. |
| Ingestion of contaminated food and water can cause infection. |
| The HAV in shellfish is from sewage-contaminated water. |
| Virus can be transmitted by food handlers, day care workers, and children. |
| **Who is at risk?** |
| People in overcrowded, unsanitary areas |
| Children: mild disease, possibly asymptomatic; day care centres: major source of spread of HAV |
| Adults: abrupt-onset hepatitis |
| Pregnant women: high mortality associated with HAV |
| **Geography/Season** |
| Virus is found worldwide. |
| There is no seasonal incidence. |
| **Means of control** |
| Good hygiene |
| HAV: passive antibody protection for contact |
| Killed vaccine |

### 9.2.4 Poliomyelitis

Poliomyelitis is primarily an infantile infection. It spreads through contaminated water. It is caused by poliovirus.

Poliomyelitis has been recognized from very early times as a paralytic disease of children. However, it was only towards the end of the $19^{th}$ century that it was characterized as a separate clinical entity capable of causing paralytic infections.

Poliomyelitis is worldwide in distribution and is an exclusively human disease. The only source of virus is man, and the patients more particularly symptomless carriers. Patients shed the virus in faeces for varying periods, about 50% for 3 wk and a small proportion for 3–4 mo. The most important spread in the community is the faecal virus however, to a lesser extent through the virus present in the throat secretion.

Though poliomyelitis is, in general, asymptomatic, the ratio of subclinical to clinical infections is 1:100. The outcome of infection is influenced by the virulence of the infecting strain, the dose of the infection and the age of the individual.

Poliovirus type-1 and 3 are responsible for most epidemics of paralytic poliomyelitis and type-2 usually causes inapparent infections in the western countries. However, type-2 causes paralytic poliomyelitis in India. The incidence of paralytic poliomyelitis in India is 20–40 for every one lakh population per year. Every two minutes, a child becomes paralytic making a total of 720 every day. The disease is equally prevalent in rural and urban areas. However, presently, the incidence of infantile paralysis has come down remarkably due to massive pulse polio programme being under taken throughout the country (India).

**Causal organism:** Poliomyelitis is caused by the poliovirus. It belongs to the viral family *Picornaviridae*, the largest family of viruses. Structurally, it consists of two parts – a capsid and genome.

1. Capsid is icosahedral in structure with 30µm diameter. The capsid has 12 pentameric vertices, each of which is composed of five protomeric units of proteins. Each protomer is made of four virion polypeptides, designated $VP_1$, $VP_2$, $VP_3$ and $VP_4$. The $VP_4$ in the virion solidifies the structure, but it is not generated until the genome is incorporated into the capsid. However, it is released on binding the virus to the cellular receptor (Fig. 88).
2. The genome consists of single strand of plus-sense RNA of approximately 7,400 bases that has a poly A at the 3' end and a small protein VPg at 5' end. It resembles mRNA. The Poly A sequence enhances the infectivity of the RNA, and VPg helps in packaging the genome into the capsid and initiating viral RNA synthesis. Some of the properties of polioviruses are precised in Table 46. The virions are naked without the surrounding envelope.

**Pathogenesis:** The virus enters the body through upper respiratory tract, the oropharynx and the intestinal tract. They are resistant to stomach acid, proteases and bile.

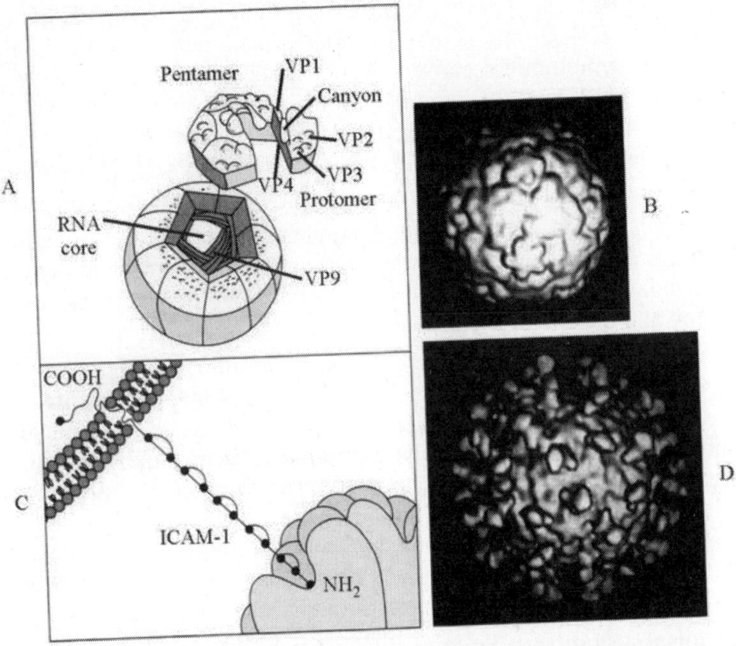


**Fig. 88.** A. Structure of the human rhinovirus and its interaction with the ICAM-1 receptor-binding canyon surrounding the star-like structure at the vertices, B. Cryoelectron microscopy computer-generated reconstruction of the human rhinovirus 16, C. Interaction of the ICAM-1 molecule with the virion, and D. Cryoelectron microscopy reconstruction of the interaction of a soluble form of ICAM-1 with human rhinovirus 16

The viruses pass to the lymphatic system of tonsils or mesenteric lymph nodes from the intestinal tract and eventually reach the blood stream. In most instances, the disease is subclinical without symptoms. In 4–6% cases, non-paralytic poliomyelitis occurs that is characterized by fever and stiffness or pain in the neck and back muscles.

In about 0.1% cases, paralytic poliomyelitis occurs. This occurs due to entry of viruses from blood stream to the CNS and cause inflammation of the gray matter of the spinal cord, especially the anterior horn, in which the cell bodies of motor neurone are located. When these cells are damaged, motor responses to the affected parts are weakened and destroyed. Paralytic poliomyelitis is not fatal, but it leads to crippling and nonfunctioning of the affected parts, mostly lower extremities.

Antibody is the major protective immune response to the poliovirus. Secretary antibodies can prevent the initial establishment of infection in the oropharynx and gastrointestinal tract while serum antibodies prevent viral spread to the target tissue and therefore, the disease does not occur (Table 46).

**Epidemiology:** Polioviruses spread mostly during the summer and autumn seasons. They are exclusively human pathogens. Capsid structure of virus is resistant to mild sewage treatment, saltwater detergents and temperature changes, which facilitates the spread of these viruses through faecal-oral routes and on hands (Fig. 89).

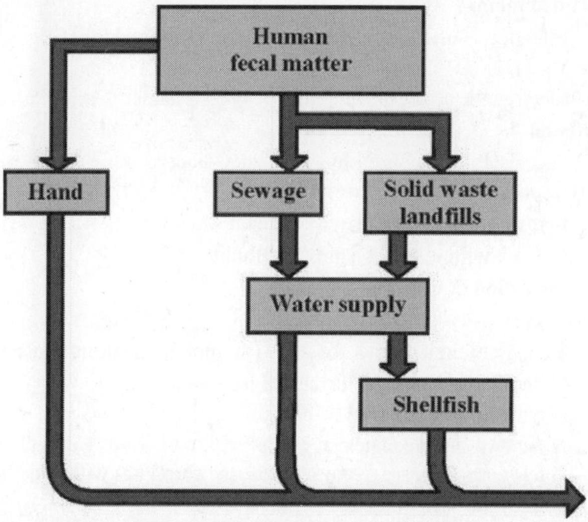


**Fig. 89** Spread of polioviruses

These viruses are spread by the faecal-oral route, shedding of viruses through faeces may continue upto one of an infected individual without any disease symptoms. The details of epidemiology are precised in (Table 47 and Table 48).

**Table 46:** Some properties of human poliovirus

1. Human poliovirus is naked, small (25–30 nm) icosahedral capsid enclosing a single-stranded positive-sense RNA genome.
2. Enteroviruses are resistant to pH 3–9, detergents, mild sewage treatment, and heat.
3. Rhinoviruses are labile at acidic pH; optimum growth temperature is 33°C.
4. Genome is mRNA.
   (i) A naked genome is sufficient for infection.
   (ii) Virus replicates in cytoplasm.
   (iii) Viral RNA is translated into polyprotein, which is then cleaved into enzymatic and structural proteins.
   (iv) Most viruses are cytolytic.

Poor sanitation, crowded living conditions and sewage contamination of water supplies may result in epidemics. Out breaks of polio disease are seen in schools and other social gatherings. Infection would produce more severe symptoms in late childhood, the adolescent years and adulthood.

**Table 47:** Epidemiology of poliovirus

**Disease/Viral factors**

- Infection is often asympatomatic, with viral shedding.
- Virion is resistant to environmental conditions (detergents, acids, drying, mild sewage treatment, and heat).

**Transmission**

- Faecal-oral route: poor hygiene, dirty diapers (especially in day care settings)
- Ingestion via contaminated food and water
- Contact with infected hands and fomites
- Inhalation of infectious aerosols

**Who is at risk?**

- Young children: at risk for polio (asymptomatic or mild disease)
- Older children and adults: at risk for polio (asymptomatic to paralytic disease)
- Newborns and neonates: at highest risk for serious Coxsackie virus and enterovirus disease. (Nature of disease correlates with specific enterovirus types and age of person.

**Geography/Season**

- Viruses have worldwide distribution; wild-type polio is virtually eradicated in developed countries because of vaccination program.
- Disease is more common in summer.

**Modes of control**

- For polio, live trivalent OPV or trivalent IPV is administered.
- For other enteroviruses, there is no vaccine; good hygiene limits spread.

**Table 48:** Disease mechanisms of picornaviruses

- Enteroviruses enter via the oropharynx, intestinal mucosa, or upper respiratory tract and infect the underlying lymphatic tissue; rhinoviruses are restricted to the upper respiratory tract.
- In the absence of serum antibody, enterovirus spreads by viraemia to cells of a receptor-bearing target tissue.
- Different picornaviruses bind to different receptors, many of which are members of the immunoglobulin super-family (i.e., ICAM-1).
- The infected target-tissue determines the subsequent disease.

- Viral, rather than immune, pathology is usually responsible for causing disease symptoms.
- The secretory antibody response is transitory, but can prevent the initiation of infection.
- Serum antibody blocks viraemic spread to target tissue, preventing symptoms.
- Enterovirus is shed in faeces for long periods.
- Infection is usually asymptomatic or causes mild influenza-like or upper respiratory tract disease.

**Symptoms:** Poliovirus may cause one of the four types of symptoms in unvaccinated individuals depending on the progression of the infection.

1. A symptomatic illness results if the viral infection is limited to the oropharynx and the gut. About 90% of poliovirus infection belongs to this type.
2. Abortive poliomyelitis: It is a non-specific febrile illness occurring in about 5% of infected people. The symptoms include fever, headache, general weakness, sore throat, and vomiting occurs within 3–4 d of exposure.
3. Non-paralytic poliomyelitis or aseptic meningitis occurs in 1–2% patients. In this case, the virus spreads to CNS and the meninges causes back pain, muscle pain and symptoms of minor illness.
4. Paralytic polio, which is the major illness, occurs in 0.1–2.0% persons and is the most severe disease. It appears 3–4 d after abortive poliomyelitis. In this disease, the virus spreads from the blood to the anterior horn cells of the spinal cord and motor cortex of the brain. The severity of the paralysis is based on the extent of the neuron infection and the neuron affected. Spinal paralysis may involve one or more limbs whereas cranial paralysis, a more severe form, may involve the muscles of the pharynx, vocal cords and respiration, and results in the death of about 75% patients.

**Diagnosis:** Isolation of poliovirus is carried out by culturing the samples from the pharynx during first few days of illness or from faeces into the kidney tissue of a monkey.

The more useful method is serological test in which estimation of specific IgM is made or by estimating the titre value of antibody. There will be four-fold increase in the antibody titre value between the time of the acute illness and the period of convalescence.

**Prevention and control:** Just like other viral diseases, there is no treatment for poliomyelitis. However, supportive therapy is made for patients with paralytic disease to assist their recovery.

The disease can be prevented by immunization through vaccination. Two types of poliovirus vaccines are available. The IPV vaccine is developed by Jonus Salk while the live attenuated OPV is developed by Albert Sabin, also called as Sabin vaccine. The advantages and disadvantages of the vaccines are precised in Table 49. Both vaccines are prepared by incorporating three strains of polioviruses. They are stable, relatively inexpensive and induce a protective antibody response. However, presently OPV vaccine is employed preferentially because of its easy delivery and its capacity to confer lifelong immunity.

Children are generally given the OPV vaccine for five times at the age of 2, 4 and 15 mo and at 4–6 y of age.

**Table 49:** Advantages and disadvantages of polio vaccines

**1. Live (OPV)**

**Advantages**

- Effectiveness
- Lifelong immunity
- Induction of secretary antibody response similar to that of natural infection
- Possibility of attenuated virus circulating in community by spread to contacts (indirect immunization) (herd immunity)
- Ease of administration
- Lack of need for repeated boosters

**Disadvantages**

- Risk of vaccine-associated poliomyelitis in vaccine recipients or contacts
- Spread of vaccine to contacts without their consent
- Unsafe administration for immunodeficient patients

**2. Inactivated (IPV)**

**Advantages**

- Effectiveness
- Good stability during transport and in storage
- Safe administration in immunodeficient patients
- No risk of vaccine-related disease

**Disadvantages**

- Lack of induction of local (gut) immunity
- Need for booster vaccine for lifelong immunity
- Fact that injection is more painful than oral administration
- Fact that higher community immunization levels are needed than with live vaccine

### 9.2.5 Amoebiasis

Several protozoa cause intestinal diseases, and are transmitted from man to man through infected food and water by flies and by direct contact. The most important of these is *Entamoeba histolytica,* which causes the disease amoebiasis that is worldwide in occurrence, sometimes leading to death. The most serious outbreak occurred in 1933, when an epidemic originated in Chicago during the century of progress worlds fair, and spread across the nation, and resulted in 1,400 amoebiasis cases and four deaths.

Amoebiasis spreads by contaminated water and food. The organism divides asexually in the human intestine where it feeds on gut cells. Man is its natural host and its life cycle is simple. Cysts are shed in infected human faeces, and new infection occurs through the ingestion of sewage-contaminated water or fruits, vegetables and other food items. The cysts open up into growing vegetative cells in the distal part of the small intestine and colon, where they feed and multiply.

The growing amoeba is highly virulent and destroys all types of human cells. It is capable of invading the epithelial and subepithelial layers of the large intestine including blood vessels causing amoebic dysentery, which is characterized by loose stools mixed with mucus and blood. Sometimes, invading-amoeba are carried to the liver and other organs of the body through blood stream where they cause amoebic abscesses, which can be life threatening.

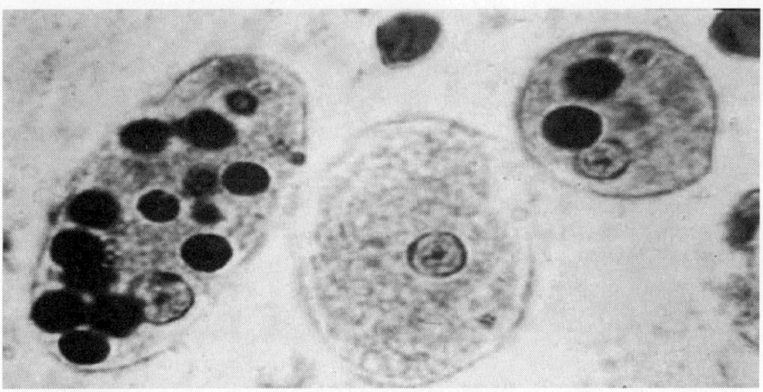

**Fig. 90** *E. histolytica* cyst form

**Causal organism:** Amoebiasis is caused by *Entamoeba histolytica.* It is a primitive unicellular microorganism belonging to the phylum protozoa. It is a human pathogen, and its life cycle consists of two stages, the active, motile, feeding stage represented by trophozoites and the quiescent, resistant, infective stage represented by cysts (Fig. 90). Replication occurs by binary fission or by the formation of trophozoites within the mature multinucleated cysts. Motility is exhibited by pseudopodia during favourable environmental conditions. Non-motile cyst-forms develop under unfavourable environmental condition.

**Pathogensis:** Trophozoites are present in flesh and cysts are formed in fecal matter (Fig. 91). After ingestion, the cysts pass through the stomach. On exposure to gastric acid, trophozoites are released from the cysts in the duodenum and their population increases due to repeated binary fission.

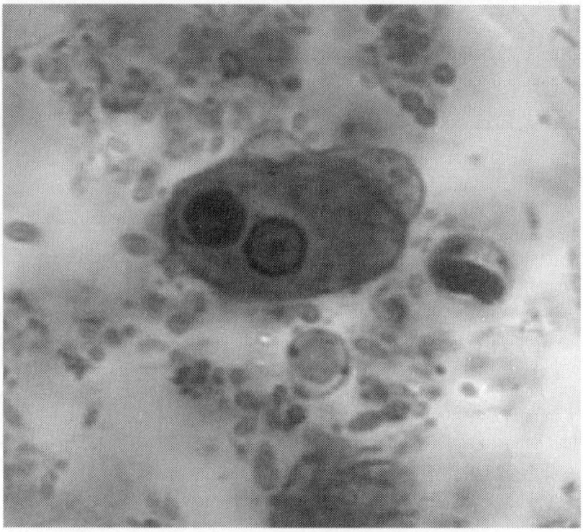

**Fig. 91** *E. histolytica* trophozoite form

Massive intestinal-tissue distruction, along with epithelial cells, neutrophils, lymphocytes and monocytes, occur due to release of a proteinaceous cytotoxin by trophozoites, which leads to local necrosis in the large intestine.

Invasion of trophozoites into the deeper mucosa, with extension into the peritoneal cavity occurs. This can lead to secondary involvement of other organs, mainly, liver, lungs, brain and heart. This results in extra-intestinal amoebiasis.

**Epidemiology:** *E. histolytica* is worldwide in distribution. Its incidence is highest in tropical and subtropical regions that have poor sanitation and contaminated water.

Many of the infected persons are asymptomatic carriers who represent reservoir for the spread of the amoeba. Hence, they constitute main source of water and food contamination.

The amoebiasis is more prevalent in mental hospitals, military and refugee camps, prisons and other crowded establishments. Flies and cockroaches can also serve as reservoirs for *E.histolytica* cysts.

Sewage containing cysts can contaminate water systems, springs and agricultural fields.

**Symptoms:** The outcome of infection may result in a carrier-state, intestinal amoebiasis and extra-intestinal amoebiasis. In a carrier state, which may be due to low virulent organism, low inoculum or patient's immune system is intact, clinical symptoms appear in the patient and they serve as carriers. In intestinal amoebiasis, patients develop abdominal pain, cramping, colitis with diarrhoea, which results due to distruction of intestinal tissue. In severe cases, there may be numerous bloody stools per day. In extra-intestinal amoebiasis, pain in the region of liver, fever and rigors may occur.

**Diagnosis:** Microscopic observation of stool sample showing presence of trophozoites or cysts of *E. histolytica* gives a positive diagnosis for amoebiasis. Extra-intestinal amoebiasis is diagnosed by abdominal scanning for the liver topography.

**Prevention and control:** Acute amoebiasis is treated with metronidazole followed by iodoquinol. A symptomatic carrier, if identified, can be treated with iodoquinol, diloxanide furoate or paromomycin.

The disease can be prevented by the introduction of adequate sanitation measures and education about the routes of pathogen transmission, chlorination and filtration of water supplies, and by boiling drinking water, and through cleaning of fruits and vegetables. These are other preventive measures to keep check on the disease.

## REVIEW QUESTIONS

### I. Answer the following questions

1. Give details of pathogenesis, clinical symptoms and treatment of cholera. How is it transmitted to humans?
2. Discuss briefly about the epidemiology, clinical symptoms and diagnosis of typhoid. Add a note on its treatment.
3. Describe the pathogenesis, clinical symptoms, diagnosis, and prevention and control of hepatitis-A viral disease.
4. Explain epidemiology, pathogenesis, clinical symptoms, and prevention and control of polio.
5. How is amoebiasis transmitted in humans? Give the details of pathogenesis, clinical symptoms, diagnosis and treatment of amoebiasis.

### II. Write short notes on

1. Causal organism of cholera

2. Causal organism of typhoid
3. Paratyphoid
4. Epidemic
5. Pandemic
6. Structure of hepatitis-A virus
7. Transmission of poliovirus in humans
8. Structure of poliovirus
9. Process of transmission of *Entamoeba histolytica*

## 9.3 INSECT-BORNE DISEASES

Insects are invertebrate animals belonging to the phylum Arthropoda. They are widely distributed all over the world. Insect-borne diseases have produced much suffering, economic loss and death of human population through the centuries, and sometimes reached pandemic proportions. Most of the insects associated with human diseases serve as vectors.

1. **Mechanical vectors:** These vectors act as mere carrier and are not involved in any stage of pathogen's life cycle. The common housefly, *Musca domestica*, is the classical example of mechanical vector. The diseases transmitted by it include salmonellosis, poliomyelitis, infectious hepatitis, amoebiosis and other enteric diseases.
2. **Biological vectors:** The other type of vectors is biological vectors in which the pathogen undergoes a period of incubation or development. The diseases in which biological vectors are involved are plague, rickettsial pox, classical typhus fever, murine typhus etc.

Majority of the bacterial diseases are caused by rickitesias, which are intracellular parasites transmitted by ticks, lice, mites or miter. The mosquitoes and ticks are the main vectors involved in the transmission of the insect-borne pathogens.

### 9.3.1 Malaria

Malaria is an insect-borne disease being transmitted by Anopheles mosquito. It is one of the most serious, global and common protozoan infections caused by one of the four species of the genus *Plasmodium,* namely *P. falciparum, P. vivax*, *P. malariae* and *P. ovale*. Forty percent of the world's population is at risk of this infection, and it is the major cause of death in many parts of tropical and subtropical parts of the world. The *P. falciparum,* specifically is more important, as it is capable of causing cerebral malaria – the most serious form of the disease – that is fatal if not treated immediately. On the other hand, *P. vivax* and *P. ovale* cause benign tertian malaria with febrile episodes (fever with chills) that typically occurs at 48 h interval. The *P .malariae* causes quartan malaria with febrile episodes occurring at 72 h intervals.

The appearance of trophozoites, one of the forms formed in the life cycle, of the four species can be illustrated in the peripheral blood films by Romonowsky staining technique.

It is reported that each year more than 300 million people suffer from malaria in the world, among which 3 million succumb to death. Though it is virtually eliminated in most of the developed countries, it is still the single greatest killer of the human race in most of the developing countries.

**Causal organism:** The disease, malaria, is caused by a protozoan parasite called *Plasmodium,* which belongs to sporozoa. Of the many species of the *Plasmodium,* four species cause malaria disease in humans and the rest in other animals. The four species that cause human disease are *P. falciparium, P. vivax, P. ovale* and *P. malariae.* The first three species cause malignant malaria and the last one causes benign malaria.

All the four species of P*lasmodium* share a common life cycle, which is complex, and requires two hosts for completion (Fig. 92):

1. The mosquito (*Anopheles*) for the sexual reproductive stages
2. Humans and other animals for the asexual reproductive stages

**1. Life cycle: Events in human host – asexual stage:** The life cycle starts in humans when an *Aanopheles* mosquito bites. Its saliva, which contains the protozoa at the sporozoite stage of its life cycle, is injected into the blood stream of the victim. The sporozoites, after entering the liver, divide and develop into multinucleated structures called schizonts. The schizonts rupture within 6–12 d and release the merozoites into the blood stream. These merozoites invade the host's red blood cells, where they grow and divide to form more schizonts. These schizonts also rupture, destroying the erythrocytes and releasing more merozoites into the blood stream to invade more red blood cells. The major symptoms of malaria are associated with rupture of the schizonts. Some of the asexual merozoites in the human stream develop into male and female gametocytes.

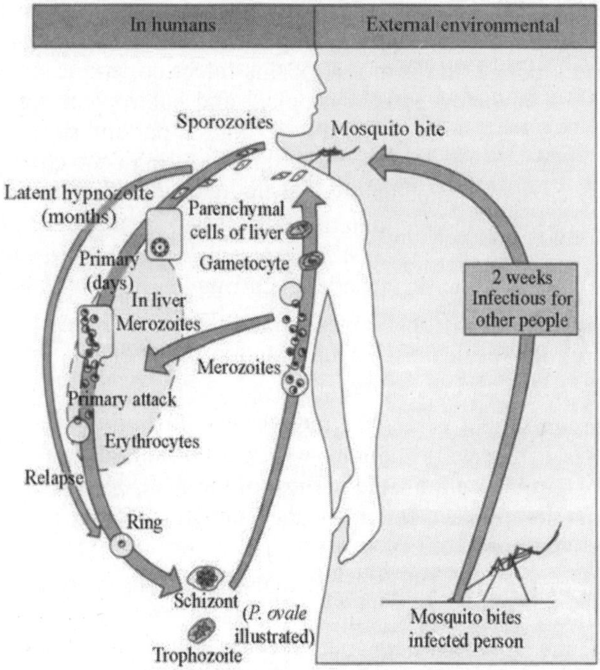


**Fig. 92** Life cycle of *Plasmodium*

**2. Events in the mosquito – sexual stage:** When a mosquito bites, the gametocytes enter into the mosquito's stomach, where they are converted into male and female gametes. Zygotes are formed in the mosquito's stomach after fertilization. These zygotes pass out of the stomach where they develop into oocytes containing many sporozoites. The mature sporozoites migrate to the salivary glands, from where they can be injected into the blood stream of another human victim, to begin the life cycle once again (Fig. 93).

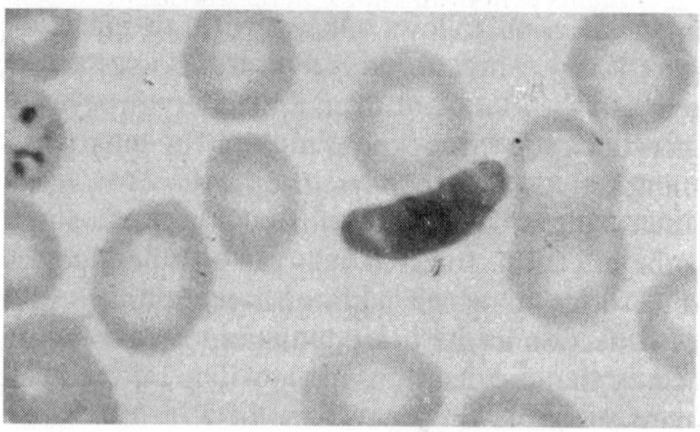

**Fig. 93** Mature gametocyte: a diagnostic feature of malaria

**Pathogenesis:** Malarial disease is caused, especially by the asexual erythrogenic cycle. The rupture of the infected erythrocytes occurs every 48 h at the completion of schizogony in *P. vivax* and *P. ovale,* and 72 h in *P. malariae* producing coincident chills and other symptoms. Synchronized or coincident schizogony, and symptoms of fever and chills are not common with *P. falciparum* infection. The release of an endogeneous pyrogene from injured cells may be the cause of the paroxysms of fever (Table 50).

**Table 50:** Human malarial parasites

| Parasite | Disease |
|---|---|
| *Plasmodium vivax* | Benign tertian malaria |
| *P. ovale* | Benign tertion or ovale malaria |
| *P. malariae* | Quarton or malarial malaria |
| *P. falciparum* | Malignant tertin malaria |

If not treated, benign malaria usually subsides spontaneously and recurs at a later stage. However, malignant malaria caused by *P. falciparum* causes the most forms of the disease. The high mortality rate of *P. falciparum* malaria is partly due to the high rate of reproduction of the asexual erythrocytic form of the parasite, clogging of the small veins and capillaries of the heart with parasitized erythrocytes, effective colonary blood flow and diminished cardiac functions.

**Epidemiology:** Malaria is known from antiquity and generally described as a single greatest killer of the human race. On a global scale, malaria is one of the most common infectious diseases of humans causing much morbility and significant mortality. Each year more than 300 million people are affected with the disease and about 3 million of the victims die of it.

**Clinical symptoms:** Clinical symptoms usually occur 10–16 d after infection by mosquitoes. Symptoms usually begin with bed shaking chills, followed by high fever, sweating, headache and severe muscular pains. However, fever cycles vary according to the species causing infection.

The symptomatic periods are usually less than 6 h. The spleen becomes enlarged and tender, eventually the patient becomes weak and exhausted, and anaemia develops.

The pattern of periodical illness alternating with periods of well-being is a characteristic of initiation of malaria caused by *P. vavix P. ovale* and *P. malariae.* In malignant malaria caused by *P. falciparum,* the fever and other symptoms are generally more persistent, and also include edema of lungs and blockage of kidney function.

**Diagnosis:** The important diagnosis for malaria is the demonstration of the malarial parasite in the blood smears from patients (Fig. 94). Serologically, the disease can be diagnosed by indirect fluorescent antibody test and indirect haemagglutination test. These antibodies can be detected in the patient after the second week of infection.

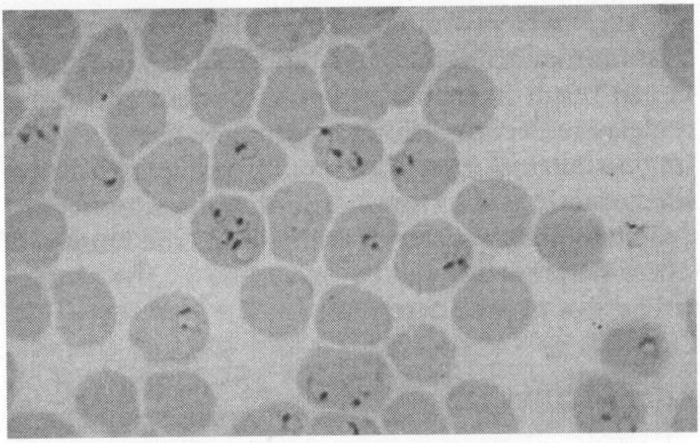

**Fig. 94** Ring form of *P. falciparum*: Note the multilple ring forms within the individual erythrocytes, which is a characterstic of this organism

**Prevention and control:** Quinine was the only drug available for over 100 years. At present, the disease is treated with a combination of two drugs, chloroquine and primaquin. Chloroquine destroys merozoites in the blood while primaquine destroys schizonts located in the liver.

However, it is now reported that *P. faciparum* has developed resistance to chloroquine and primaquine. In such cases, treatment is supplemented with sulphonamides and pyrimethamine. The malaria can be controlled on social front by the destruction of the anopheles mosquito vector, its larval and breeding areas. At the individual level of control, the disease can be controlled by using mosquito curtains around sleeping areas, by spraying insecticides, important of them is the dichlorodiphenyltrichloroethane (DDT), and using mosquito repellents in houses. Presently, there is no vaccine available against malaria.

### 9.3.2 Filariasis

It is an insect-borne human disease caused by *Wuchereria bancrofti,* which belongs to the group of nematodes of the phylum Nematoda. It is transmitted to humans by the bite of various species of mosquito (especially culex mosquito) throughout the tropical regions of the world. It is reported that over 100 million people are affected annually over the world. The larvae invade lymphatic vessels, usually of the lower limbs, where they develop into adult worms. Presence of adult worms cause blockage of lymph vessels leading to gross lymph edema, which sometimes leads to the ugly deformities associated with elephantiasis.

Embryonic forms (microfilaria) are liberated into the blood stream. They retain the elastic egg membrane as a sheath, which cover the whole larva. Microfilaria remains in the pulmonary circulation during the day, emerging into the peripheral circulation during night, to coincide with the biting habit of the insect vector. The physiological basis of this nocturnal periodicity is not understood, but can be reversed by altered sleep patterns e.g., night shift workers. Due to this nocturnal periodicity, blood of the organism is collected for examination during night, especially between 12.00 midnight–2.00 am. However, strains of *W.bancrofti* of Pacific islands do not exhibit this nocturnal periodicity.

**Causal organism:** The disease filariasis is caused by two structurally-similar round worms *Wuchereria bancrofti* and *Brugia malayi.* In India, most cases of filariasis are due to *Brugia malayi,* which are structurally long, slender, unsegmented round worms living as parasites in human blood, lymph, subcutaneous and connective tissue. There are no animal reservoirs for these worms (Fig. 95).

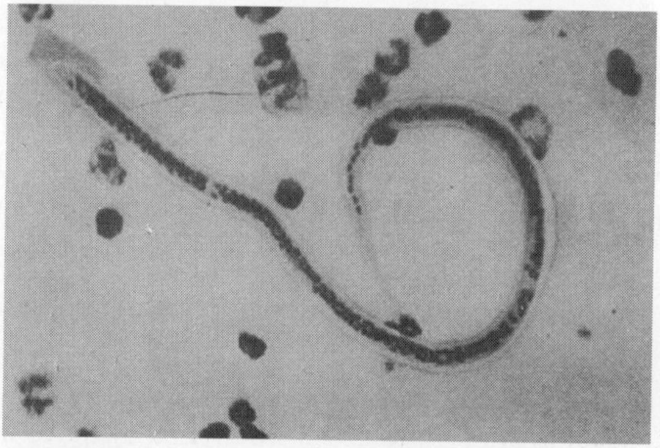

**Fig. 95** *W. bancrofti* microfilaria in blood smears

**Life cycle:** Two hosts, a mosquito and human, are required for the completion of the life cycle (Fig. 96). Various species of mosquito such as *Anopheles, Aedes* and *culex* serve as vectors for filariasis, more particularly species of *Culex.*

**1. Events in human:** The life cycle starts with the bite of the specific mosquito due to which infective larvae are introduced into the human body. The larvae migrate from the region of the bite to the lymphatic system primarily in the arms, legs or groin, where larvae grow into adult worms and can persist as long as 10 y.

Fertilization occurs between male and female worms from 3–12 mo. Then sheathed larval microfilaria are produced from the oocytes, which enter into the blood circulation.

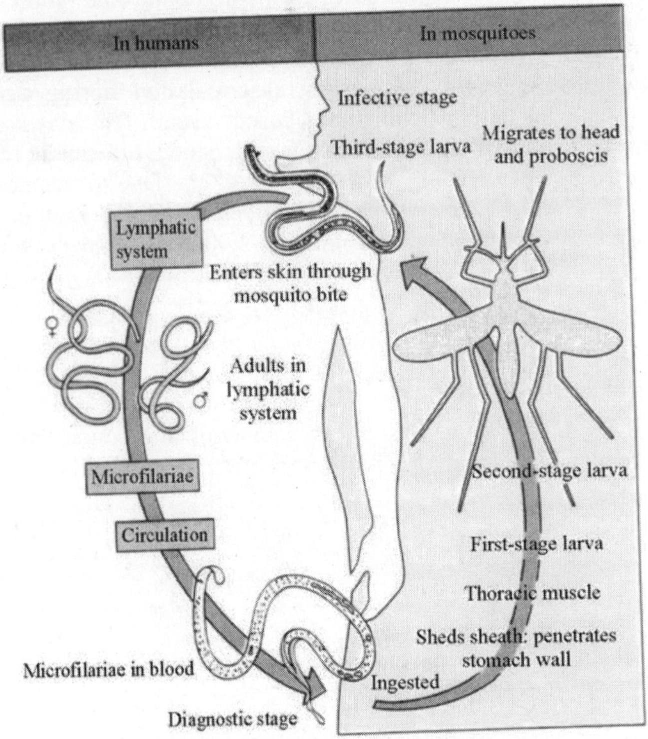


**Fig. 96** Life cycle of *W. bancrofti*

**2. Events in mosquito:** The microfilarial larvae move into the stomach of mosquito after a bite. From stomach, they move into thoracic muscles. During this moving stage, they develop into first stage and second stage larvae, and finally migrate into the proboscis where they develop into third stage larvae. They are transferred from mosquito into the humans at this stage.

**Epidemiology:** Both the species of round worms occur in tropical and subtropical areas. *W. brancrofti* is endemic to Central Africa along the mediterranean coast, and in many parts of Asia including China, Korea, Japan and Philippines.

*Brugia malayi* is found primarily in Malaysia, India, Thailand, and Vietnam, in parts of China, Korea, Japan and many Pacific Islands.

**Clinical symptoms:** In some patients, there is no sign of disease even though many microfilariae are present in the blood. In other patients, the acute symptoms include fever, lymphangitis and lymphadenitis with chills, and recurrent febrile attack. These symptoms are due to inflammatory response to the presence of molting, dead or dying adult worms in the lymph vessels.

As the infection progresses, lymphnodes in many parts of the body enlarge including the legs and hands, the scrotum and the testes with occasional abscess formation. This enlargement of lymph nodes followed by lymph vessels is due to obstruction of lymph flow caused by the presence of adult worms.

In severe cases, due to obstruction of lymph, there may be thickening and hypertrophy of tissue with abnormal swelling of extremities leading to filarial elephantiasis. Filaria of this type is a chronic, debilitating and disfiguring disease.

**Diagnosis:** The disease can be diagnosed by the demonstration of microfilariae in the patient's blood. The blood sample is collected between 10.00 pm–4.00 am because during this period more number of active microfilariae will be present in the patient's blood. The collected blood sample is stained with Giemsa stain to have clear visibility of microfilariae in the blood sample. It can also be diagnosed by the detection of antigens by serological tests.

**Prevention and control:** The drug of choice for the treatment of filariasis is diethylcarbamazine. However, treatment is of little benefit in most cases of chronic lymphatic filariasis.

Supportive and surgical therapy for lymphatic obstruction may be of some cosmetic help. Prevention, rather than treatment, is more beneficial in controlling the disease. The filariasis can be effectively prevented by eradication of mosquito vectors, use of protective clothing, use of mosquito curtains in sleeping areas and use of insecticides.

### 9.3.3 Dengue Fever

Dengue fever is an insect-borne disease caused to humans by dengue virus, which belongs to the group of arboviruses. Dengue virus is widely distributed throughout the tropic and subtropic regions of the world. Dengue fever, which occurs intermittently in large epidemics, is usually similar to the fever caused by chickungunya virus. Four types of dengue viruses exist, which include type-1 - first isolated from Hawai in 1944, type-2 from New Guinea in 1944, and types 3 and 4 from the Phillippines in 1956.

The dengue virus is transmitted from man to man through *Aedes aegypti* mosquitoes. No vertebrate host other than man has been identified. There is, however, some evidence of dengue infection in monkeys, squirrels and other forest animals.

Dengue occurs typically as a fever of sudden onset with headache, pain at the back of eyes, and in the back and limbs, lymphadenopathy and rash on the skin. The fever typically lasts for 5–7 d followed by recovery. In the more serious form, there may be haemorrhagic manifestation such as Dengue Haemorrhagic Fever (DHF) or Dengue Shock Syndrome (DSS). These symptoms have been first recognized in Thailand, and since then recorded in many countries in Southeast Asia and Western Pacific. These symptoms are believed to represent hypersensitivity response to sequential dengue virus infection in person already sensitized by prior exposure to other serotypes of the virus.

Haemorrhagic dengue is seen rarely in India. However, dengue is common along the coast of India and has frequently caused highly wide-spread epidemics sometimes along with chikungunya virus. All the four types of dengue viruses have been isolated in India. Occasionally, more than one type of dengue virus has been isolated from the same patient.

**Causal organism:** Dengue fever is caused by a virus, which belongs to the viral family *Flaviviridae*. Hence, it is generally called as flavivirus. They range in size from 37–50 μm. The virion has three parts – an envelope, capsid and genome (Fig. 97).

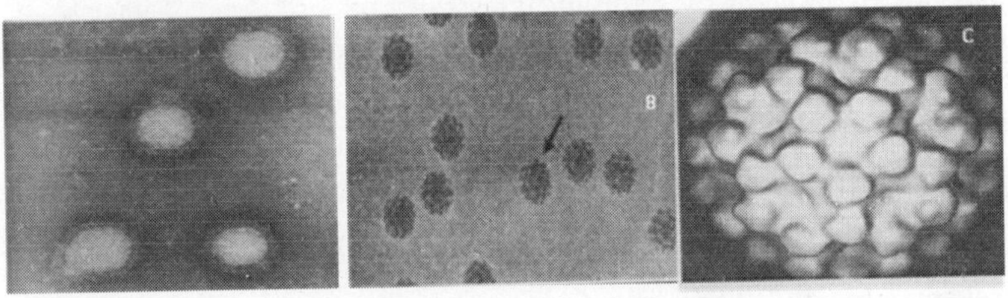


**Fig. 97** Alpha virus morphology: A. The outline of the envelope and the glycoprotein spikes of the dengue virus are visualized by negative staining, B. More details on the morphology of the virion electron micrograph, and C. Surface representation of the negative virus

The envelope consists of lipids and GP spikes, which protrude from the surface of the virion. Each spike is made up of 2–3 GPs. These GPs are also called as envelope proteins, which exhibit haemagglutination activity and are the targets of neutralizing antibodies. The capsid is icosahedral in shape and contains capsid proteins.

The virus genome consists of one single-stranded, positive-sense RNA molecule that is 1,000–12,000 nucleotides in length. The genome does not contain poly-A tail at the 3' terminus. Two-third of the 5' end genome encodes three structural protein genes while the remaining one-third of the genome encodes seven non-structural protein genes (designated as NS1, NS2A, NS2B, NS3, NS4A, NS4B and NS5) of which NS5 encodes the RNA-dependent-RNA polymerase enzyme, which catalyses the synthesis of new genomes during replication.

The flaviviruses replicate in the cytoplasm of the host cell. During replication, the single-stranded positive-sense RNA synthesizes polyprotein precursor that is rapidly processed into structural and non-structural proteins by viral and host cellular proteins, which help in the viral replication.

**Pathogensis:** Dengue virus requires two hosts for the completion of the disease cycle, a vertebrate (humans) and an invertebrate host (mostly *Culex* or *Aedes* female mosquito) (Fig. 98).

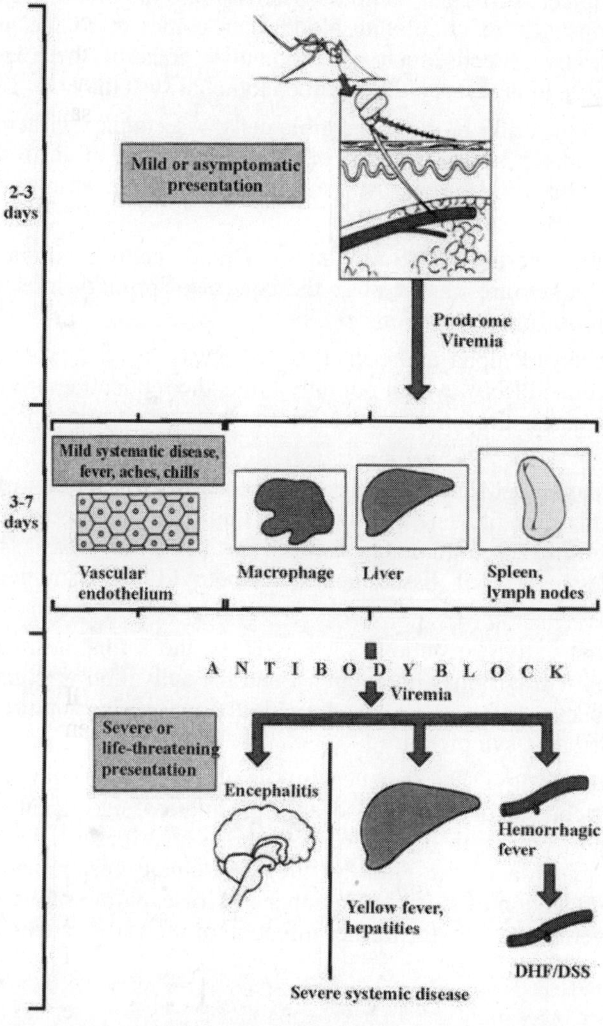


**Fig. 98** Life cycle of dengue virus

**1. Events in the mosquito:** Female mosquitoes acquire the viruses by sucking blood from the human host. The virus then infects the epithelial cells of the midgut of the mosquito, spreads through the wall of the midgut to the blood circulation and finally reaches the salivary glands of the mosquito. The viruses replicate repeatedly in the cells of the salivary glands. The newly-formed viruses then migrate into the saliva.

**2. Events in the vertebrate host:** On biting the human host, the mosquito injects sufficient number of viruses into the blood stream. The virus, along with the circulating blood, then comes in contact with the susceptible target cells such as endothelial cells of the capillaries, macrophages, monocytes and reticuloendothelial system.

After replication in the susceptible cells, a secondary viraemia may develop resulting in the formation of sufficient number of viruses, which can infect the target organs such as the brain, liver, skin and blood vessels.

**Immune response:** Both humoral and cellular immunity is developed. They are important in the control of primary infection and prevention of further infection.

Induction of alpha and beta IFNs is caused by the production of double-stranded RNA, which not only limits the replication of virus, but also stimulates the immune response that results in the disease symptoms.

Circulating IgM and IgG are produced within the 6 d of infection. These antibodies prevent the spread of the virus and infection of the tissue. Cell-mediated immunity to the virus sometimes may also result in the destruction of tissue and contribute to the pathogenesis of encephalitis.

Hypersensitivity reactions such as DTH, the formation of immune complexes with virions and antigens, and the activation of complement can also occur. This can weaken the vasculature causing rupture leading to haemorrhagic symptoms.

**Epidemiology:** The dengue virus is able to infect both vertebrate and invertebrate hosts. It is also able to intiate a significant viraemia (presence of viruses in blood) in a vertebrate host for sufficient time to allow acquisition of the virus by the invertebrate vector; and able to initiate production of sufficient number of viruses in the salivary glands of the invertebrate host for further infection of the vertebrate host (Table 51).

**Table 51:** Epidemiology of dengue Infection

**Disease viral factors**

- Enveloped virus must stay wet and can be inactivated by drying, soap, and detergents.
- Virus can infect mammals, birds, reptiles, and insects.
- Asymptomatic or nonspecific (flu-like fever or chills), encephalitis, haemorrhagic fever, or arthritis

**Transmission**

- Togaviruses and flaviviruses: specific arthropods characteristic of each virus (zoonosis: arbovirus)

**Who is at risk?**

- People who enter ecological niche of arthropod: arboviruses

**Geography/Season**

- Endemic regions for each arbovirus are determined by habitat of mosquito or other vector.
- Aedes mosquito, which causes dengue and yellow fever, is found in urban areas and in pools of water.
- Culex mosquito, which causes St. Louis encephalitis, is found in forest and urban areas.
- Disease is more common in summer.

**Modes of control**

- Mosquito breeding sites and mosquitoes should be eliminated.
- Live attenuated vaccines are available for yellow fever virus and Japanese encephalitis virus.

---

Humans, generally, remain as dead-end hosts because they cannot spread the virus back to the vector as sufficient numbers of viruses are not maintained. In such cases, a full cycle of the disease occurs when the virus is transmitted by the arthropod vector and is amplified in susceptible new (reservoirs) host that allows the new infection of other arthropod.

Birds, reptiles, amphibians and small mammals act as reservoir hosts. The viruses are found in most tropical parts of the world including Asia, Africa, and Central and South America.

**Clinical symptoms:** Dengue fever is caused by four serologically-related viruses called dengue-1, dengue-2, dengue-3 and dengue-4.

Dengue fever is of two types: acute dengue fever and DHF. The acute dengue fever is characterized by the symptoms including chills, headache, high fever, rash, back and bone pains accompanied by nausea or vomiting, which lasts for 2–7 d in most cases. These symptoms subside after 7 d. Afterwards the disease may relapse with severe pains in the bones, muscles and joints; hence, the dengue fever is commonly called as break bone fever. However, there will be complete recovery (Fig. 99).

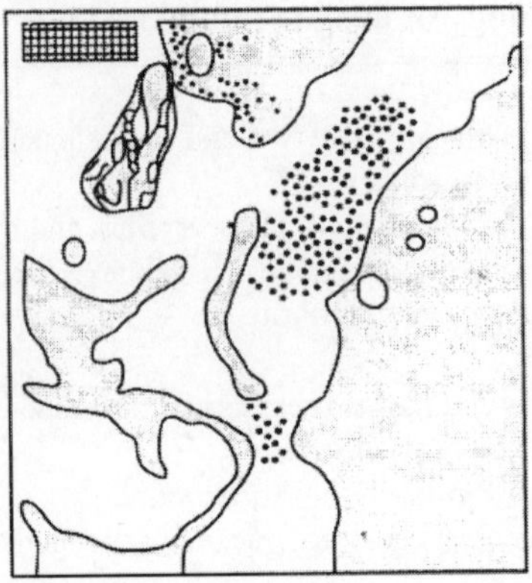

**Fig. 99** Clinical symptoms of dengue fever

The DHF is caused by another four related strains during or after acute dengue fever. Due to immune response of the entry of viruses into muscles of blood vessels, the vessels become weak and rupture, which leads to internal bleeding and loss of plasma. All these changes result in shock symptoms, which may culminate in death if immediate medical care and treatment is not given.

**Diagnosis:** Laboratory diagnosis consists of:

- Isolation of the virus from blood, CSF or tissue
- Detection of specific viral RNA in blood, CSF or tissue. After isolation, the viruses can be identified by cytopathological studies which include:
  - Immunofluorescence technique
  - RNA finger printing of the genomic RNA
  - Monoclonal antibody technique is used for the detection of individual strains and species.
- Serological tests: When compared to other tests, the infection can be identified quickly by serological tests, which include:
- Haemagglutination inhibition test
- ELISA, latex agglutination test

## REVIEW QUESTIONS

### I. Answer the following questions:

1. Describe the life cycle of malarial parasite.
2. Discuss briefly pathogenesis, clinical symptoms, and diagnosis and treatment of the disease malaria.

3. Give the details of the life cycle of filarial parasite.
4. Discuss pathogenesis and epidemiology of filariasis.
5. Describe pathogenesis, immune response and clinical symptoms of dengue fever.

**II. Write short notes on the following:**

1. Nocturnal periodicity of *Wuchereria bancrofti*
2. Describe different species, which cause malaria in humans
3. Elephantiasis
4. Prevention and control of filariasis
5. Structure of virion of dengue fever
6. Diagnosis of dengue

## 9.4 CONTACT DISEASES

A number of diseases are spread primarily by direct contact with an infected person or by contact with blood or excreta of the infected person. Many of the respiratory tract diseases are spread from person-to-person through direct contact with the infected individuals. The diseases include *Staphylococcus* infections, infection by *Helicobacter pylori,* gastric ulcers, diseases caused by hepatitis viruses.

The STDs also spread by person-to-person direct contact. The STDs, also called as venereal diseases, are caused by a wide variety of bacteria, viruses, protozoa and even fungi. Unlike respiratory pathogens that are constantly shed in large numbers by infected individual, sexually-transmitted pathogens are generally present in the body fluids exchanged during sexual activity. This is because sexually-transmitted pathogens are very sensitive to drying and other environmental stresses such as heat and light. Their habitat, that is, the human genitourinary tract is a protected and moist environment. Therefore, these organisms preferentially, sometimes exclusively, colonize the genitourinary tract.

Some pathogens are transmitted by direct contact with the tissues of infected animal/man. Diseases caused by such pathogens often have an occupational incidence, being contacted mainly by veterinarians, butchers, slaughterhouse workers, etc. Examples are anthrax, brucellosis.

### 9.4.1 Syphilis

Syphilis is a sexually-transmitted human disease caused by the bacterium, *Treponema pallidum*. It was discovered by Schaudinn and Hoffmann in 1905 in the chancres and inguinal lymph nodes of syphilis patients.

The origin of syphilis is not definitely known. Towards the end of the 15$^{th}$ century, syphilis spread widely throughout Europe in a virulent form. The name 'syphilis' was derived from a poem written by Fracaustorias of Verona, describing the legend of a shepherd's name – syphilis – who had been struck with the disease. It was widely held that syphilis was a new disease brought from America by Columbus's crew. Since then, the natural history of the disease has undergone alterations, but syphilis continues to be one of the most important and wide spread diseases of man.

Syphilis is worldwide in distribution. During the last five centuries, it has been recorded and studied extensively, and the disease has undergone much variation in its natural history and clinical symptoms. As originally described, it was a highly virulent disease with florid cutaneous manifestation. With the discovery of the dramatic therapeutic response to penicillin, it was hoped that it may even be possible to eradicate syphilis as the disease has no extra reservoir. However, it was not possible to eliminate the disease, but there has been even an increase in incidence due to the changing customs, habits and values in societies.

**Causal organism:** Syphilis is one of the venereal diseases caused by a bacterium *Tryponema pallidum* subspecies *pallidum*, but generally it is referred as *Tryponema pallidum*. It belongs to the order spirochaetales. It is strictly a human pathogen.

The spirochete is thin, coiled, with 8–14 coils, gram-negative, motile, aerobic or anaerobic. It is so thin that it cannot be seen under a light microscope in specimens stained with Giemsa stain (Fig. 100).

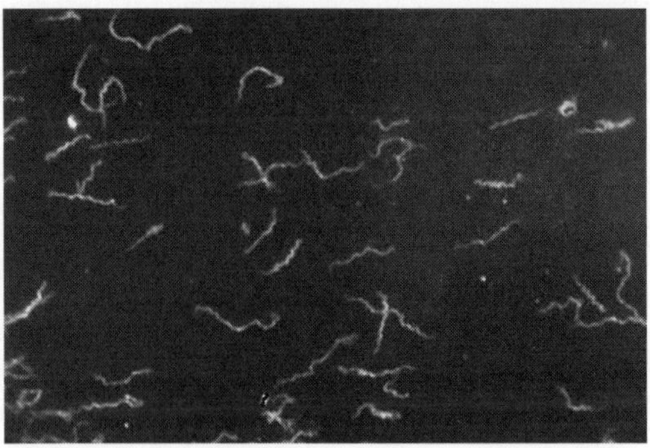

**Fig. 100** *Tryponema pallidum* in exudates (Dark-field microscopy)

**Pathogenesis:** The inability of this pathogen to grow in large number *in vitro* has limited the detection of specific virulence factors. However, by cloning experiments, several genes and their products have been isolated, and their virulent properties are characterized. The analysis has revealed that some of the gene products are associated with virulent strains.

The outer membrane proteins are involved in the adherence to the surface of the host cells. Virulent strains produce hyaluronidase that facilitates perivascular infiltration. Virulent strains are also coated with host cell fibronectins, which can protect against phagocytosis. The details are precised in Table 52.

**Table 52:** Precise information on *Treponema* infections

**Bacterial factors**

- Diseases: venereal syphilis (*T. pallidum* subspecies *pallidum)*, endemic syphilis or bejel *(T. pallidum* subspecies *endemicum*), *yows (T. pallidum* subspecies *pertenue*), pinta (*T. carateum)*
- Virulence factors: outer membrane proteins, hyaluronidase, coating of fibronectin
- Tissue destruction primarily results from host's immune response to infection.

**Transmission**

- Humans are the only host.
- Sexual contact or congenitally
- Contact of mucous membranes with infectious lesions; congenital infections rare.

**Who is at risk?**

- Sexually-active adolescents and adults; children borne of mothers having active disease
- Children or adults in contact with infectious lesions

**Geography/Season**

- It is worldwide.
- Endemic syphilis (bejel) occurs in desert and temperate regions of North Africa, Middle East, and Northern Australia.
- Yaws occurs in tropical or desert regions of Africa, South America and Indonesia.
- Pinta occurs in tropical areas of Central and South America.
- No seasonal incidence.

**Modes of control**

- Penicillin is a drug of choice; tetracycline, erythromycin, chloramphenicol are administered if patient is allergic to penicillin.
- Safe sex practices should be emphasized and infected sexual partners need to be treated.
- Endemic syphilis (yaws and pinta) can be eliminated through organized public-health measures. However, these efforts have been inconsistently applied.

The tissue destruction and lesions observed in syphilis are primarily the consequence of the patient's immune response to infection. The clinical course of syphilis evolves through three phases, the details of which have been given under clinical symptoms.

**Epidemiology:** Syphilis is worldwide and is the third most commonly sexually-transmitted bacterial disease after gonnorrhoea and *Chlamydia* infection. It occurs only in humans and is transmitted by direct sexual contact of untreated patients. About 25% exhibit a spontaneous cure; about 40% develop symptoms of tertiary syphilis. It is also transmitted to the foetus during first 4 mo of pregnancy through the placentum of an infected mother. This is called congenital syphilis. In 80% of the congenital syphilis cases, the child is born asymptomatic, but develops the disease symptoms several weeks or months later.

**Clinical symptoms:** As stated above, an untreated syphilis passes through three stages. The patient is able to transmit the infection to others only in the first two stages.

**1. Primary stage:** This stage develops after incubation period of 10–90 d in which the tryponemas attach to host cells, multiply locally, invade the lymphatic system and blood, and get distributed throughout the body. The first symptom of the disease is the formation of chancre, which is a painless ulcer with a hard margin. It develops on genitals or other parts of the body. Dark field microscopy reveals treponemas in this lesion. During this stage, the patient does not feel ill and the chancre heals within 25–40 d.

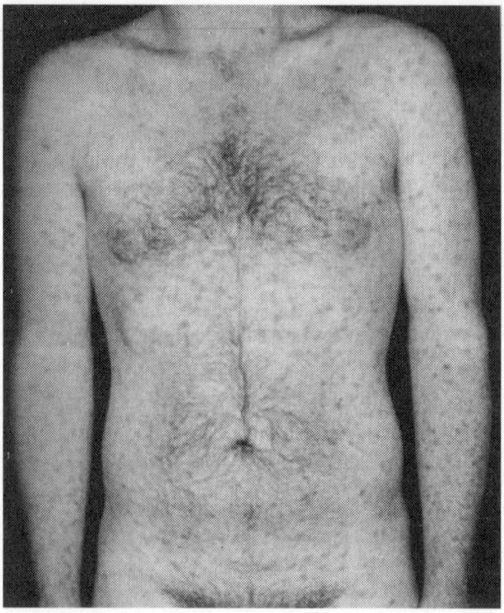

**Fig. 101** Development of rash in secondary syphilis

**2. Secondary stage:** This stage develops after 2–6 mo. A generalized eruption appears on the skin and mucous membranes of the body (Fig. 101). The patient may have swollen lymphnodes (lymphadenopathy), a vague feeling of bodily discomfort (malaise) and a slight fever or headache. The treponemas may be present in the lesions, but it is often difficult to demonstrate. The lesions on the skin and mucous membranes disappear after 3 wk to 3 mo. A latent stage of 3–30 y may intervene between the secondary and tertiary stages.

**3. Tertiary stage:** About 75% of untreated patients will progress to this stage, which is also called as late stage, and may exhibit symptoms of mild or severe syphilis. In about 5–6 y, patient develops mild syphilis due to which there may be blindness, difficulty in hearing and other symptoms of premature oldness. In severe syphilis, brain damage may occur resulting in paralysis along with decreased memory, personality changes, insomnia, headache and delusions. The patient becomes severely incapacitated and requires hospitalization. Tabes dorsalis results from damage to the spinal cord; the patient is uncoordinated and has loss of reflexes, uncontrollable urination, loss of perception and sensation, impotency, severe muscular pains. In addition to these symptoms, there may be damage to blood vessels such as aorta and those nourishing the CNS. Gummas, which are disfiguring granulomatous lesions, may appear on or within various parts of the body (Fig. 102).

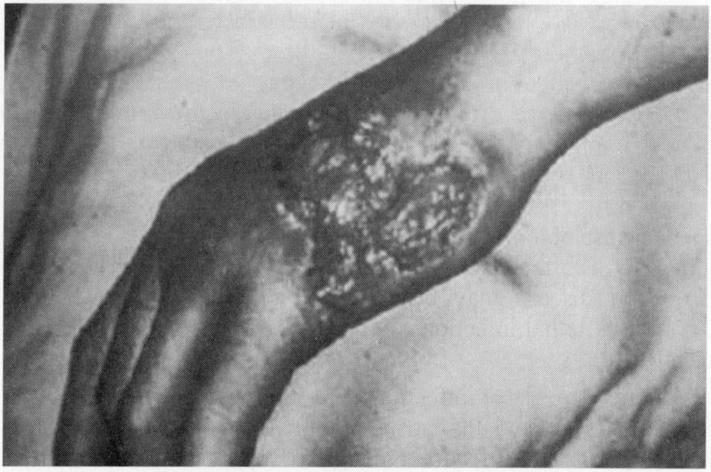

**Fig. 102** Inn Tritary or late syphilis lesions called Gammas rupture and results in ulcers on forehand

**Diagnosis:** As *Treponema pallidum* cannot be cultivated on laboratory media, the laboratory diagnosis of syphilis does not depend on its isolation from the infected samples. However, diagnosis is mostly based on the demonstration of antibodies, which are formed in response to the infection.

Two serological tests, which are more commonly performed, are VDRL and RPR (Rapid Plasma Reagin). However, these tests are not specific because non-treponemal antibodies, cardiolipins, are used to detect Wassermann test or Wassermann reaction (WR) antibodies (anticardiolipin antibodies), which occur in the serum of syphilitic patient. However, these antibodies are also formed in certain other diseases such as malaria, rheumatic arthritis etc. Therefore, certain specific tests, in which antitreponemal antibodies are used, are also performed. Two widely-used specific tests are FTA (Fluorescent Treponemal Antibody) and TPH (*Treponema pallidum* Haemagglutination) tests.

**Prevention and control:** Penicillin is the most effective drug for the treatment of syphilis. Long acting benzathine penicillin is used for the early stages of syphilis, and penicillin G for congenital and late syphilis. Penicillin-sensitive patients are treated with tetralycline or doxycycline.

Immunization of syphilis is not possible because of the absence of effective vaccine. The disease can be effectively controlled by following safe sex methods.

### 9.4.2 Gonorrhoea

Gonorrhoea is a venereal disease, which has been known from ancient times. The name 'gonorrhea' (meaning - flow of seed) was first employed by Galen in 130 AD. It is one of the STDs caused by *Neisseria gonorrhea,* a bacterium belonging to the family *Neisseriaceae*. It is commonly called gonococcus, which was first described in gonorrheal pus by Neisser (1870). Bunn, in 1885, cultured the coccus and proved its pathogenicity by inoculating to human volunteers. Gonococci resemble meningococci in many properties. It is exclusively a human disease, their being no natural infection in animals. However, experimental disease may be produced in chimpanzees by urethral inoculation. A lethal infection can be produced in mice by intracerebral inoculation.

The only source of infection is a human patient or carrier. The existence of asymptomatic carriage in females makes them a reservoir serving to perpetuate infection among their male contacts. The mode of infection is exclusively through sexual contact. Any contaminated article (fomite) does not play any significant role as the cocci rapidly die outside the human body. The only non-venereal infection is ophthalmia neonatorum (infection occurring to the newly born child during birth), which was so common previously, has now been controlled by the practice of instilling silver nitrate solution into the eyes of all newborn babies.

When sulphonamides and later penicillin were found effective for the treatment of gonorrhea, it was hoped that the disease could be eradicated. However, after a temporary decline, the incidence of the disease has been rising steeply. In 1970, the global incidence of new cases was estimated at 16 million, making it one of the commonest infectious diseases. In some areas, gonorrhoea has reached epidemic proportions, especially in adolescents and young adults. The reasons for the increase in gonorrhoea infection are largely social and cultural. A higher incidence of gonorrhoea has been observed in persons belonging to blood group B. The basis for this is not known.

**Causal organism:** The disease gonorrhoea in humans is caused by *Neisseria gonnorrhoeae*. It is aerobic, gram-negative coccus, typically arranged in pairs (diplococcus) with the adjacent sides flattened together. It is non-motile, nonendosporous encapsulated, oxidase positive and produces catalase. It is a fastidious organism and requires complex medium for growth, which makes recovery of the organism from clinical specimen difficult (Fig. 103).

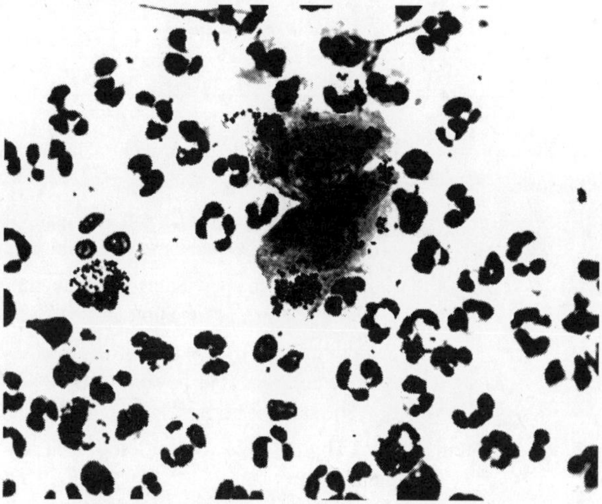

**Fig. 103** *N. gonococci* in the leukocytes

Capsule helps to prevent phagocytosis. The surface of the bacterium is covered with pili, which contain pilin protein and act as virulence factor. The pili help in adherence of the bacterium to host cells.

The cell wall is in the form of peptidoglycan layer, which is between the cytoplasm and outer membrane. The outer membrane consists of three types of protein:

1. Porin (Pore) proteins that form pores or channels in the outer membrane
2. Opacity protein (Opa) proteins that help to bind with host cells
3. RMP proteins that are closely associated with Pore proteins.

The cell wall also contains Lipooligosaccharides (LOS), which act as antigens along with proteases and betalactamases. Proteases are capable of cleaving IgA antibodies and betalactamases degrades penicillin by hydrolysis.

**Pathogenesis:** Gonococci enter the human body through urethra. They attach first to urethral mucosa cells, penetrate into the cells and then pass through the cells into the subepithelial space where infection is established.

The initial attachment to mucosal cells is mediated by pili. Opa proteins help in tighter association and migration into the epithelial cells. Pore proteins protect the phagocytosed bacteria from intracellular killing by inhibiting phagolysosome fusion. The gonococcal LOS stimulates the inflammatory response, which is responsible for most of the symptoms of gonococcal disease (Table 53).

**Table 53:** Virulence factors in *N. gonorrhoeae*

| Virulence factor | Biological effect |
|---|---|
| Capsule | Antiphagocytic |
| Pilin | Protein that mediates initial attachment to human cells |
| Pore (protein I) | Porin protein – promotes intracellular survival by preventing phagolysosome fusion in neutrophils |
| Opa (protein II) | Opacity protein – mediates firm attachment to epithelial cells and invasion into cells |
| Rmp (protein III) | Reduction-modifiable protein – protects other surface antigens (Par protein LOS) from bactericidal antibodies |
| Transferring-binding protein (Tbp) 1, Tbp2 | Two Tbp mediate acquisition of iron for bacterial metabolism |
| Lps | LOS – has endotoxin activity |
| IgA, protease | Destroys IgA (its role in virulence is unknown) |
| β-lactamase | Hydrolyzes beta-lactam ring of penicillin |

**Immune response:** $IgG_3$ is the predominant IgG antibody formed in response to gonococcal infection. Although the antibody response to Pore protein is minimal, serum antibodies to pilin, opa protein and LOS are readily formed, and can be detected. Antibodies to LOS can activate complement releasing C5a factor, which has chemotactic effect on neutrophils. However, IgG and secretary IgA1 antibodies directed against Rmp protein can block this bactericidal antibody response. People with inherited complement deficiencies are at greater risk for acquiring gonorrheal disease.

**Epidemiology:** Gonorrhoea occurs only in humans. *N. gonorrhoeae* is transmitted primarily by sexual contact. Hence, it is called as venereal disease. In males, the primary site of infection is the urethra and in females, it is the cervix. The disease is generally more prominent in males. The infection may extend to the prostate and epididymitis (Table 54).

**Table 54:** Premise information on *N. gonorrhoeae* infections

**Bacterial factors**

- Diseases: see Table 2
- Virulence factors: see Table 1

**Transmission**

- Primarily by sexual contact
- Asymptomatic carriage major reservoir

**Who is at risk?**

- Humans are only natural hosts
- Patients with multiple sexual encounters
- Patients with late complement deficiencies

**Geography/Season**

- Worldwide distribution
- Disease corresponds to sexual activity

**Modes of control**

- Ceftriaxone, cefixime or ciprofloxacin can be administered in uncomplicated cases. Penicillin should be avoided because resistance is common.
- Deoxycycline or azithromycin should be added for infections complicated by Chlamydia.
- For neonates, prophylaxis with 1% silver nitrate; ophthalmia neonatarum is treated with ceftriaxone.
- Prevention includes patients' education, use of condoms or spermicides with nonoxynol 9 (only partially effective) and aggressive follow-up of sexual partners of infected patients.
- Effective vaccines are not available.

Usually 20–80% of the females are asymptomatic, and those who are symptomatic, may have painful urination, vaginal discharge, fever and abdominal pain. In severe cases, the disease may spread to genital tract, and in about 10% of cases, the fallopian tubes become infected leading to Pelvic Inflammatory Disease (PID), which is a major cause of sterility in females. Other complications may include arthritis, endocarditis and meningitis (Table 55).

**Table 55:** Different species of Neisseria causing diseases in man

| Organisms | Disease |
|---|---|
| *N. gonorrhoeae* | Urethritis, cervicitis salpingitis, Pelvic Inflammatory Diseases (PID), proctitis, bacteraemia, arthritis, conjuctivitis, pharyngitis |
| *N. meningitides* | Meningitis, meningoencephalitis, bacteraemia, pheumonia, arthritis, urethritis |
| Other *Neisseria* species | Opportunistic infections |

**Clinical symptoms:** Acute gonorrheal infections are restricted only to genital organs, both in males and females. In men, the primary site of infection is the urethra. A purulent yellowish urethral discharge and dysuria develop after 2–5 d of incubation period. Approximately, 95% of all infected men have acute symptoms. In severe cases, there may be development of epididymitis, prostatitis, arthritis and periurethral abscesses.

In females, the primary site of infection is the cervix because the bacteria grow well in the endocervical columnar epithelial cells of cervix. Symptomatic patients generally experience vaginal discharge, dysuria and abdominal pain. Ascending genital infection including salpingitis (sterility) tuboovarian abscesses and pelvic inflammatory diseases (PID) are observed in 10–20% of women. Clinical symptoms of disseminated disease occur in about 1–3% women, which include fever, migratory arthralgis, suppurative arthritis in the wrists, knees and erythomatous base over the extremites but not on the head and trunk, ankles, and paricular rash on the extremities.

**Diagnosis:** Laboratory diagnosis is primarily made by demonstrating the *N. gonorrheae* inside neutrophils of the smear from urethral exudates (Fig. 104).

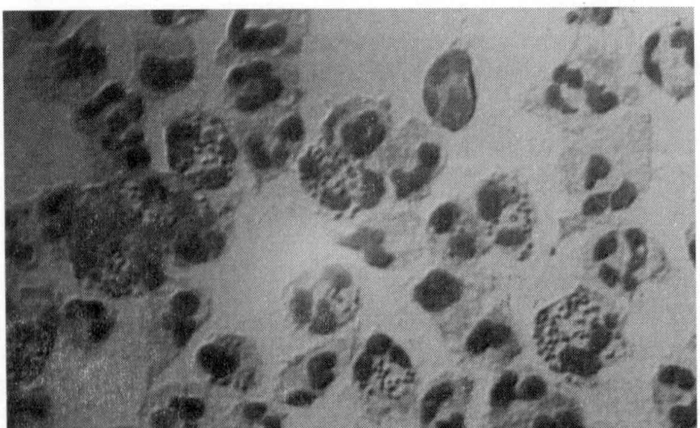

**Fig. 104** *N. gonorrheae* in urethral exudates

Isolation of the bacteria is carried out on Thayer martin medium or chocolate blood agar medium. *Neisseria* colonies are oxidase positive, which is indicated by the development of reddish-purple colour in the colonies when they are flooded with 1% tetramethyl-p-phenylenediamine reagent. Samples for the diagnosis of gonorrhoea can be collected from different body parts of a patient (Fig. 105). Species identification is made on the basis of biochemical and serological tests.

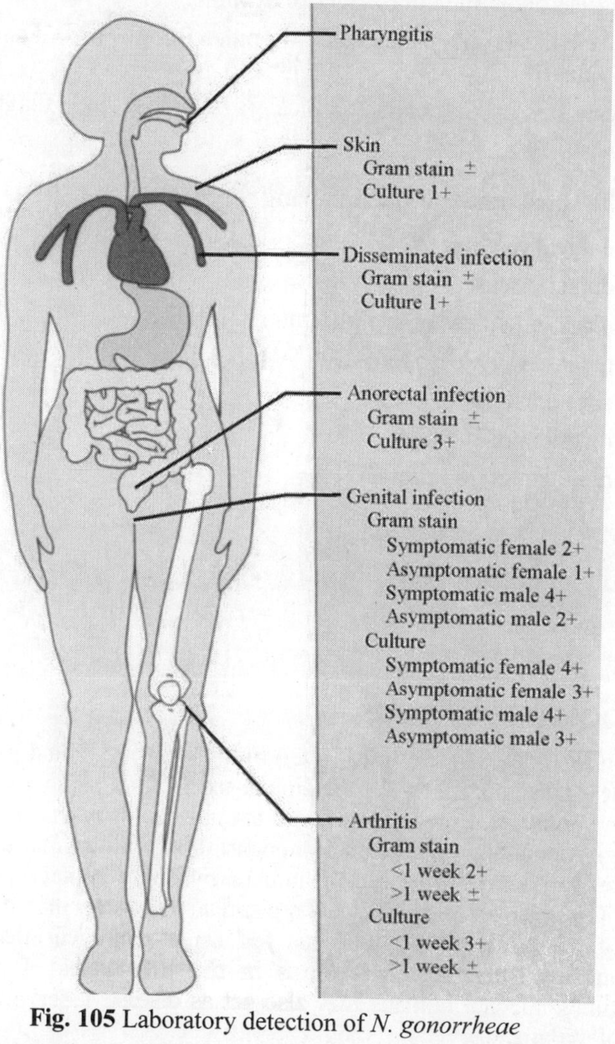


**Fig. 105** Laboratory detection of *N. gonorrheae*

**Prevention and control:** The disease can be treated with penicillin. Streptomycin is used for treating those cases, which are caused by penicillin-resistant strains. A vaccine is also developed for controlling the disease, which inhibits the disease by preventing pillary attachment of the bacteria to the host cells.

## REVIEW QUESTIONS

**I. Answer the following questions**

1. Discuss about the three stages of syphilis.
2. Give details of pathogenesis, epidemiology and diagnosis of syphilis.
3. Describe pathogenesis, clinical symptoms and diagnosis of gonorrhea.

**II. Write short notes on the following**

1. Venereal disease
2. Contact disease
3. Causal organisms of syphilis and gonorrhoea
4. Epidemiology of gonorrhoea
5. Tabes dorsalis
6. Treatment of syphilis
7. Treatment of gonorrhoea

## 9.5 ZOONOTIC DISEASES

Human diseases that are transmitted by animals are called as zoonotic diseases. The natural host for the animal-transmitted pathogens is a non-human vertebrate. Population of wild animals act as reservoirs for these pathogens making eradication impossible. The animal-transmitted diseases are generally spread to humans by direct contact, aerosols or bites. The vector-borne pathogens are spread to healthy man through the bite of an insect vector that last fed on a pathogen-infected host. Humans are often accidental hosts in the life cycles of pathogens, sometimes infected humans may also act as disease reservoir as in the case of malaria.

Animals contract a number of infectious diseases, some of which can be passed to humans. Fortunately, the use of effective immunization practices and good veterinary care prevents most of the infectious diseases to humans. However, the situation is different in the wild animal populations; they cannot be routinely immunized and do not receive veterinary care. These animal-transmistted diseases of man occur in wild animal population, usually on a periodic, cyclic basis. The important zoonotic diseases are rabies, antrax, hantavirus pulmonary syndrome and lyme disease. However, only rabies and anthrax are described in this chapter.

### 9.5.1 Rabies

The name 'rabies' comes from the Latin word meaning, 'mad', derived from the Sanskrit root 'rabhas', means to do violence. Reference to rabies has been recorded in the Mesopotamian Laws of Eshnunna (circa 1800 BC). Rabies is one of the contact human diseases transmitted by the bite of a mad dog. Rabies is caused to humans by rabies virus, which belongs to the rhabdoviruses groups.

The disease in the dog was traditionally associated with the appearance of the Dog Star, Sirius, during summer days, when dogs were considered to be prone to spells of madness. The disease is also called as hydrophobia because the patient exhibits fear of water, being incapable of drinking though subjected to intolerable thirst. Rabies in animals is not called hydrophobia because they do not have this peculiar character.

The causative agent of rabies has been associated with the saliva of rabid dog, but it was only in 1804 when Zinke gave proof by transmitting the disease to normal dogs by inoculation of the saliva from rabid dog. In 1821, Magendie and Brischet convincingly proved that rabies in man and animals is caused by the same pathogen, the rabies virus, by infecting healthy dogs with saliva from human patient. In a series of studies dating from 1881, Louis Pasteur established that rabies virus was present in the brain of infected animals. He obtained the fixed virus by serial intracerebral passage in rabbits and demonstrated that the dog could be rendered immune by a series of injections of fixed virus of great infectivity. This vaccine was prepared by drying pieces of spinal cord from rabies infected with fixed virus for various periods. In July 1885, Joseph Meister, a 9-year old boy, severly bitten by a rabid dog and in grave risk of developing rabies, was given a course of 13 inoculations of the infected cord vaccine by Pasteur. The boy survived. This dramatic event was marked as a mile stone in the development of modern medicine.

Most cases of human rabies are caused due to dog bite, but in endemic areas, almost any animal – domestic or wild – can transmit rabies. In India, anti-rabies treatment is indicated after the bite of any animal except the rat. Though rabies caused due to domestic animals has been controlled, the majority of rabies by wild animals will remain uncontrolled.

Rabies is endemic in India. It has been estimated that more than 15,000 people die of rabies every year and more than 7,00,000 receive anti-rabies vaccine. Human rabies can be checked bycontrol of rabies in domestic animals by killing, licensing and vaccination of the pets, and destruction of the stray animals. Rabies can be eradicated only if the wild animals are controlled. Eradication of rabies from India may not be possible, thanks to the abundant wild life.

**Causal organism:** The rabies in humans is caused by rabies virus. It belongs to the genus *Lissavirus,* which is included in the family *Rhabdoviridae.* Apart from humans, the virus also causes disease in many mammalian vertebrates (Table 56).

**Table 56:** Characters of rabies virus

| |
|---|
| (i) Bullet-shaped, enveloped, negative-sense single-stranded RNA that encodes five proteins |
| (ii) Prototype for replication of negative-sense enveloped viruses |
| (iii) Replication in the cytoplasm |

Rabies virus is typically bullet-shaped and measures 75 X 180 μm. It consists of two parts, a central core and a peripheral envelope. There are many spikes on the surface of the envelope. Each spike is composed of a trimer of the GP, which helps in the attachment to the host tissue and also generates neutralizing antibodies.

There is helical nucleocapsid in the core, which remains coiled symmetrically into a cylindrical structure, giving it the appearance of striations. The nucleocapsid consists of RNA and NPs. The RNA is single-stranded, negative-sense RNA. Two types of NPs surround the RNA, the structural proteins (N), and the Large and Non-structural (LNS) proteins. The N proteins are the major structural proteins of the virus, which protect the RNA from host cell ribonuclease digestion and maintain the RNA in a configuration suitable for replication. The LNS proteins constitute the RNA-dependent RNA polymerase (Fig. 106). The matrix (M) protein lies between the envelope and the nucleocapsid.

**Pathogenesis:** Rabies virus infects several mammalian vertebrates. Based on their habituation, they are of two types – wild and domestic animals. The former is called as sylvatic type that includes wolfs, foxes, bats etc. and the latter called as urban type that includes dogs, cats etc.

The rabies virus in wild animals is localized in CNS, cerebellum, cerebrum, cortex and spinal cord. The virus is transmitted to domestic animals when they are bitten by wild animals and from domestic animals to humans. Therefore, the rabies virus moves in a cyclic manner among the wild animals, domestic animals and humans. Infected animals are called as rabid animals, which show aggressive behaviour or madness. The dog is the principal agent for the transfer of rabies virus to humans.

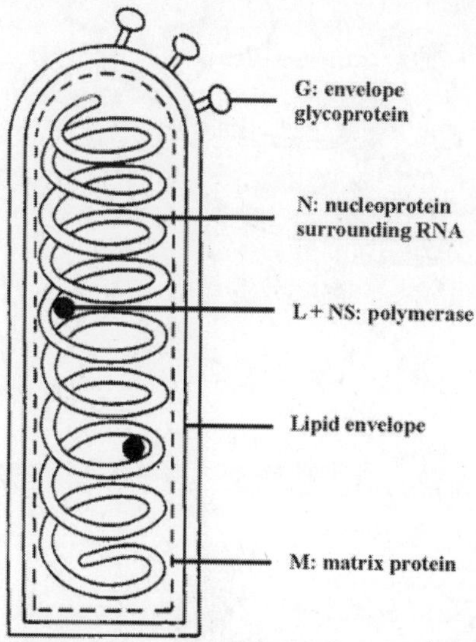


**Fig. 106** Diagramatic structure of rabies virus

A human contracts the rabies virus when a mad dog bites him. The saliva of a mad dog or any other animal contains more number of rabies virions. The virus can also be transmitted through inhalation of aerosols with virus, or through transplanted infected tissue (e.g., cornea) or by inoculation through intact mucosal membranes.

After entering into the human body, the virus binds to the muscle cells at the site of bite where it may remain for days to months before progressing into CNS. Once the virus gains an access to the spinal cord, several parts of the brain get rapidly infected. The virus then spreads from the CNS through afferent neurons to other sites of the body such as skin of the head and neck, adrenal medulla, renal parenchyma and pancreatic cells. Encephalitis develops once the virus infects the brain (Fig. 107).

Rabies becomes fatal once the clinical symptoms appear in the patient. The length of the incubation period is determined by the concentration of the virus in the inoculum, the proximity of the wound to the brain, the severity of the wound, the host age and immune status (Table 57).

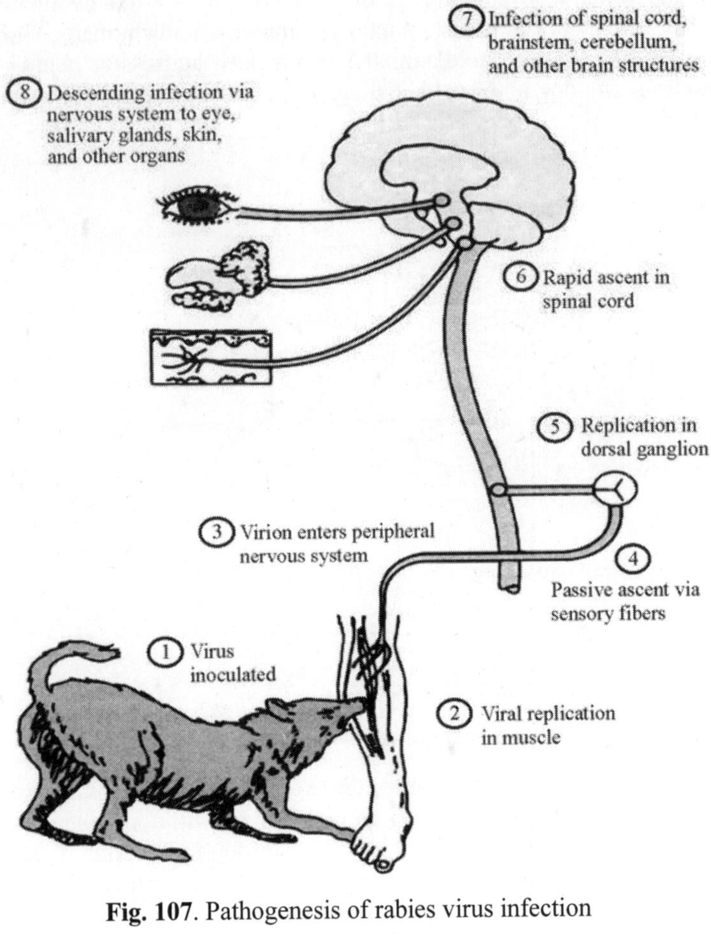


**Fig. 107**. Pathogenesis of rabies virus infection

**Immune response:** Neutralizing antibodies are not apparent until the clinical symptoms are well established, probably because the infection remains hidden from the immune response. Cell-mediated immunity appears to play little or no role in protection against rabies virus infection.

Therefore, the antibodies are to be administered or generated to stop the spread of the virus to the CNS and brain. These antibodies are generally produced through vaccination to control and prevent the disease.

**Epidemiology:** Rabies is a classic zoonotic infection that spreads from animals to human as explained above. It is endemic worldwide in a variety of animals. Dog is the principal reservoir of rabies in most parts of the world. It is estimated that rabies accounts for 25,000 deaths per year in India, in which the virus is transmitted by dogs in 90% of the cases, whereas in the US, the incidence of human's rabies is approximately 1% because of effective dog-vaccination programs (Table 58).

**Table 57: Disease mechanisms of rabies virus**

Rabies is usually transmitted in saliva and is acquired from the bite of a rabid animal.

Rabies virus is not very cytolytic and seems to remain cell associated.

Virus replicates in the muscle at the site of the bite with minimal or no symptoms (incubation phase).

The length of the incubation phase is determined by the infectious dose, and the proximity of the infection site to the CNS and brain.

After weeks to months, the virus infects the peripheral nerves and travels up the CNS to the brain (prodrome phase).

Infection of the brain causes classic symptoms, coma and death (neurological phase).

During the neurological phase, the virus spreads to the glands, skin, and other body parts including the salivary glands, from where it is transmitted.

Rabies infection does not elicit an antibody response until the late stages of the disease, when the virus has spread from the CNS to other sites.

Antibody can block the progression of the virus.

The long incubation period allows active immunization as a post-exposure treatment.

**Table 58:** Epidemiology of Rabies virus

**Disease/Viral factors**

- Virus-induced aggressive behaviour in animals promotes the spread.
- Disease has long, asymptomatic incubation period.

**Transmission**

**Zoonosis:**

- Reservoir: wild animals
- Vector: wild animals, and unvaccinated dogs and cats

**Source of virus:**

- Major: saliva in bite of rabid animal
- Minor: aerosols in bat caves containing rabid bats

**Who is at risk?**

- Veterinarians and animals handlers
- Person bitten by a rabid animal
- Inhabitants of countries with absence of pet vaccination programmes

**Geography/Season**

- Virus is found worldwide, except in some island nations.
- There is no seasonal incidence, summer is comparatively more favourable.

**Modes of control**

- Vaccination program is available for pets.
- Vaccination is available for high-risk personnel.
- Future vaccination program is needed for control of rabies in forest mammals.

**Clinical symptoms:** Rabies is almost always a fatal disease unless prevented by vaccination. In humans, the incubation period from the time of infection varies from 6 d to 1 y, but is usually about 3–8 wk (Table 59). The development of symptoms and the length of incubation period in untreated cases depend on the severity and location of bite. It has been estimated that only 5–15% of the persons bitten by a rabid animal contract rabies.

**Table 59:** Progress of rabies disease in human beings

| Disease phase | Symptoms | Time | Viral status | Immunological status |
|---|---|---|---|---|
| Incubation phase | Asymptomatic | 60–365 d after bite | Low titre, virus in muscle | ----- |
| Prodrome phase | Fever, nausea, vomiting, loss of appetite, headache, lethargy, pain at the site of bite | 2–10 d | Low titre, virus in CNS and brain | ------ |
| Neurological phase | Hydrophobia, pharyngeal spasms, hyperactivity, anxiety, depression; CNS symptoms: loss of coordin-ation, paralysis, confusion, delirium | 2–7 d | High titre, virus in brain and other sites | Detectable antibody in serum and CNS |
| Coma | Coma; cardiac arrest, hypotension, hypoventilation, secondary infections | 0–14 d | High titre, virus in brain and other sites | ----- |
| Death | ---- | ---- | ---- | ---- |

Symptoms in humans include fever, headache, pain or itching at the site of bite, gastrointestinal symptoms, fatigue and aneroxia. These symptoms persist for two to several days, after which the neurological symptoms specific to rabies appear. The most important of these symptoms is hydrophobia (fear of water), which occurs in 20–50% patients. It is characterized by the patients who desire to drink water, but cannot drink it because of the intensive pain in the throat while swallowing water. Apart from hydrophobia, generalized seizures and hallucinations are common during the neurological phase. Death usually follows paralysis or convulsive seizures. The mortality rate from rabies is nearly 100%.

**Diagnosis:** Diagnosis of rabies is not possible during preliminary stage of the disease because evidence of infection, including symptoms and the detection of antibody, does not occur until the disease progresses to advanced neurological phase. However, laboratory tests are generally conducted to assess whether the clinical symptoms are due to rabies or not. For this, diagnostic tests are made by detection of viral antigen in the CNS or skin isolation of the virus and serological findings.

The most important finding is the detection of intracytoplasmic inclusion bodies, called Negri bodies, in the affected neurons. Immunofluorescence tests are also conducted to detect the antigen for diagnosing rabies.

**Prevention and control:** As with most of the viral diseases, there is no treatment for rabies. The only way available for preventing the virus from reaching the CNS of a bitten person is to follow the immunization process. Immunization process consists of administration of a combination of passive and active vaccines. Passive vaccine consists of horse globulin, which provides an immediate source of antibodies against the virus, but it lasts for only about 14 d. Active vaccine consists of rabies vaccine, which is a killed-virus vaccine prepared by the chemical inactivation of rabies-infected tissue-culture of Human Diploid Cells Vaccine (HDCV). The vaccine is given intramuscularly on 0, 3, 7, 14 and 28 d after exposure. Prevention of human rabies also requires the control of rabies in domestic and wild animals.

### 9.5.2 Anthrax

Anthrax is a zoonotic disease of man being transmitted by animals to humans. Animals are infected by ingestion of the spores present in the soil. Direct spread from animal to animal is rare. The disease generally results in septicaemia, but sometimes may be localized resembling the skin disease in man. Infected animals shed large number of bacilli through discharges from the mouth, nose and rectum, which sporulate in soil and serve as the source of infection.

Human anthrax is contracted from animals directly or indirectly. The disease may be cutaneous, pulmonary or intestinal; all types result in fatal septicaemia.

Anthrax is primarily an animal disease in India. The number of animals infected annually runs into lakhs. The disease is rare in some countries such as Britain where infection occurs through contaminated hides, bone meal fertilizer and other animal products. The spread of anthrax in man is not clear, but about 10,000 cases are reported to occur annually throughout the world mostly in rural areas. Anthrax in man provides permanent immunity and second attack is extremely rare.

**Causal organism:** Anthrax is caused by *Bacillus anthracis,* a bacterium belonging to the family *Bacillaceae*. Of the known 51 species of *Bacillus,* only *Bacillus anthracis* is strictly pathogenic to humans and cattle.

*Bacillus anthracis* is a large, 1 x 3–5 µm in size, organism arranged as single or paired bacilli in clinical specimen, and as long serpentine chains and clumps in culture. It is aerobic or facultative anaerobic, gram-positive or gram variable, spore forming, motile and catalase positive. Its ability to form spores allows it to survive in soil for years (Fig. 108). Capsule is clearly visible in clinical specimen, but is not produced *in vitro* cultures unless special growth conditions are provided. The capsule is made up of a polypeptide consisting of glutamic acid.

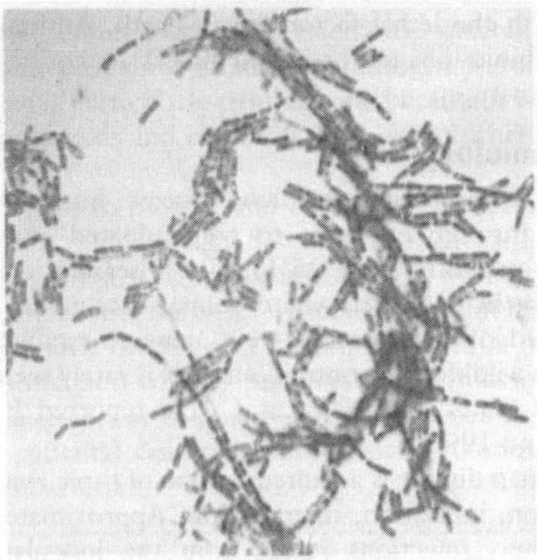

**Fig. 108** Spores of *Bacillus anthracis*

The cell wall is made up of a polysaccharide, which acts as an antigen. The organism also produces a toxin called anthrax toxin. The toxin consists of three antigenically-distinct heat-labile components:

1. Protective antigen
2. Lethal factor
3. Edema factor

All the three components together exhibit toxic properties.

**Pathogenesis:** The capsule and the toxin are the two major factors responsible for the virulence of *B.anthracis*. The capsule is antiphagocytic, and antibodies directed against the capsule are not protective. The combination of anthrax toxin-protective antigen and the edema factor cause edema in the affected persons and animals. The combination of protective antigen with the lethal factor causes death. Anthrax toxin also inhibits phagocytosis and blocks oxidative activity.

**Epidemiology:** Anthrax is primarily a cattle disease. Humans are infected through exposure to contaminated animals or animal products. Human disease is acquired by one of the three routes:

1. Inhalation anthrax is also called wool-sorter disease, which results from inhalation of *B.anthracis* spores during the processing of goat hair.
2. About 95% of anthrax infection results from the inoculation of *Bacillus* spores through exposed skin, either from contaminated soil or infected animal products such as hides, goat hair or wool.
3. Ingestion anthrax is very rare in humans, but ingestion is a common route of infection in many cattles. The organism is not highly contagious and person-to-person transmission does not occur (Table 60)

**Table 60:** Characteristic features of *Bacillus anthracis* infections

**Bacterial factors**

- Spores can survive in soil for years.
- Virulence factors include polypeptide capsule and anthrax toxin (lethal factor, edema factor, and protective antigen).

**Transmission**

- Inoculation (most common)
- Inhalation (rare in humans, more common in herbivores)

**Who is at risk?**

- People in endemic areas in contact with infected animals or contaminated soil
- People who work with animal materials (hides, fur, wool, hair) imported from endemic areas

**Geography/Season**

- Prevalent in impoverished areas where vaccination is not practiced
- No seasonal incidence

**Modes of control**

- Penicillin is the drug of choice; erythromycin or chloramphencol is administered for patients allergic to penicillin.
- Vaccination of animal herds and people in endemic area is preferred.
- Disposal of animals that die of anthrax.

**Clinical symptoms:** Generally, cutaneous anthrax starts with the development of a painless papule at the site of inoculation that rapidly progresses to an ulcer surrounded by vesicles and then to a necrotic structure. This structure may develop into a massive edema due to anthrax toxin. The mortality rate in untreated cutaneous anthrax is about 20%.

Symptoms of inhalation anthrax initially are similar to viral respiratory infection. It may rapidly progress to scattered pulmonary involvement leading to respiratory failure. The mortality rate is high even in suitably-treated cases because the disease is usually not apparent till advanced stage.

Gastrointestinal anthrax is very rare in humans with varied clinical symptoms including edema of mesentry and intestinal haemorrhage. The mortality rate is very high.

**Diagnosis:** Laboratory diagnosis of *B.anthracis* can be readily made by the microscopic observation of cultures made from cutaneous papules or ulcers. It can also be detected by the absence of haemolysis, the sticky consistency of the colonies and the microscopic appearance of serpentine chains of bacilli. It can also be diagnosed by chemical and serological tests.

**Prevention and control:** The anthrax disease can be effectively controlled with penicillin. Erythromycin or chloremphenicol is used for the patients allergic to penicillin. As the disease spreads through animals, control of human disease requires control of animal disease, which can be achieved by vaccination of animal herds in endemic region and the burning or burial of animals that die of anthrax.

Vaccination of humans who live in endemic areas and who work with animal products such as hides, fur, bone meal, wool and hairs is necessary. However, complete eradication of anthrax is not possible because the spores of the organism can exist for many years in the soil.

## REVIEW QUESTIONS

### I. Answer the following questions:

1. Discuss about the pathogenesis, clinical symptoms and diagnosis of rabies.

2. Explain the causal organism of rabies.
3. Describe the pathogenesis, epidemiology, diagnosis, and prevention and control of anthrax.

**II. Write short notes on the following:**

1. Preparation of rabies vaccine by Louis Pasteur
2. Situation of rabies in India
3. Structure of anthrax pathogen
4. Prevention and control of rabies
5. Immune response in host against rabies
6. Zoonotic disease

## 9.6 BLOOD-BORNE DISEASES

Some microbial pathogens are transmitted from person to person by transfusion of blood or blood products, or by contact with traces of blood or blood serum on hypodermic syringes, syringe needles, contaminated razors and singular items that have been used by more than one person without being properly sterilized after each use. A classical example is viral Hep B, a disease that is common among drug addicts, blood bank workers, dentists and surgeons, who because of their occupation can come into contact with infected blood or blood products.

### 9.6.1 Serum Hepatitis

It is also called as hep B disease. The term 'hepatitis' refers to a primary infection of the liver by hepatitis virus type-A or type-B, and other hepatitis viruses presently designated as non-A and non-B viral infection transmitted mainly by inoculation. Persons receiving serum inoculation or blood transfusion are called as Hep B. It is also called as homologous serum jaundice or serum hepatitis.

Blumberg and coworkers reported a protein antigen in the serum of an Australian aborigin, which gave a clear precipitation with sera from two haemophiliacs who had received multiple transfusions. This antigen was called Australian antigen. By 1968, the Australian antigen was shown to be associated with serum hepatitis. It was subsequently shown to be the surface component of HBV (HBsAg).

The disease occurs throughout the world. It has no seasonal distribution. The incidence of infection is more in adults than in children, and in urban than in rural areas. The virus is maintained in nature by chronic viraemia carriers who are estimated up to a number over 200 million, of which 35 millions are in India. The reported prevalence of carriers in different populations varies widely, from 0.1% in the advanced countries to 20% in the developing countries. Carrier rate is higher in tropical than in temperate regions and in males than in females. Though acute and chronic infections can be established experimentally in chimpanzees, no natural animal reservoir is known for the virus.

**Causal organism:** The disease serum hepatitis is caused by HBV, which belongs to the *Hepadnaviruses* family (Table 61).

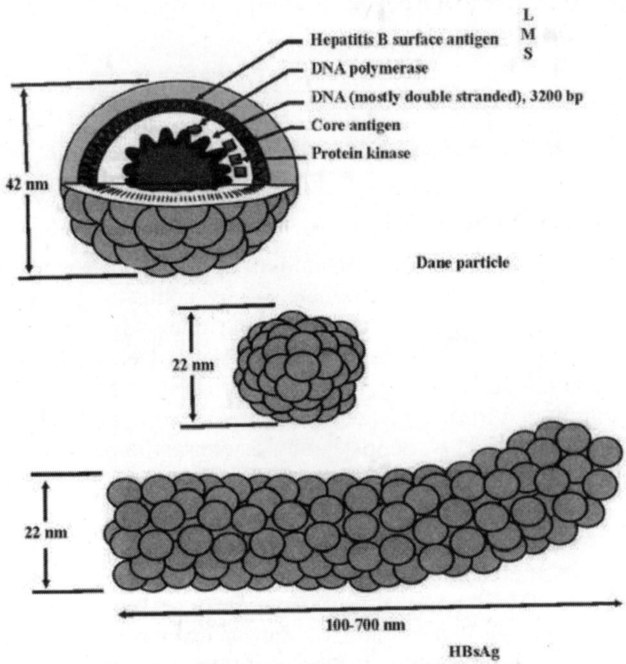


**Fig. 109** Structure of HBV

The HBV is circular and 42 μm in diameter. It consists of peripheral envelope and centrally-located genome. The genome is a small, circular, partly double-stranded DNA of 3,200 bases. Although HBV is a DNA virus, it encodes reverse transcriptase. The virion is also called Dane particle.

The HBV virion contains ribonuclease. Its activity is determined by a P protein attached to the genome, which is surrounded by the HBcAg and an envelope containing the GP HBsAg. It also contains HBcAg antigen protein as a minor component.

**Table 61:** Characteristic features of *Hepadnaviruses*

- Virus has enveloped virion containing partially double-stranded, circular DNA genome.
- Replication is through a circular RNA intermediate.
- Virus encodes and carries a reverse transcriptase.
- Virus encodes several proteins (HBsAg [L, M, S], HBe/ Hepatitis B core antigen [HBc]) that share genetic sequences, but with different in-frame start codons (AUG).
- The HBV has a strict tissue tropism to the liver.
- The HBV-infected cells produce and release large amounts of HBsAg particles lacking DNA.
- The HBV genome can integrate into the host chromosome.

HBsAg-containing particles are released into the serum and their number is more than the actual virions. These particles can be spherical, but smaller than the Dane particles or filaments. They are immunogenic and were processed into the first commercial vaccine against HBV (Fig. 109). The HBsAg includes three GPs – L, M and S – encoded by the same gene, and read in the same frame, but translated into protein from different AUG start codons.

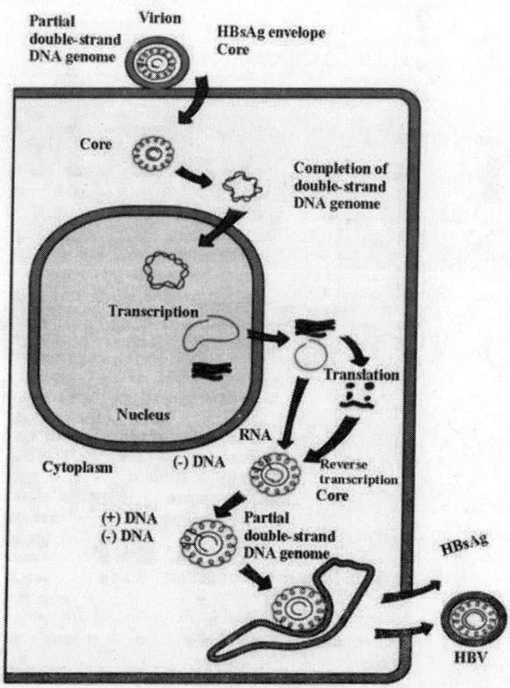


**Fig. 110** Proposed pathway of replication of HBV

**Pathogenesis:** The HBV can cause acute or chronic symptomatic or asymptomatic disease based on the person's immune response to the infection. Presence of HBsAg and HBcAg components of the virion in the blood indicates ongoing active infection (Fig. 110). The major source of infection is the transfusion of infected blood or sexual contact or infection to a newly born child during birth from infected mother.

The virus starts to replicate within 3 d of its acquisition. However, symptoms may not be observed for 45 d or longer, depending on the infection dose, the route of infection and the person. The virus replicates in hepatocytes with minimal harm, after which, the infection proceeds for a relatively long time without causing liver damage or symptoms. Presence of intracellular filamentous form of HBsAg can produce hepatocyte cytopathology characteristic of HBV infection.

Cell-mediated immunity and inflammation are responsible for causing the symptomology, and effecting resolution of the HBV infection. An insufficient T-cell response to the infection generally results in the occurrence of mild symptoms, an inability to resolve the infection and the development of chronic hepatitis (Fig. 112).

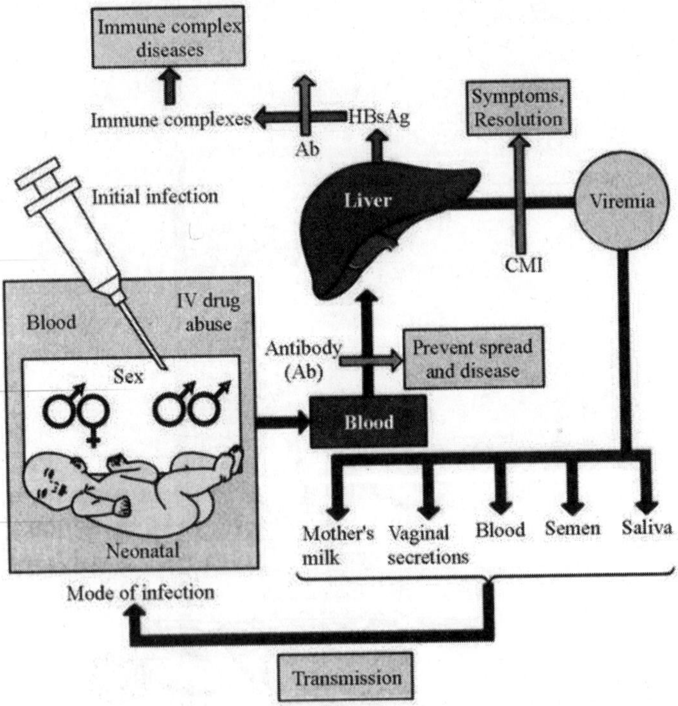


**Fig. 111** Spread of HBV in the body

Antibody, (as generated by vaccination), can protect against the infection. However, serum HBsAg binds to and blocks the action of neutralizing antibody, which limits the antibodies' capacity to resolve an infection. Immune complexes formed between HBsAg and anti-HBsAg contribute to the development of hypersensitivity reactions leading to problems such as vasculitis, arthralgia, rash and kidney damage.

Fulminant infections, the activation of chronic infections or co-infection with the delta agent, can lead to permanent damage of liver causing cirrhosis. Infants and young children have an immature cell-mediated immune response, and exhibit less tissue damage and milder symptoms. When infected perinatally, as many as 90% of the infants become chronic carriers. Viral replication persists in these people for long periods (Table 62).

**Table 62:** High-risk groups for HBV infection

- People from endemic regions
- Babies of mothers with chronic HBV
- Intravenous drug abusers
- People with multiple sex partners: homosexual and heterosexual
- Haemophiliacs and other patients, requiring blood and blood product treatments
- Healthcare personnel who have contact with blood
- Residents and staff members of institutions for the mentally retarded

**Epidemiology:** There are three routes for transmission of HBV (Table 63).

**Table 63:** Epidemiology of HBV, HCV, and HDV

**Disease/Viral factors**

- Enveloped virus is labile to drying. The HBV is less sensitive to detergents than other enveloped viruses.
- Virus is shed during asymptomatic periods.
- Virus causes chronic disease with potential shedding.

**Transmission**

- In blood, semen and vaginal secretions (HBV: saliva and mother's milk).
- Via transfusion: needle-stick injury, shared drug paraphernalia, sex and breast-feeding.

**Who is at risk?**

- Children: mild asymptomatic disease with establishment of chronic infection
- Adults: insidious onset of hepatitis

- HBV-infected people co-infected or superinfected with HDV: abrupt, more severe symptoms with possible fulminant disease.
- Adults with chronic HBV: at high risk for primary hepatocellular carcinoma

**Geography/Season**

- Viruses are found worldwide.
- There is no seasonal incidence.

**Modes of control**

- Avoidance of high-risk behaviour
- HBV: vaccine and screening of blood supply

**1. Exposure to contaminated blood:** This can occur through a number of mechanisms, most commonly by blood transfusion. However, this can be prevented by screening all blood donors for evidence of HBV. An exposure to contaminated blood can occur in much more subtle ways such as sharing of contaminated needles by drug users or by sharing tooth brushes or razors with a chronic carrier. In hospitals, healthcare workers come in contact with HBV by accidental needle stick. (Fig. 111).

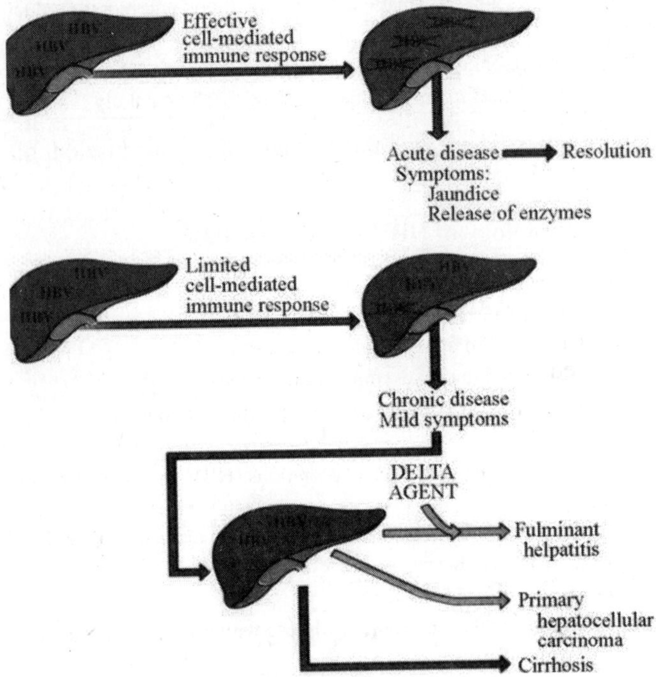


**Fig. 112** Major determinants of acute and chronic HBV infection

**2. Through sexual contact:** This occurs due to presence of virus in genital tract secretions of the chronic carriers. In most of the European countries, HBV is primarily a STD. Sexual workers are at high risk of contracting HBV infection by this route.

**3. From mother to baby:** Most commonly this occurs at the time of birth (perinatally) through exposure to a contaminated birth canal and maternal blood. This results in high rate of carriage in the offsprings. In countries where transmission is through this route, upto 20% of the population may be HBV carriers and some of them even develop hepatoma (cancer tumor). Globally, this is the most important route. As per the estimation of WHO, there are 350 million HBV carriers in the world. They are not evenly distributed around the globe because there are some countries with very high carriage rate and others with very low carriage rate.

**Diagnosis:** The HBV infection results by exposure of the host immune system with a number of antigens to which an antibody response may or may not be there. The laboratory diagnosis mainly consists of estimation or detection of HBsAg / anti-HBs, antiHBc and HBeAg / antiHBe. Presence of HBsAg and anti-HBc in the serum indicates the HBV infection as acute/chronic.

Presence of HBeAg in a serum sample indicates a high degree of infection and the patient at increased risk of serious liver disease. Presence of anti-HBe at greater concentration in a serum sample indicates that the patient is at much lower-risk of serious liver disease.

**Prevention and control:** Chronic HBV carriers can be treated with IFN-a, which acts as an immunomodulatory agent. This drug empowers the host immune system to eliminate the majority of the HBV-infected hepatocytes, which results in seroconversion from HBeAg to anti-HBe. However, IFN is not effective for those carriers who are co-infected with HIV or for those who acquired an infection at birth, because the immune system of these patients is tolerant to HBV. The HBV infection can be prevented by vaccination programme.

A subunit vaccine, consisting of entirely HBsAg, is effective in inducing protective immunity against HBV infection. The standard method consists of three doses at 0, 1 and 6 mo. The same vaccine is also used routinely in many countries to prevent HBV infection in newly born children.

For non-vaccinated individuals or vaccine nonresponders, protection after an acute infection can be given by passive immunization, in which preformed anti-HBs is used, which is also called as HBIg) that is collected from the blood donors.

### 9.6.2 AIDS

AIDS is a blood-borne disease transmitted to man through blood. It is caused by HIV. It belongs to *Retroviridae* family of viruses. Because of its pandemic nature, AIDS is a greatest challenge to public health in modern times. The full consequence of AIDS could not be realized for several years because of the silent spread and slow evolution of this infection. It is truly global in its magnitude and can affect any continent, nation, race or political systems.

The first indication of this new disease began in the summer of 1981 by reports from New York and California (USA) of a sudden increase in the incidence of two very rare diseases, Kaposi's sarcoma pneumocystis casini, a type of pneumonia, in young adults who were homosexuals or addicted to heroin or other injected narcotics. Further investigations revealed that they were also suffering from fatal infections of virulent microorganisms as well as lymphoid and other malignancies because of failure of immune competency. Due to this nature of occurrence of simultaneous multiple infections, it is called as Acquired Immune Deficiency Syndrome.

Subsequently, it is demonstrated that AIDS is transmitted to humans through sexual contact, transfusion of blood or other fluids and infected mother to newborn. With the rapid increase in the number of AIDS patients, alarm about AIDS mounted and investigation about its causative agent assumed the highest priority.

Isolation of causative agent of AIDS was first reported by Lue Montagmies (1983) and his colleagues from the Pasteur Institute, Paris. They isolated a retrovirus from a West African patient with persistent generalized lymphadenopathy, which is a major symptom of AIDS and called it Lymphadenopathy-associated Virus (LAV). It produced lytic infection in fresh peripheral blood lymphocytes, but could not be established in permanent cell lines.

In 1984, Robert Gallo and his colleagues from National Institute of Health, USA isolated a retrovirus from an AIDS patient and called it Human T Lymphotrophic Virus-III (HTLV-III). It can be grown in continuous culture in T-cell leukaemia cell lines resulting in the formation of sufficient antigens for serological tests. Other similar isolates were reported from AIDS patients under different names as AIDS-related Virus (ARV) etc.

Serological and molecular cloning studies have established common origin of these viruses though they show differences in antigenic properties. Therefore, International committee on virus nomenclature, in 1986, decided to name all these different isolates under the generic name HIV.

The first serological test became available for detection of anti-HIV antibodies in 1984. This made possible a more realistic estimation of the extent of HIV infection. Till then, the infection could be recognized only when patients developed the characteristic clinical features such as opportunistic infections or malignancies. Serological screening of high-risk groups, blood donors and others, revealed a very large and increasing reservoir of HIV in patients and carriers in different parts of the world.

By the end of 1998, over 10 million persons are believed to have been infected with HIV. Some 1.5 lakhs of them, spread over 142 countries representing all continents, have developed clinical AIDS. These figures indicate the dimension of the problem, which is bound to increase in succeeding years.

In India, HIV infection has arrived late and has spread slowly. From the available information, transmission has been mainly in heterosexuals, involving prostitutes. Infection through blood transfusion has also been reported. The first case of AIDS identified in India had acquired the infection from abroad. However, indigenous cases since then are on the increase.

**Table 64:** Characteristic features of retroviruses

- Enveloped, spherical virion that is 80–12 nm in diameter and that encloses a capsid containing two copies of the positive-strand RNA genome (approximately 9 kilobases for HIV and TLV).
- RNA-dependent DNA polymerase (reverse transcriptase) and integrase enzymes are carried in the virion.
- Virus receptor is the initial determinant of tissue tropism.
- Replication proceeds through a DNA intermediate, termed as the provirus.
- The provirus integrates randomly into the host chromosome and becomes a cellular gene.
- Transcription of the genome is regulated by the interaction of host transcription factors with promoter and enhancer elements in the Long Terminal Repeat (LTR) portion of the genome.
- Simple retroviruses encode gag, pol, and env genes. Complex viruses also encode accessory genes (e.g., tat, rev, nef, vif, vpu) for HIV.
- Virus assembles and buds from the plasma membrane.
- Final morphogenesis of HIV requires protease cleavage of gag and gag-pol polypeptides after envelopment.

**Causal organisms:** The AIDS is a human disease caused by the HIV. Two types of viruses are identified HIV-1 and HIV-2. However, they are identical in virological properties and nature of infection. Therefore, both of them are referred simply as HIV (Table 64). It belongs to the retroviruses group. Baltimore and Temin feel that retroviruses encode RNA-dependent DNA polymerase, also called reverse transcriptase, and replicate through a DNA intermediate. The HIV is roughly spherical with a diameter of 80–120 nm. Structurally, it consists of a peripheral envelope and centrally-located core. The envelope is a double-membraned structure. Both membranes are made up of GPs. Glycoproteins remain around the core and consists of glycoproteinaceous spikes (gp 120) and fusion proteins (gp 41).

The core consists of two identical copies of the positive, single-stranded RNA genome. The core also contains 10–50 copies of reverse transcriptase and integrase enzymes, and two cellular t-RNAs. These t-RNAs are base-paired to each of the two genomic RNAs and act as primers for the reverse transcriptase. The morphology of the core differs for different viruses and therefore, is used as a basis for the classification of retroviruses (Fig. 113).

Although the genome resembles an m-RNA, it is not infectious because it does not encode a polymerase that can directly generate more m-RNA.

**Fig. 113** Structure of AIDS virus

The genome of HIV consists of three major genes that encode polyproteins for the following enzymatic and structural proteins of the virus. They are gag - capsid proteins, which are group-specific proteins, pol - polymerase, protease and integrase, and env - envelope proteins. At each end of the core RNA, there are LTR sequence containing promoters, enhancers and other genes, which help in binding different cellular transcription factors.

The viral glycol proteins are produced by proteolytic cleavage of the polyprotein encoded by the env gene. The size of the GP differs within the virion. For example, there are larger GPs (gp 120) and smaller GPs (gp41) in the envelope of HIV. Both of these GPs prevent the immune clearance of the virus. Detection of these GPs is useful in the diagnosis of HIV infection.

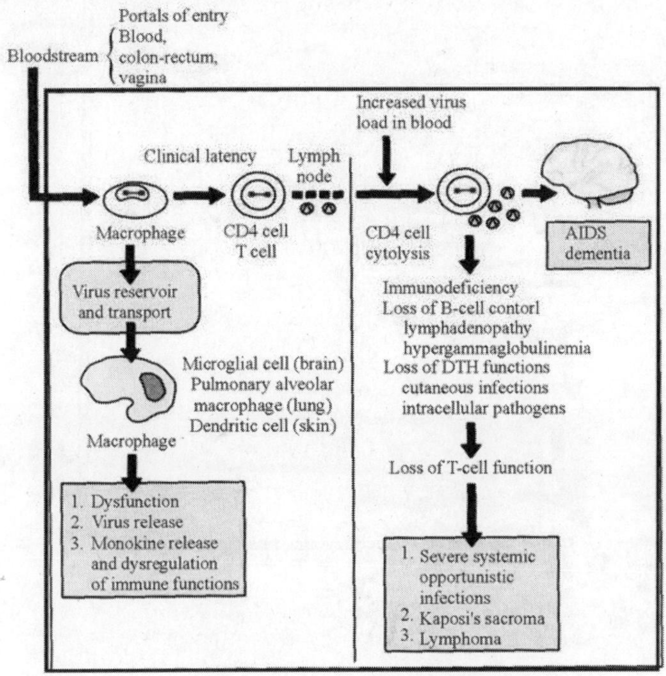


**Fig. 114** Pathogenesis of HIV

**Pathogenesis:** Initially the infection occurs in macrophages when the virus adheres to CD4 molecules and CCCKR5 chemokine receptors, which are small peptide molecules involved in promoting inflammatory reactions. People who are genetically deficient of CCCKR5 co-receptors become resistant to HIV infection (Fig. 113).

1. Knob of protein gp 120 anchored to another Protein gp 41 (marked 2 in figure)
2. P.17/18 viral protein

3. P.24/25 core protein
4. Virus core
5. P9/P7 viral protein
6. Reverse transcriptase
7. Lipid bilayer

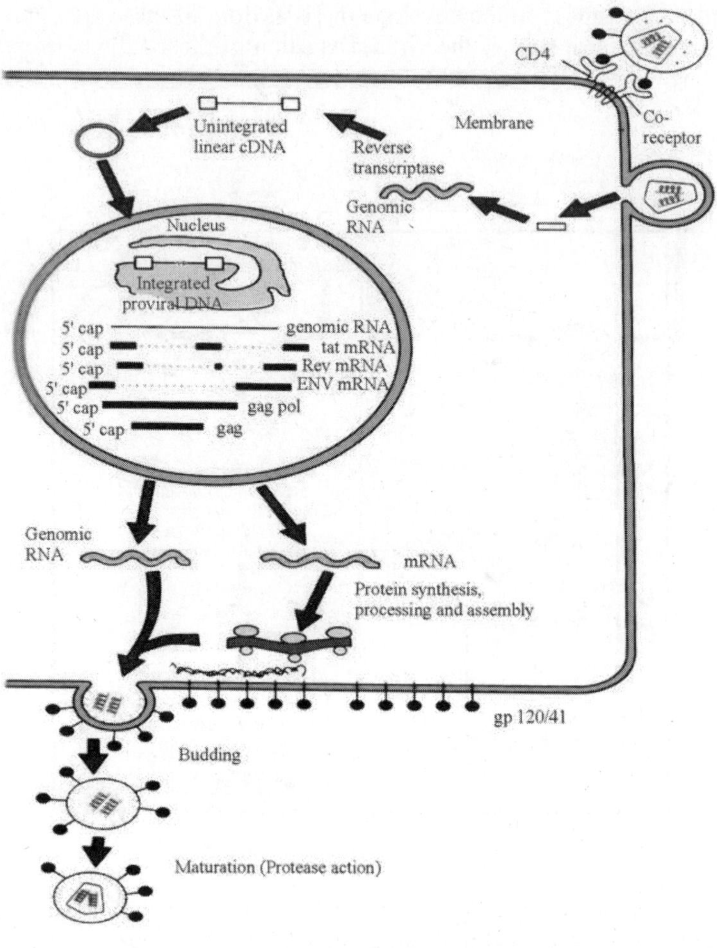


**Fig. 115** Life cycle of HIV

As the disease progresses, there is decrease in the CD4 T-cell number and increase in the viral population in the blood. Later in the disease, CD4 levels are remarkably decreased, the structure of lymphnodes are destroyed and the patient becomes immunosuppressed.

Further spread of the virus in the body occurs on filtration through the follicular-dendritic cell network of the germinal centre of the lymph nodes, macrophage cell lineage, which become infected. Infected macrophages and monocytes are the major reservoirs and means of distribution of HIV (Table 65).

**Table 65:** Disease mechanisms of HIV

- Primarily, infects CD4 T-cells and cells of the macrophage lineage
- Causes lytic infection of CD4 T-cells and persistent low-level productive infection of macrophage lineage cells
- Causes syncytia formation, with cells expressing large amounts of CD4 antigen (T-cells) with subsequent lysis of the cells
- Alters T-cell and macrophage cell function

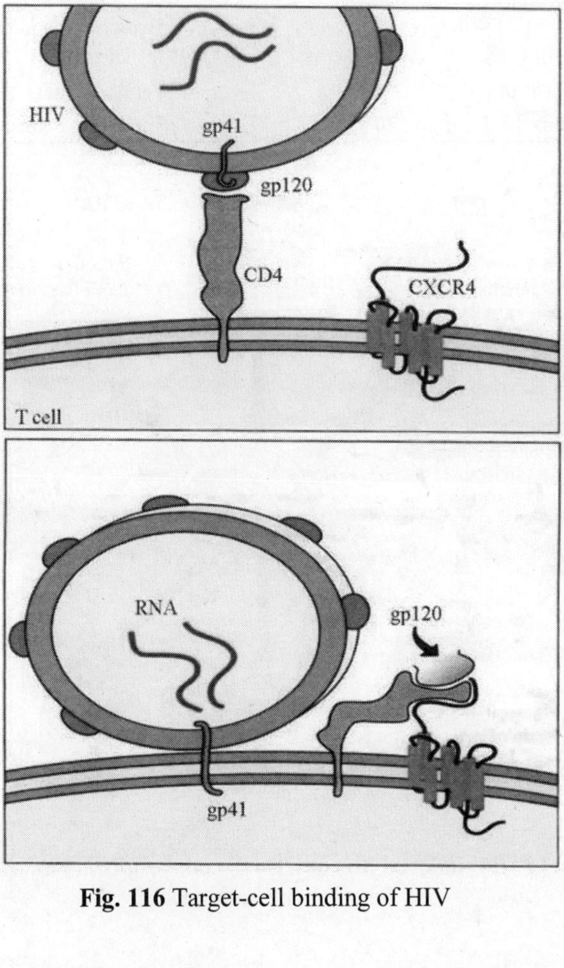


**Fig. 116** Target-cell binding of HIV

Circulating macrophages, microglial cells of the brain, pulmonary alveolar macrophages, dendritic cells and other cells of the monocyte-macrophage lineage can spread the virus, and potentially contribute to the clinical manifestation of HIV disease (Fig. 114 and Fig. 115).

Infact, the important factor in the pathogenesis and HIV disease is the tendency of virus to grow on CD4-expressing T-cells and macrophages. HIV-induced immune suppression (AIDS) results from a reduction in the number of CD4 T-cells, which destroys the helper and DTH functions of the immune response (Table 66).

**Table 66:** Means of HIV escape from the immune system

| Characteristics | Function |
|---|---|
| Infection of lymphocytes and macrophages | Inactivation of key element of immune defence |
| Inactivation of CD4 helper cells | Loss of activator of the immune system and DTH response |
| Antigenic drift of gp 120 | Evasion of antibody detection |
| Heavy glycosylation of gp 120 | Evasion of antibody detection |

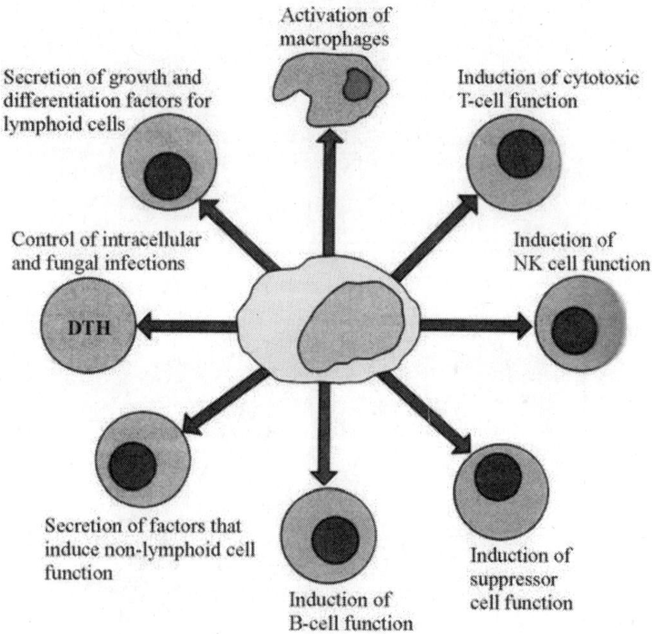


**Fig. 117** HIV-induced loss of CD4 T-cells results in loss of several functions including DTH

The accessory proteins of HIV are important for replication and virulence. The nef proteins appear to be essential for promoting the progression of HIV infection to AIDS. Individuals infected with natural mutant of nef proteins, which lacks nef, have lived beyond their expected life times and primates with the Simian Immunodeficiency Virus (SIV). Activation of CD4 T-cells is one of the first steps in the initiation of normal immune response. The TH-cells then release lymphokines and the IFN gamma required for the activation of macrophages, other T-cells, B-cells and NK cells. When CD4 T-cells are absent or not functioning, antigen-specific immune responses are incapitated and humoral responses also become uncontrollable. The absence of CD4 T-cells, which also induce DTH, permits the manifestation of many intracellular infections that are characteristic of AIDS (Fig. 117).

**Epidemiology:** AIDS was first noticed in male homosexuals in the US, but later on spread in epidemic proportions throughout the world. As per the 1996 statistics, there are 28 million people affected with AIDS infection.

**Table 67:** Characteristic features of HIV infections

**Disease / Viral factors**

- Enveloped virus is easily inactivated and must be transmitted in body fluids.
- Disease has long prodromal period.
- Virus can be shed before development of identifiable symptoms.

**Transmission**

- Virus is present in blood, semen and vaginal secretions.

**See Table 68 for modes of transmission**

**Who is at risk?**

- Intravenous drug abusers; sexually-active people with multiple partners (homosexual and heterosexual); prostitutes; newborns of HIV-positive mothers
- Blood and organ transplant recipients, and haemophiliacs: before 1985

**Geography/Season**

- There is an expanding epidemic worldwide.
- There is no seasonal incidence.

**Modes of control**

- Antiviral drugs limit progression of disease.
- Vaccines for prevention and treatment are in trials.
- Safe, monogamous sex helps limit the spread.
- Sterile injection needles should be used.
- Large-scale screening programs have been developed for blood transfusion, organs for transplants, and clotting factors used by haemophiliacs.

The HIV is thought to be derived from a Simian Immunodeficiency Virus (SIV). The HIV-2 is, in fact, similar to SIV. It is reported that initial human infection may have occurred in Africa, but it went unnoticed. The virus then spread to other areas of the world with the migration of infected people of Africa. Promotion of prostitution leads to its transmission throughout the world (Table 67).

The virus is transmitted through three important routes – sexual intercourse with infected people, from mother to baby and exposure to blood or blood products such as blood transfusion or sharing of contaminated needles (Table 68).

**Table 68: Transmission of HIV infection**

| Routes | Specific transmission |
|---|---|
| Inoculation in blood | • Transfusion of blood and blood products<br>• Needle sharing among intravenous drug abusers<br>• Needlestick, open wound, and mucous membrane exposure in healthcare workers.<br>• Tattoo needles |
| • Sexual transmission | • Anal and vaginal intercourse |
| • Perinatal transmission | • Intrauterine transmission<br>• Peripartum transmission<br>• Breast milk |
| Routes not involved in transmission | |
| Close personal contact | • Household members<br>• Healthcare workers not exposed to blood |

However, HIV is not transmitted by casual contact, touching, hugging, kissing, coughing, sneezing, insect bites, water, food, utensils, toilets, swimming pools, etc. Sexually-active people (homosexuals and heterosexuals), intravenous drug abusers and their sexual partners, and the newborns of HIV-positive mother are at high risk for HIV infection.

**Clinical symptoms:** The AIDS is one of the most devastating epidemics ever recorded. Most of the HIV-infected people will become symptomatic and finally succumb to the disease. Though rare, there are cases of survivors, which result from those strains of HIV that lack functional nef protein. Host resistance to the virus results into a lack of chemokine co-receptors' expression, which helps in binding of the virus to TH-cells.

Acute infection of HIV occurs after 6–8 wk of acquisition of the virus. Symptoms of acute infection are fever, pharyngitis, generalized lymphadenopathy (swelling of lymph glands). After an acute infection, patients enter into a stage of asymptomatic carriage, where they are infected, but the damage to the immune system is not yet sufficient to cause clinical symptoms.

During this stage, HIV-induced cell death of CD4-positive TH-cells is compensated by increased production of these cells. However, eventually, cell loss becomes more than cell replenishment, which results in the decline of peripheral blood CD4 count.

The loss of CD4 cells results in impaired functioning of the host immune system, rendering the individual susceptible to a wide range of infections and life-threatening complication (Table 69). Once a patient is infected with these pathogens, he is said to have AIDS.

**Table 69:** Indicator diseases of AIDS

| **Infection** | **Disease** |
|---|---|
| Opportunistic infections | |
| Protozoal | Toxoplasmasis of the brain |
| | Cryptosporidisis with diarrhoea |
| | Isosporiasis with diarrhoea |
| Fungal | Candidiasis of the oesophagus, trachea, and lungs |
| | *Pneumocystis carinii* pneumonia |
| | Cryptococcosis (extrapulmonary) |
| | Histoplamosis (disseminated) |
| | Coccidioidomycosis (disseminated) |
| Viral | CMV disease |
| | HSV infection (persisting or disseminated) |
| | Progressive multifocal leukoencephalopathy |
| | Hairy leukoplakia caused by Epstein-Barr virus |
| Bacterial | *Mycobacterium avium*-intracellulare complex (disseminated) |
| | Any 'atypical' mycobacterial disease |
| | Extrapulmonary tuberculosis |
| | *Salmonella septicemia* (recurrent) |
| | Pyogenic bacterial infections (multiple or recurrent) |
| Opportunistic neoplasias | Kaposi's sarcoma |
| | Primary lymphoma of the brain |
| | Other non-Hodgkin's lymphomas |
| Others | HIV wasting syndrome |
| | HIV encephalopathy |
| | Lymphoid interstitial pneumonia |

The nature of clinical symptoms varies according to the opportunistic disease. However, the onset of these clinical symptoms, generally, is correlated with reduction in the number of CD4 T-cells to less than 450/μl, and increased levels of virus and proteins p24 in the blood. Full-blown AIDS occurs when the CD4 T-cell count is less than 200/μl and involves the onset of more significant diseases including HIV wasting syndrome, characterized by weight loss and diarrhoea for months together.

AIDS may manifest in one of several different ways including lymphadenopathy and fever, opportunistic infections (Table 70), malignancies and AIDS-related impairment of mental powers (dementia).

1. **Lymphadenopathy and fever:** Combination of lymphadenopathy and fever is called as AIDS-related Complex (ARC) characterized by weight loss and uneasiness, which may persist indefinitely. There may also be symptoms of opportunistic infections such as diarrhoea, night sweats and fatigue.
2. **Opportunistic infections:** These are caused by several agents such as yeast and fungi, DNA viruses, parasites and intracellular bacteria. However, pneumonia is the most common opportunistic infection in AIDS patients (Table 71).
3. **Malignancies:** Kaposi's sarcoma is the most notable cancer (malignancy) that develops in AIDS patient, which spreads involving visceral organs in immunodeficiency patients. Other malignancies include Epstein-Barr virus-related lymphoma and Hodkin's lymphoma.
4. **AIDS-related dementia:** This results from HIV infection of microglial cells and neurons of the brain, which result in slow deterioration of intellectual abilities and other signs of neurological discorder.

**Diagnosis:** Diagnostic tests for HIV infection are primarily for the identification of the following:

- Those with the infection so as to facilitate antiviral drug treatment can be initiated
- To identify carriers, specifically blood and organ donors, pregnant women and sex partners, who may transfer the disease to others
- To confirm the diagnosis of AIDS

Generally, serological, immunological tests are conducted for diagnosis of the AIDS disease.

### Serological tests

- The chronic nature of AIDS infection can be identified by the serological tests, but these tests fail to identify AIDS of recent origin.

- Serological tests are conducted with western blot analysis, which confirms the seropositive results. This test detrmines the presence of antibody to the viral antigens (P24 or P31), and GPs gp 41 and gp 120/160 (Table 70).

**Table 70:** Laboratory analysis of HIV

| Test | Purpose |
|---|---|
| Serology | |
| ELISA | Initial screening |
| Latex agglutination | Initial screening |
| Western blot analysis | Confirmation test |
| Immunofluorescence | Confirmation test |
| virion RNA RT-PCR | Detection of virus in blood |
| P24 antigen | Early marker of infection |
| Isolation of virus | Test not readily available |
| CD4:CD8 T-cell ratio | Correlation of HIV disease |

- **Immunological tests:** These tests determine the number or concentration of CD4 T-cells. The particular concentration of CD4 lymphocytes indicates the stage of AIDS.

**Table 71:** Antiviral therapies for HIV infection

**Nucleoside analogue reverse transcriptase inhibitors**

- AZT (Zidovudine)
- ddC (Zalcitabine)
- ddI (Didonosine)
- d4T (Stavudine)
- 3TC (Lamivudine)

**Non-nucleoside reverse transcriptase inhibitors**

- Nevirapine
- Delavirdine

**Protease inhibitors**

- Squinavir
- Ritonavir
- Indinavir
- Nelfinavir

**Combination therapy**

- Indinavir/AZT/3TC
- Ritonavir/AZT/3TC
- Nelfinavir/AZT/3TC
- Nevirapine/AZT/ddI
- Nevirapine/indinavir/3TC

**Prevention and control:** The important methods of treatment of AIDS are detailed in Table 71. A range of antiretroviral drugs is available for the treatment of HIV-infected individuals. These include nucleoside analogues, non-nucleoside analogues, reverse transcriptase inhibitors, protease inhibitors and fusion inhibitors.

The nucleotide analogues such as AZT, dideoxyinosine (ddI) and dideoxycytidine (ddC) inhibit the reverse transcriptase activity. Non-nucleoside reverse transcriptase inhibitors such as nevirapine, inhibit the activity of the enzyme by other mechanisms. Protease inhibitors such as retonavir, block the morphogenesis of virion by inhibiting the cleavage of gag and gap polyprotein (Table 71).

A mixture of several antiviral drugs such as AZT, 3TC, protease inhibitors, each with a different mechanism of action, is presently adopted in the treatment due to a high rate of drug resistance developed by HIV to the treatment by a single type of drug. This approach prevents development of resistance by HIV.

Though several types of vaccines are at various stages of clinical trials, none of them, so far, have shown any evidence of inducing useful protection against HIV infection. Therefore, prevention strategies are currently given utmost importance that includes:

- Educating the people regarding the methods of transmission and the measures that may prevent viral spread.
- Screening of donated blood and organs.
- Behavioural modification to reduce the risk of person-to-person spread e.g., safe sex practices, needle exchange schemes for injecting drug users.
- The use of post-exposure prophylaxis, a 4 wk course of a mixture of antiretroviral agents taken immediately after a known exposure. This is relevant to healthcare workers after a needle stick injury or following sexual exchange.
- Prevention of mother to baby transmission of HIV, which can be achieved by appropriate use of drugs in pregnancy, delivery by caesarian section and avoidance of breast feeding.

## REVIEW QUESTIONS

**I. Answer the following questions:**

1. How are blood-borne diseases transmitted in humans?
2. Discuss briefly about pathogenesis and diagnosis of hepatitis-B disease.
3. How is hepatitis-B-virus transmitted? Discuss.

4. Explain the structure of HIV.
5. Discuss about pathogenesis of AIDS.
6. Explain epidemiology and clinical symptoms of AIDS.

**II. Write short notes on the following:**

1. Discovery of hepatitis-B-virus
2. Structure of hepatitis-B-virus
3. Prevention and control of hepatitis-B virus
4. Syndrome
5. Opportunistic infections associated with AIDS
6. Prevention and control of AIDS
7. Diagnosis of AIDS

## 9.7 NOSOCOMIAL INFECTIONS

Nosocomial infections (Greek Nosos = disease, and Komeion = to take care of) result from pathogens that develop within a hospital and other type of clinical care institutions, and are acquired by patients while they are in the hospital. Nosocomial infections, besides affecting people, can also affect physicians, nurses, aides, visitors and anyone who has contact with hospital. Most of these infections become clinically apparent while patients are still in the hospitals. However, infections that are in incubation stage when patients are admitted to a hospital are not nosocomial. They are commonly considered as acquired. As such infections can serve as a ready source or reservoir of pathogen for other patients or personnel. They are also considered in the total epidemiology of nosocomial infections.

According to an estimate, 5–10% of all the hospital patients acquire some type of nosocomial infection. Of the 40 million people that are admitted to hospitals annually throughout the world, about 2–4 million people may develop an infection that was not there while entering the hospital. Thus, nosocomial infections represent a significant proporition of all infectious diseases acquired by humans.

These infections are caused by all types of microorganisms, bacteria, most of which are non-invasive and part of the normal microbial flora, viruses, protozoa and fungi.

### 9.7.1 Hospital Environmental Factors Promoting Nosocomial Infections

Nosocomial infections spread early and rapidly in hospital environments because of the following reasons:

- Decreased resistance in patients to the infectious diseases because of their illness
- Patients admitted to a hospital may be reservoirs of highly-virulent pathogens.
- Increased chances of cross infection among the group of patients admitted into rooms and wards.
- Increased probability of transfer of pathogens due to movement of hospital personnel from patient to patient.
- Many hospital procedures such as catheterization, hypodermic injections, spinal puncture and removal of tissue samples (biopsy) or fluids (drawing blood) breach the skin barriers, and carry with them the risk of introducing pathogens to the patient.
- Newborn infants, in maternity wards of hospitals, are usually susceptible to certain kinds of infections because of lack of well-developed defence mechanism.
- Surgery procedures expose internal organs to source of contamination and the stress of surgery commonly diminishes the resistance of the patients towards infection.
- Certain therapeutic drugs such as steroids that are used for controlling inflammation increase susceptibility to infection.
- Use of antibiotics to control infection may lead to the development of antibiotic resistance in the pathogenic organisms.

### 9.7.2 Sources

There are three sources of nosocomial pathogens:

1. Endogenous source
2. Exogenous source
3. Autogenous source

A patient's own microbial flora constitutes endogenous source while microbial flora other than patient becomes the exogenous source. A pathogen whose source cannot be determined precisely whether endogenous or exogenous, will constitute autogenous source.

Pathogens of endogenous source are either brought into the hospital by the patient or are acquired when the patient gets colonized after admission. In either case, the pathogen colonizing the patient may subsequently cause a nosocomial disease. For example, when the pathogen is transported to another part of the body or when the patient's resistance drops.

There are many potential exogenous sources in hospitals. Animate sources are the hospital staff, other patients and visitors. Some examples of inanimate exogenous sources are food, computer keyboard, urinary catheters, intravenous and respiratory therapy equipments, and water systems such as softeners, dialysis units and hydrotherapy equipments.

### 9.7.3 Nosocomial Pathogens

Nosocomial pathogens generally cause infection to urinary tract, respiratory tract and to the blood. However, their number is relatively less (Fig. 118).

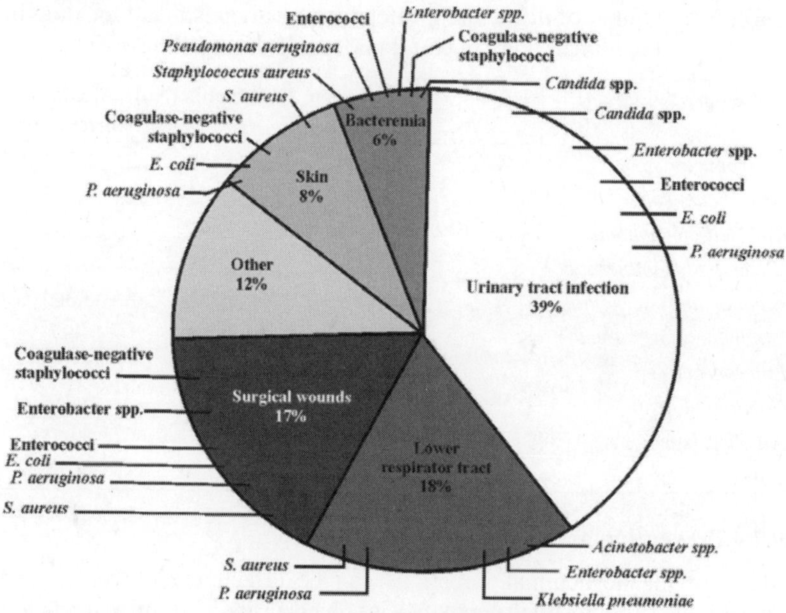


**Fig. 118** Nosocomial infection

*Staphylococcus aureus* is the most important and widespread hospital pathogen. It is the most common cause of pneumonia and blood infections, and is also problematic in nurseries. Many strains are unusually virulent and are also resistant to common antibiotics making their treatment very difficult. In addition, *S. aureus* and other *Staphylococcus* species are now considered as collective cause of hospital-acquired blood infections and wound infections. Most members of *Staphylococcus* are found in the upper respiratory tract or on the skin where they are a part of the normal flora in many individuals including hospital patients and personnel.

*Escherichia coli* is another most common cause of urinary tract infection in hospital, but *Enterococcus* species, *Pseudomonos aeruginosa, Candida albicans* and *Klebsiella pneumoniae* infections are also common while *Enterococcus*, *E. coli* and *K. pneumoniae* are normally found only in human body. Species of *Candida* and *Pseudomonos* are good examples of opportunistic pathogens and are commonly found in the environment and cause disease only in individuals whose defences are somehow weakened. Isolates of *Pseudomonos aeruginosa* from hospital infections are often resistant to multiple antibiotics and complicates the treatment. *E. coli*, species of *Staphylococcus* and *Enterococcus* also have the potential for multiple drug resistance (Table 72).

**Table 72:** Number of nosocomial infections by organism and location in the human body in certain hospitals during 1992-1999

| Pathogen | Blood stream (No) | Pneumonia (No) | Urinary tract (No) |
|---|---|---|---|
| *Enterobacter* sp | 1,083 | 4,444 | 1,560 |
| *Ercherichia coli* | 514 | 1,725 | 5,393 |
| *Klebsiella pneumoniae* | 735 | 2,865 | 1,891 |
| *Haemophilus influenzae* | - | 1,738 | - |
| *Pseudomonas aeruginosa* | 841 | 6,752 | 3,365 |
| *Staphylococcus aureus* | 2,758 | 7,205 | 497 |
| *Enterococcus* sp | 2,967 | 682 | 4,226 |
| *Candida albicans* | 1,090 | 1,862 | 4,856 |
| All other pathogens | 3,774 | 1,2537 | 8,075 |
| **Total** | **21,943** | **39,810** | **30,701** |

### 9.7.4 Prevention and Control

Many of the nosocomial infections can be effectively prevented and controlled if the hospital personnel involved in the care of patients are trained in the basics of infection. Control measures such as aseptic techniques, proper handling of equipment supplies, food and excreta, and surgical wound care and dressing. Patients should be monitored with respect to the frequency, distribution, symptomology and other characteristics common to nosocomial infections. A dynamic control and surveillance programme will be invaluable in preventing many nosocomial infections. Patient's discomfort, extended stays, and further expansion should be avoided.

## REVIEW QUESTIONS

### I. Answer the following questions

1. What are nosocomial infections? Explain.
2. Discuss about the factors, which promote nosocomial infection.

### II. Write short notes on the following

1. Nosocomial pathogens
2. Source of nosocomial infections
3. Prevention and control of nosocomial infections

## 9.8 FURTHER READING


Ananthanarayan R and Jayaram Paniker CK 2000 *Text book of Microbiology*, Chennai: Orient Longman Limited

Baron EJ, Peterson LR, and Finegold SM 1994 *Bailey and Scott's Diagnostic Microbiology*, London: Mosby

Bhatia R and Icchpujami RL 1999 *Essentials of Medical Microbiology*, New Delhi: Jaypee Publications

Brubaker RR 1991 Factors promiting acute and chronic diseases caused byYersiniae, *Clinical Microbiology Reviews* **4** 309–324

Bumann D, Hueck C, Aebischer T, and Meyer TP 2000 Recombinant live Salmonella spp. for human vaccination against heterogenous pathogens, *FEMS Immunology and Medical Microbiology* **27** 357–364

Collee JG, Fraser AG, Marimion BP, Simmons AM, and McCartneys 1996 *Practical Medical Microbiology*, Edingburh: Churchill Livingstone

Roy C III 2002 Bacterial infectious disease control by Vaccine development, *The Journal of Clinical Investigation* **110** 1061–1066

Destrosiers R 2004 Prospects for an AIDS vaccine, *Nature Medicine* **10** 221–223

Elliot T, Hasting M, and Dessel berger U 1996 *Lecturer notes on Medical Microbiology*, Oxford: Blackwell Sciences

Falows DA and Goff AP 1996 Hepadnavirus: Current models of RNA encapsulation and reverse transcription, *Advances in Virus Research* **46** 167–196

Fields BS, Benson RF, and Besser RE 2002 Legionella and Legionnaires' disease: 25 years of investigation, *Clinical Microbiology Reviews* **15** 506–526

Frosner G 1991 Hepatitis A virus, in *Textbook of Human Virology* (ed) RB Belshe, St. Louis: Mosby, 498–515 pages

Garavelli OL and Corti E 1992 Chloroquinine resistance in Plasmodium vivax in Braz., Transactions of the Royal Society of Tropical Medicine and Hygiene **86** 128

GeeJ et al 1992 *Oxford Textbook of Pathology,* Oxford University Press

Golden MR and Monhart FF 2005 Innovative approaches to the prevention and control of bacteria sexually transmitted infections, Infectious Disease Clinics of North America **19** 513–540

Gorbach S et al 1992 *Infectious Diseases,* Philadelphia: WB Saunders

Greenwood DR, Slalk CB, and Peuther FJ 2007 *Medical Microbiology,* Elsevier Science Ltd. Churchill Livingstone

Haldy WG and Klontz KC 1996 The epidemiologies of vibrio infections in Florida 1981-1993, *Journal of Infectious Diseases* **173** 1176–1183

Kiran TJ, Jude BA, and Tayler RK 2005 A colonization factor links *Vibrio cholera* environmental survival and human infection, *Nature* **438** 863–866

Liu LX and Weller PF 1996 Antiparasitic drug, *The New England Journal of Medicine* **334** 1178–1184

Mandell GJ et al 1995 Mandell, Douglas and Benett's Principles and Practices of Infectious Diseases, New York: Churchill Livingstone

Markel EK, Voce M, and John DT 1992 *Medical Pharmocology*, Philadelphia: WB Saunders

Moaeri M and Leppla SH 2004 The role of Anthrax toxin in Pathogenesis, *Current Opinion in Microbiology* **7** 19–24

Murray PR, Baron EJ, Pfaller MA, Tenover FC, and Yolken RH 1999 *Manual of Clinical Microbiology,* Washington, DC: ASM

Murray PR 1996 ASM Pocket Guide to Clinical Microbiology, Washington, DC: ASM

Murray PR, Rosenthal KS, Kabayashi GS, and Pfaller MA 1998 *Medical Microbiology*, St.Louis: Mosby

Nagoba BS 2005 *ClinicalMicrobiology,* New Delhi: Bi Publications Pvt. Ltd.

Ottesen EA 1993 Filarial Infections, Infectious Disease Clinics of North America **7** 619–633

Rolfs R 1993 Treatment of Syphilis, *Clinical Infectious Diseases* **22** S 23–28

Salyers AA and Whitt DD 1994 *Bacterial Pathogenesis: a molecular approach*, Washington, DC: ASM

Sepkowitz KA et al 1995 Tuberculosis in AIDS era, *Clinical Microbiology Reviews* **8** 180–199

Shanson DC 1999 *Microbiology in Clinical Practice*, Oxford: Butter worth Heinemann

Waehsrmuth LK et al (eds) 1994 Vibrio cholera and chlora: molecular to gloval prospectives, Washington, DC: ASM

World Health Organisation (WHO) 1994 Control of salmonella infections in animals and prevention of human food borne Salmonella infections *Bulletin of the World Health Organization* **72** 831–833

Zuker JR and Campbell CC 1993 Malaria: Principles of prevention and treatment, *Infectious Disease Clinics of North America* **7** 547–567

# 10

# History and Development of Immunology

It had been observed from very early days that persons surviving an attack of small pox did not develop this disease when exposed to the infection. Subsequently, this observation had been applied for the prevention of the disease by producing a mild form of small pox *in vitro*. This process is called as variolation, which was prevalent in India, China and other ancient civilizations. It was established in England by LMW Montague (1718). Edward Jenner (1796) introduced the technique of variolation using cowpox material by observing the immunity of milkmaids to small pox who were liable to occupational cowpox infection. This was the first successful instance of scientific immunization, which stood the test of time, and paved the way for vaccination and the ultimate eradication of smallpox.

Immunology is a broad branch of Biomedical Sciences, which covers the study of all aspects of immune system. Immunology deals with the physiological function of immune system in states of both health and diseases, and malfunctions of the immune systems in immunological disorders. The different areas of immunology include classical immunology, development immunology, immunotheraphy, diagnostic immunology, evolutionary immunology and reproductive immunology. Some of the milestones in the development of immunology are precised in Table 73.

**Table 73:** Discoveries in the development of immunology

| Scientist | | Contributions |
|---|---|---|
| Edward Jenner | 1798 | Proposed cowpox vaccination for smallpox |
| Elie Metchonikoff | 1880 | Phagocytic theory |
| Emil Von Behring and Kitasato | 1890 | Passive immunization, Nobel prize |
| E Von Behring | 1901 | Serum antitoxin as basis of immunity |

| Scientist | | Contributions |
|---|---|---|
| Portier and Charles Richert | 1902 | Described hypersensitivity |
| Almroth Wright and Stewart Donglas | 1903 | Discovered opsonin |
| Wright et al | 1903 | Demonstrated antibodies in the blood of immunized animal |
| Robert Koch | 1905 | Demonstrated cellular immunity to tuberculosis |
| Paul Ehrlich | 1906 | Antitoxin in acquired immunity |
| Metchonikoff and Paul Ehrlich | 1908 | Nobel prize for understanding inflammation |
| Metchonikoff | 1908 | Role of phagocytosis in acquired immunity |
| Charles Richert | 1913 | Nobel prize for proposition of anaphylaxis |
| Behring, Nuttall and Nissen | 1916 | Humoral factors in disease prevention |
| J. Border | 1919 | Complement-mediated bacteriolysis |
| Jules Bordet | 1920 | Antigen and antibody reactions |
| Ferdinated Widal | 1920 | Widal test for diagnosis of typhoid fever |
| Carl Land Stainer | 1930 | Blood groups |
| Macfarlane Burnet and Peter Medawar | 1940 | Clonal selection of antibody formation |
| Max Theiler | 1951 | Development yellow fever vaccine |
| D. Bover | 1957 | Discovery of antihistamine |
| F.M. Barnet | 1960 | Discovery of acquired immunological tolerance |
| Wu and Kabat | 1970 | Demonstrated the presence of variable region in the antibody |
| Rodeny R. Porter and Gerald M.Edleman | 1972 | Antibody structure |
| Geoges Kohler and Cesar and Milstein | 1975 | Nobel prize in 1984, Monoclonal antibodies |
| R.R.Yalow | 1977 | Development of radioimmunoassay |
| Tonegawa | 1978 | Nobel prize, found production of antibodies |
| G. Snell, J. Dasset and B. Benalerraf | 1980 | Discovery of major histocompatibility complex |
| Niels Jerney | 1984 | Clonal concept, Nobel prize |
| Hozumi and Onegawa,S | 1987 | Mechanism of antibody production |
| ED Thomas and J Murray | 1991 | Contribution to transplantation immunology |
| PC Doherty and R.M. Zinkernagel | 1996 | Role of major histocompatibility in a antigen recognization by T-cells |

In the process of evolution of individual organisms developed, the property of distinguishing self is accepted and nonself (foreign) is rejected or degraded. For this purpose, body has developed a versatile defence system called immune system that operates via different mechanisms, which is called as immunity or resistance. The term 'immunity', in English, is derived from a Latin term 'immunis', meaning exempt. The immunity operates against non-self agents and is called immune response. Immunity is of two types:

1. Innate (Non-specific or natural)
2. Acquired (Specific or adaptive)

Immunity is more highly-evolved system of specific responses, and it recognizes and selectively eliminates specific microbial pathogens or foreign substances. Substances that are recognized by the immune system as foreign and provoke immune responses are called antigens. The special substances (proteins) that are generated by an antigen are called as antibodies (immunoglobulin). These proteins bind to the specific cells that generate antibodies. There are also some other cells of immune system that destroy the foreign invaders. Both these immunities work together and provide protection to the body repeatedly.

## 10.1 DEVELOPMENT OF VACCINATION

Louis Pasteur, in 1880, isolated the bacterium responsible for chicken cholera and grew it in pure culture. During the course of experiments, Pasteur noticed that chicken cholera bacterium lost its disease-causing ability or virulence when the bacterial culture was incubated for longer interval between transfers. Such bacteria are called as attenuated bacteria. When such attenuated culture was inoculated to the healthy chicken, it did not develop disease symptoms; instead exhibited ability to resist the pathogen.

These experiments of Louis Pasteur clearly established that most of the animals possess a mechanism by which they are able to resist pathogens. Pasteur called the attenuated cultures as vaccines in honour of Edward Jenner because many years before the Pasteur experiments, Jenner had used material from cowpox lessions for vaccination to protect the people against smallpox. Shortly after these experiments, Pasteur and Chamberland developed an attenuated vaccine for the control of anthrax disease.

Pasteur also developed a dead vaccine for rabies. He attenuated rabies pathogen by growing it in the rabbit brain. The brain and spinal cord of the infected rabbit was excised and dried for several days, pulverized and mixed with glycerin. This vaccine was given to Joseph Meister, a nine-year old boy, who had been bitten by a rabid dog repeatedly for 13 times over a period of 10 d with increasingly virulent preparations of the attenuated culture, Joseph Meister survived. Pasteur was as surprised as anyone else. Joseph Meister did not die even after several weeks of crucial trials. Having convinced people all over the world contributed liberally for the construction of the Research Institute in Paris, France in honour of Pasteur for the development of vaccines.

Pastuer's development of vaccines for cholera, anthrax and rabies though successful, the mechanism of protection provided by them remained obscure and different persons interpreted the involved mechanism differently. Nuttal (1988) observed that defabrinated blood had a bactericidal effect. Buchner (1989) observed involvement of a heat-labile bactericidal factor called alexine, while Behring and Kitasato (1890) explained the involvement of humoral factor or antibody. Elie Metchnikoff (1880) observed that certain leukocytes present in the blood could swallow disease-causing bacteria. He called these cells as phagocytes and the phenomenon as phagocytosis. Wright (1903) demonstrated the combined action of phagocytes and antibodies in neutralizing the antigens. Pfeffer (1893) demonstrated lysis of live cholera Vibrios in the blood of guinea pigs that were previously injected with killed vibrios. Bordet (1895) explained the participation of two components in the lysis of live cholera vibrios; the first being heat stable and found in immune sera (later called as antibody), and the second being heat labile and identical with Buchner's alexine, which was named subsequently as complement. Further studies reported other processes in which antibodies react with antigens such as agglutination, precipitation, complement fixation and neutralization.

## 10.2 DISCOVERY OF HUMORAL IMMUNITY

Emil Von Behring and Shibasaburo Kitasato (1890) injected inactivated diphtheria toxin into rabbits stimulating them to produce antitoxin in the blood cells. This antitoxin would inactivate the toxin produced by the diphtherial bacterium and would protect the animal from diphtheria disease. They have also developed tetanus antitoxin. These toxins are used even now in the prevention of diphtheria and tetanus respectively.

The antitoxin discovery by Behring and Kitasato proved that disease resistance or immunity could be developed from certain soluble substances produced in the blood, which are now called as antibodies. This has resulted in the development of concept of humoral immunity mediated by non-cellualr body fluids for which Behring was awarded Nobel prize in 1901. However, humoral immunity was fully realized only by the work of Elvin Kabat and other workers during 1930 who demonstrated that a fraction of serum, called gamma globulin, was responsible for immunity acquired following immunization. Elie Metchnikoff (1940) discovered that certain leukocytes present in the blood could swallow disease-causing bacteria. He called these cells as phagocytes and the phenomenon as phagocytosis. He further observed that these phagocytes were more active in immunized animal and contributed to the immune state. Metchonikoff hypothesized that the cells were the major effector of immunity and he gave the concept of cell-mediated immunity.

A great deal of controversy developed between the supporters of cell-mediated immunity and of humoral immunity. However, concepts of humoral immunity got more acceptance due to ready availability of blood while it was difficult to study the activities of immune cells. With the development of tissue culture techniques, cell-mediated immunity also got acceptance during 1940. Subsequent studies during 1950 proved that both the types of immunity are interrelated. The identification of lymphocytes as the responsible cells for both cell-mediated and humoral immunity resulted in the end of controversy. Bruce Glkick (1950) demonstaretd two types of lymphocyets; B-lymphocytes that are involved in humoral immunity, while T-lympocytes play an important role in cell-mediated immunity.

## 10.3 IMMUNITY AGAINST NON-PATHOGENIC SUBSTRATES

Jules Border demonstared that immunity can also be caused by any non-pathogenic organisms or organic substances from other animals. Karl Landstainer and others have demonstrated that inoculating an animal with almost any organic substance induces production of antibodies. The discovery of blood groups (ABO) by Landstainer (1930) led to the successful blood transfusion in humans. This sensational discovery won Landteiner a Nobel prize in 1930.

## 10.4 SELECTIVE THEORY

Paul Ehrlich (1900) proposed the concept of selective theory, which states that cells in the blood possess variety of receptors on their surface. He called these receptors as side-chain receptors that react specifically with pathogenic organisms and kill them. In addition, such cells also produce more receptors with same specificity. Thus, he demonstrated specificity between antigen and the receptors. This concept was further refined by subsequent workers such as David Talmadge and Macfarlane Burret during 1950, and named it as clonal selection theory. According to this concept, an individual lymphocyte binds to an antigen on its specific receptor site resulting in the proliferation of clone of cells, which have same immunological specificity as the parent cell. The clonal selection theory has been widely accepted and in recognization. Burnet was awarded the Nobel prize along with Sir Peter Medawar. Subsequently, Neiljerne (1984) was also awarded a Nobel prize for his contribution to the development of clonal expression and evaluation of the idiotype network in the regulation of immune responses.

## 10.5 STRUCTURE OF AN ANTIBODY

Rodrey Porter and Gerald M. Edleman (1972) elucidated the structure of antibody (immunoglobulin) with the help of enzymes. The immunoglobulin polypeptide chain can be cleaved into 3–4 pieces i.e., two antibody fragments (Fab) and one crystallizable fragment (FC). These people were awarded for Nobel prize in 1972 for this work. Wu and Kabat (1972) demonstrated the hypervariable region in antibody molecule.

Peter Gorer proposed a concept of rejection of a foreign tissue in responses to cell surface molecules regulated by the genes. These molecules were identified to be four groups of genes (group I to IV) that encoded blood cell antigens. They further confirmed that group II genes participated in the rejection of transplanted tumours and other tissue. Snell called these genes as histocompatible genes. For this work, Snell along with Dausset and Benacerraf, who worked on Human Leucocyte Antigen (HLA) complexes, were awarded Nobel prize in 1980.

## 10.6 SOMATIC HYBRIDIZATION

George Kohler, and Ceaser and Milstein (1975) devised a technique called somatic hybradization by which homozygous monoclonal antibodies could be developed. Later this technique got popularity by the name 'hybridoma technology'. Kohler and Milstein were awarded Nobel prize in 1984 for this achievement.

Invasion by a microorganism or presence of an antigen did not always result in the beneficial effect of immunity or protection. Sometimes, it caused the opposite effect. Koch (1890) had noticed that when the tubercle bacillus or its protein was injected into guinea pig, an abnormal reaction took place, known as Koch phenomenon. This was later on known as hypersensitivity reaction. Ricket and Portier (1902) made the strange observation while studying the effect of toxic extract of sea anemones in dogs. According to their observation, dogs, which had prior contact with the toxins, were abnormally sensitive even to minute quantities of the toxin. Subsequently, this phenomenon was termed as 'anaphylaxis'. Later, many similar reactions were observed, both experimentally and in nature of injury, disease or even death resulting from repeated contacts with antigen. The importance of this phenomenon in the pathogensis of many human diseases led to the development of the discipline of allergy.

## 10.7 SCOPE OF IMMUNITY

Till recently, a generalized concept of immunity prevailed. It was considered as a protective mechanism intended to defend the body against invading microorganisms. Based on the original suggestion of Thomas (1954), Burnet (1967) developed the concept of immunological surveillance. According to this concept, the primary function of the immune system is to preserve the integrity of the body by identifying and destroying all foreign antigens whether autogenous (internal) or external in origin. Malignancy was visualized as a failure of immunological surveillance, and the scope of immunity was enlarged to include natural defence against cancer.

Another aspect of the role of immunity is in the rejection of homograft. It was possible to understand the immunological basis of organ transplantation by the works of Medawar and Burnet. This concept was made possible with the elective immunosuppression and proper selection of donors based on histocompatibility. The history of transplantation, thus, parallel to the history of blood transfusion, which was unsuccessful and even fatal before the discovery of blood groups by Landstainer (1900).

The last two decades have witnessed an explosion of conceptual and technical advances in immunology. Immunological processes in health and diseases are now better understood following the identification of the two components of immunity i.e., the humoral or antibody-mediated processes and the cellular or cell-mediated processes, which are developed and manifested in separate pathways.

Possibility of culturing of different types of cells *in vitro* and to clone these cells to get a population of cells with identical genetic makeup helped to investigate cellular interactions and gene inactivations. The role of a number of gene products has been elucidated by producing transgenic animals and studying the effect of the introduced genes.

Many new techniques developed by immunologists such as monoclonal antibodies revolutionized immunoassays and detection systems that employ antibodies. Potentials of these substances are infinite in the treatment of infectious diseases, cancers, organ transplant, etc. In addition to antibodies, other immunologically-important molecules have been produced and are being developed as therapeutic agents. The flow cytometry has revolutionized the analysis of the populations, and PCR has increased the sensitivity of the detection of microorganism.

The interaction between the cells and molecules of the immune system is extremely complex, and intricacies of immune recognition are far from the understanding. Some molecules play different functions depending on the location and presence of other molecules, which need to be exploited not only in the elimination of pathogens, but also the cancer cells. It may also enable us to treat the immunodiseases, allergies and tumours. It will also help us to develop new strategies to reduce transplant rejection.

Tak W. Mak and Mark M. Davis (1984) characterized T-cell receptor. They further cloned and sequenced, a gene expressed and rearranged in T-cells, but not in B-cells. They further revealed that many of the T-cell receptor sequences were homologous to those of immunoglobulin genes. Peter Dobety and Rolf M. Zinkernagel (1955) reported that these T-cells from mice infected with meningitis virus were destroyed. Virus infected cells only from the same strain of mice showed that T-cells must recognize two signals on an infected cell; one from the virus and other from the cell's own antigen to destroy it. For these observations, Mak and Davis, and Dobety and Zinkernagel shared Nobel Prize in 1996. Thus, various developments in the field of immunology led to rapid progress in different fields of Medical Microbiology such as organ and tissue transplantation, vaccinology and molecular biology.

Knowledge of the immunological process underlying the reactions of the body to infectious agents, tumours, and transplanted tissues and organs has advanced significantly by the use of modern techniques, which have enabled the identification of gene codings for molecules such as cell receptors and MHC molecules. The genes coding for immunologically-important molecule have been cloned and relatively large amounts of recombinant proteins have been produced.

The field of immunology continues to fascinate and frustrate the researchers all over the world, and may lead to the better understanding of tumour immunology, autoimmunity and responses to immunization. It could also help to develop strategies for protection against many crippling diseases. Immunity is a powerful tool created by nature to ensure the continued survival of the human beings and could perhaps be manipulated to yield a better quality of life.

## REVIEW QUESTIONS

### I. Ansewer the following questions

1. Define immunity. Discuss different types of immune responses.
2. Trace important events in the understanding of resistance of human beings to different diseases.
3. Critically analyse the impact of host-pathogen interaction over the control of human diseases.
4. What is immunity? In what way it is different from immune response? Give concise account of scope of immunology.
5. Give detailed account of developments in immunology.
6. Discuss impact of combination of Edward Jenner, E von Behring, Lantsainer and Metchnikoff in development of immunology.

### II. Wrire short notes on the following

1. Antigen
2. Antibody
3. Attenuation
4. Vaccine
5. Receptors
6. Phagocytosis
7. Hypersensitivity
8. Immune response
9. Clonal selection theory
10. Somatic hybradization

## 10.8 FURTHER READING

Abbas AK, Lichtman AH, and Probe JS 2000 *Cellular and Molecular Immunology,* Philadelphia: Saunders

Canard DA and Jensen HB 1996 New and improved vaccine, *Postgraduate Medical Journal* **100** 113–127

Janeway CA, Travers P, Walport M, and Shlomchik 2005 Immunology: The immune system in health and Disease *Current Biology*, London

Male D et al 1996 *Advanced Immunology*, St. Louis: Mosby

Plolkin SA and Mortiner EA 1988 *Vaccines*, Philadelphia: WB Saunders

Roitt I, Broshoff J, and Male D 1996 *Immunology*, St. Louis: Mosby

Weir DM and Stewart J 1997 *Immunology*, Edinburgh: Churchill Livingstone

# 11

# Scope of Immunity

## 11.1 INTRODUCTION

The adaptive or specific immunity of most of the vertibrates possess three important functions:

1. To recognize anything that is foreign to the body
2. To respond to the foreign material / pathogen
3. To remember the foreign pathogen / invader

The immune system is capable of differentiating one pathogen from another to recognize the body's own or self proteins and cells from non-self proteins, cells and organs, and to identify the cancer cells. After recognition of an invader or pathogen, the specific immune system responds by employing its defence molecules and cells to destroy the invader or pathogen. This phenomenon is called as effect or response, which either eliminates the foreign material or renders it harmless to the host, thus, inhibiting the disease. If the invader attacks second time, at a later time the immune system remembers it, and sets up a more intense and rapid memory, which results in the elimination of the invader or pathogen once again and protects the host from the development of disease. Adaptive (specific) immunity differs from non-specific or innate immunity by the following characteristics:

- **Specificity:** Immunity is specific i.e., it is directed against a particular pathogen or foreign material. Immunity developed against a pathogen or foreign material is not effective against other pathogen or foreign material.
- **Memory:** The immune system possesses specific memory, that is, the body reacts so rapidly that there is no manifestation of pathogenesis when it is re-exposed to the same pathogen or foreign material. In contrast, the reaction time for non-specific or innate defence mechanism is the same for the initial and subsequent attacks e.g., the inflammation.
- **Diversity:** The immune system is capable of producing large quantities of diversified molecules such as antibodies that recognize several different antigens.

- **Discrimination between self and non-self:** The specific immune system generally responds only to non-self foreign substances or antigens and does not destroy the organism, which is protecting.

## 11.2 TYPES OF IMMUNITY

There are two types of immunity in a vertebrate animal: 1. Innate immunity or natural immunity 2. Specific or adaptive immunity

1. **Innate immunity or natural immunity:** It is also called as non-specific immunity or non-specific response. It acts as a first line of defence. It offers resistance to any microorganism or foreign substance that attacks a vertebrate host. It includes general mechanisms inherited as a part of the innate structure and function of each animal. This system lacks immunological memory i.e., non-specific response occurs to the same extent each time a pathogen or invader is encountered.
2. **Specific immunity:** It is also called as acquired or adaptive immunity. It confers resistance against a foreign agent such as pathogens, bacteria, viruses or toxins. Substances that are identified as foreign and stimulate an immune response are called antigens. Specific cells are generated in response to the antigens, which produce specific proteins called antibodies. Antibodies bind to the antigens and thereby, inactivate them, which results in the non-manifestation of the disease.

Both these immune systems – innate immunity and specific immunity – usually work together to destroy pathogenic micro-organisms and other foreign agents. Two types of specific immunity or adaptive immunity are recognized in the vertebrate body: 1. Cellular immunity or cell-mediated immunity 2. Humoral immunity or antibody-mediated immunity

**1. Cellular immunity:** This type of immunity occurs with the help of special types of defence cells called T-lymphocytes that directly destroy cells infected with bacteria, viruses, parasites or transplanted cells or organs and cancer cells. The T-lymphocytes destroy these cells either by lysis or by releasing certain chemicals called cytokines or by stimulating non-specific defence mechanisms (Fig. 119).

**2. Humoral immunity:** Fluids of the body such as blood and lymph are called as humors. This type of immunity is based on the action of soluble proteins called antibodies, which are present in the blood and lymph as well as on the surface of B-lymphocytes. The antibodies bind to the surface of bacterial cells, viruses and toxic chemicals, and neutralize them directly or mark them for destruction by mechanisms such as opsonization, immune complex formation, complementation, etc.

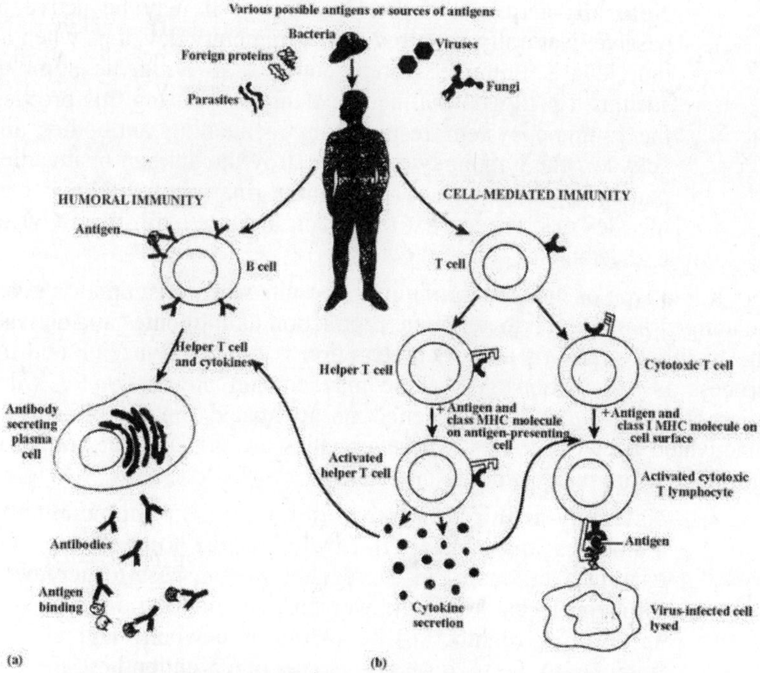


**Fig. 119** Humoral and cell-mediated branches of immunity

a. **Kinds of acquired immunity:** It includes types of specific immunity of a host that develops after entry of a suitable antigen or foreign material or after transfer of antibodies or lymphocytes from an immune donor. Acquired immunity can develop in a host by natural or artificial means and actively or passively (Fig. 120).

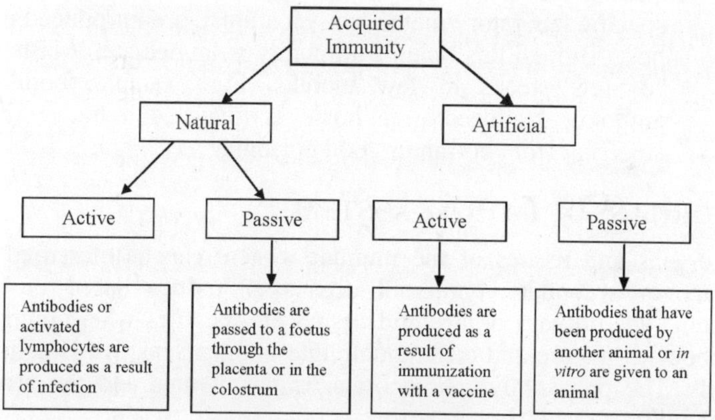


**Fig. 120** Types of acquired immunity

- **Naturally-acquired active immunity:** It may be active or passive. Naturally-acquired active immunity develops when an individual's immune system contacts an antigenic stimulus such as a pathogen that causes a disease. During this process, the immune system responds by generating antibodies and activates the lymphocytes that destroy the antigen or invading pathogen. This type of immunity may last lifelong e.g., measles or chicken pox, or remain in force only for few years e.g., tetanus.

It is a type of induced immunity. It results when an animal is given an antigen preparation to stimulate production of antibodies and activate the lymphocytes. This antigen preparation is called as vaccine and the process as vaccination. A vaccine may contain preparation of killed microorganisms, live or weakened or attenuated microorganisms or inactivated bacterial toxins. These vaccines are given to an animal to stimulate an immune response artificially.

- **Naturally-acquired passive immunity:** In this immunity, antibodies are transferred from one host to the other. For example, antibodies of a pregnant woman pass to her foetus through placentum. If the woman is immune to diseases such as polio or diphtheria, the foetus or newborn will also be immune to these diseases. Some other antibodies are also transferred from mother to her child through the first milk called colostrums secreted from the mammary glands. These antibodies transferred from mother to her child are essential for conferring immunity till the child develops its own immunity. However, this type of immunity is short time immunity and lasts only for certain time i.e., weeks or months.
- **Passive immunity:** Artificially-acquired immunity develops when antibodies that have been produced, artificially or by specific laboratory methods in an animal, are introduced into a host. Though this type of immunity is immediate, it lasts only for few weeks to few months. For example, botulinum antitoxin produced in a horse is given to a human being suffering from botilinum food poisoning.

## 11.3 ORGANS OF IMMUNE SYSTEM

The organs and tissues of the immune system can be classified into primary or secondary lymphoid organs or tissues based on their function. The primary organs and tissues are the sites where immature lymphocytes mature and differentiate into antigen sensitive B-cells and T-cells. The primary lymphoid organ is the thymus and the primary lymphoid tissue is the bone marrow.

The secondary organs and tissues of the immune system provide sites where lymphocytes may encounter and bind antigen, and then proliferate and differentiate into fully-mature antigenic receptor cells. The secondary lymphoid organ is the spleen and the secondary lymphoid tissue includes lymph nodes and mucosal associated tissue such as GALT and SALT.

### 11.3.1 Primary Lymphoid Organs and Tissues

**The thymus:** It is a lymphoid organ located above the heart. Immature undifferentiated lymphocytes from the bone marrow move into the outer cortex of the thymus where they divide. As they mature and acquire T-cell surface markers, they migrate into the inner cortex. Here, approximately 90% of them die due to acquisition of immune tolerance. The remaining 10% migrate into the medulla and migrate into T-cells, which carry immune functions after activation.

**Bursa of fabricius:** This is a lymphoepithelial organ arising as a pouch from the dorsal part of the cloaca in birds. Its development, structure and functions are similar to the thymus. It originates from an epithelial organ, becomes a lymphoid organ by about 15$^{th}$ day of embryo development, assumes full functional ability near hatching and starts involuting by 7–13 wk of age, which is the age of puberty in birds. This organ is a site of lymphocytic proliferation and differentiation. Stem cells from the yolk sac, fetal liver and bone marrow enter this organ, proliferate and develop into immunocompetent bursal lymphocytes. These cells then migrate and reside in selective areas, the peripheral lymphoid organs such as the mantle, germinal follicle, and perifollicular regions of the spleen, thc far cortical areas and medullar cords of lymph nodes. These areas are called bursa dependent and thymus independent areas. Following a suitable antigenic stimulation, bursal lymphocytes transform into plasma cells and secrete antibodies. They play an important role in humoral immunity. Immunocompetence is conferred on the lymphocytes in birds by the bursa in stages. Competence for IgM production is acquired early about 14$^{th}$ day of embryo development, and for IgG, it is 21$^{st}$ day of embryo development. Birds bursectomised on the 18$^{th}$–20$^{th}$ day synthesize IgM, but not IgG. The mechanism of differentiation of bursal lymphocytes in the bursa is not clear.

### 11.3.2 Secondary Lymphoid Organs and Tissues

The important and highly-organized secondary lymphoid organ is the spleen and the tissue is the lymph node.

**The spleen:** It is located in the abdominal cavity. It helps in the filteration of blood, and trapped blood-borne pathogens and antigens. Once pathogens are trapped in the blood by macrophages and dendritic cells, they are swallowed, and antigens are presented to B-cells and T-cells, which get activated to carry out the destruction of the pathogens (Fig. 121).

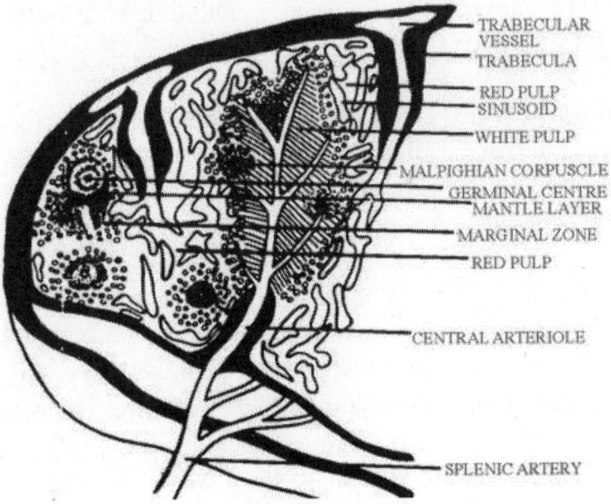


**Fig. 121** Schematic diagram of spleen architecture

**Lymphnodes:** They are situated at the junctions of lymphatic vessels where they filter out harmful pathogens and antigens from the lymph (Fig. 121a). The pathogen and antigen are trapped by phagocytic and dendritic cells. Fixed macrophages then swallow the foreign material. The B-cells divide and become antibody-secreting plasma cells within the lymph nodes, which also contain T-cells and dendritic cells. The former cells promote the B-cell immune response while the later serve as Antigen Presenting Cells (APCs) (Fig. 121 a).

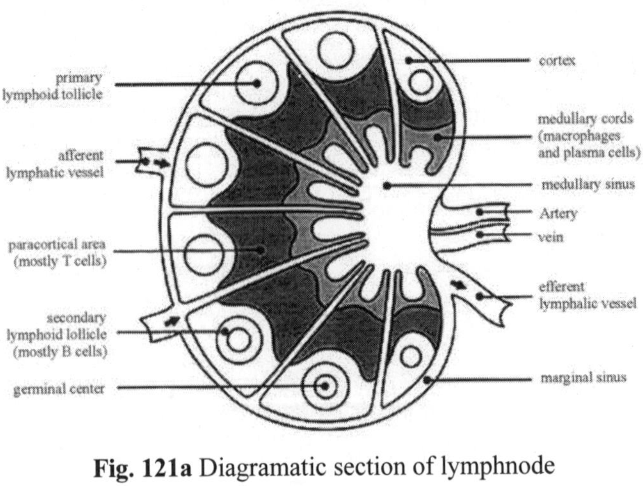


**Fig. 121a** Diagramatic section of lymphnode

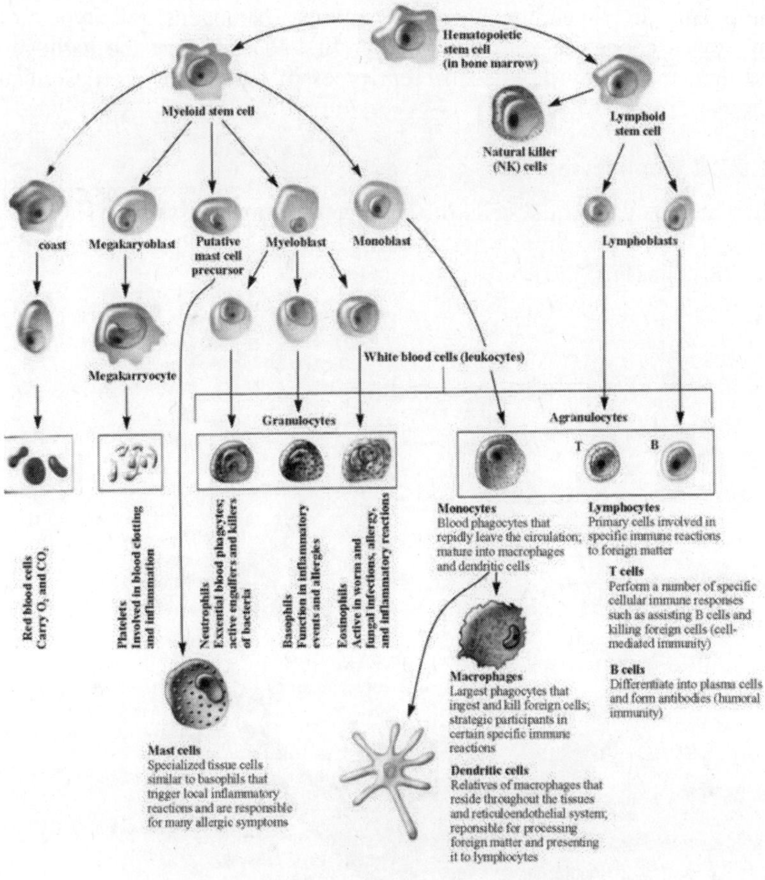


**Fig. 122** The different types of human blood cells

## 11.4 CELLS OF IMMUNE SYSTEM

Immune system consists of molecules, cells, tissues and organs with a specific role of defending the host against invading pathogens, cancer cells and non-self proteins. The cells capable of conferring both non-specific and specific immunity to the host are the leukocytes or white blood cells. Pluripotent stem cells present in the fetal liver and in the bone marrow of the animal body produce leukocytes. After their formation in these sites, they move to other body sites, undergo further development and carry out their several functions. These leukocytes are present throughout the animal body (Fig. 122). However, some of them become localized in certain specific tissue, where they respond to the local trauma and protect the tissue while others remain in blood and lymph circulation, and reach the site of infection to destroy the pathogens. In defending the host against pathogens, all types of leukocytes cooperate with one another first to recognize the pathogen and then to destroy it. The different types of leukocytes are described below:

### 11.4.1. Lymphocytes

They are the prominent cells of the specific immune system. They can be categorized into three types: T-cells, B-cells and null cells, which are the NK cells (Fig. 123).

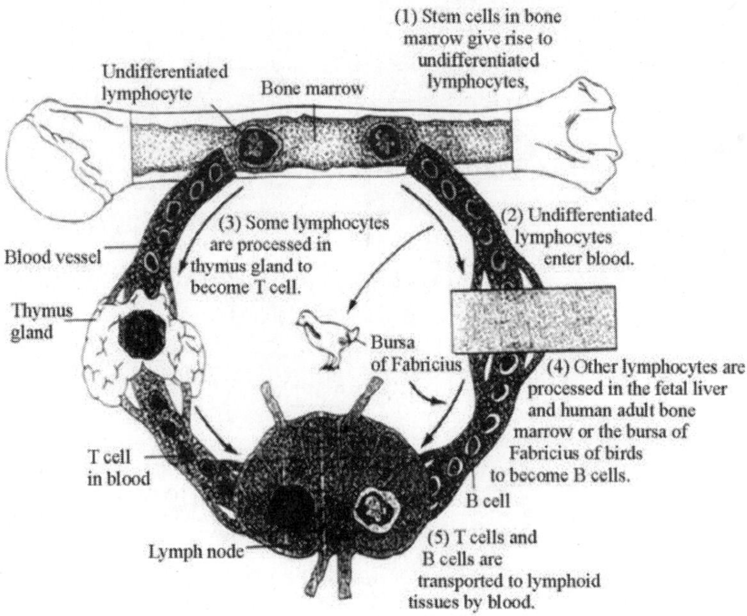


**Fig. 123** Schematic representation of lymphocyte development

#### T-cells

The T-cells or T-lymphocytes mature in thymus gland. They may remain within the thymus or circulate in the blood or present in the lymphoid organs such as the lymph nodes or spleen. The B-cells or B-lymphocytes mature within the bone marrow, circulate in the blood and also settle down in various lymphoid organs. The null cells or NK cells kill the cells infected with either viruses or intracellular pathogens, and destroy the cancer cells.

T-cell receptors (TCRs): The T-cells possess surface TCRs on their plasma membrane for antigens. The receptor site consists of two parts: 1. an alpha polypeptide chain and 2. a beta polypeptide chain. Each chain is stabilized by disulphide bonds (Fig. 125). The receptor is embedded in the plasma membrane, and parts of alpha and beta chains extend into the cytoplasm. The recognition site of the TCR extends out from the plasma membrane and possesses a terminal variable part complementary to antigen fragments. The T-cells respond to the antigen fragments exposed on the surface of APCs most of which are macrophages, dendritic cells and B-cells. These cells remove foreign materials from the blood and tissues, and later digest them. Inside the cells, small pieces of the antigens bind to a special class of proteins called MHC class-II molecules. The MHC antigen complex is then transported to the surface of the presenting cell and displayed to passing TH-cells.

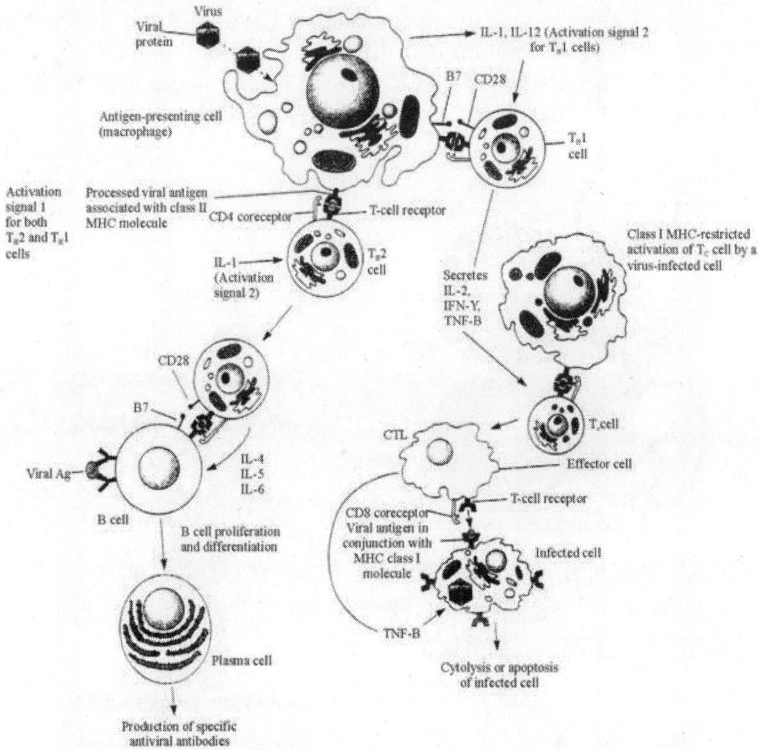


**Fig. 124** Helper T-cells responses in cell-mediated imuunity

**Types of T-cells:** There are several classes of T-cells of which effector cells and regulatory T-cells are very important classes.

1. **Effector cells:** They are also called as Cytotoxic-T (Tc) cells or CD8T cells, which cause cytolysis and death of the target cells. In order to make them participate in an immune

response, they are first activated in a complex process that involves atleast three signals. They attach to viral antigen class 1 MHCs present on the surface of virus-infected cells by TCR proteins. A constimulatory signal CD28-B7 is also involved during this attachment. Finally, the T-cells are exposed to IL-2 from TH1 cells. The Tc-cells then undergo proliferation and differentiation to form active Cytotoxic T-lymphocytes (CTLs), which can attack viral-infected cells (Fig. 124). The CTLs and TH1-cells release cytokines such as gamma IFN, and both macrophages and T-cells release TNF-alpha. These cytokines limit not only viral replication, but also activate macrophages and other phagocytes that can destroy the infected cells.

The TH2-cells secrete cytokines such as IL-4, IL-5 and IL-6, which stimulate B-cell proliferation and differentiation into antibody-forming plasma cells. This activation of TH cells by macrophages is required for the formation of both CTLs and antibodies (Fig. 124).

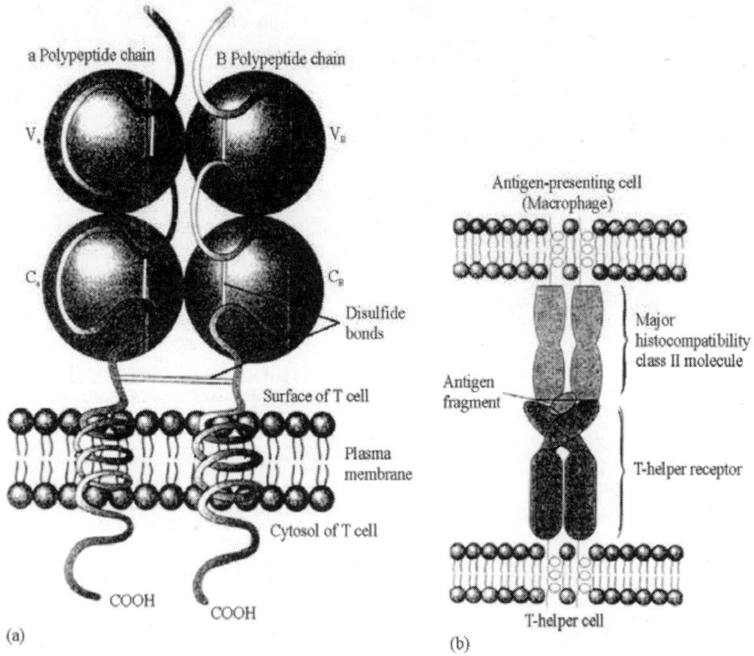


**Fig. 125** Role of the T-cell receptor protein in T-helper cell activation: a. Structure of the antigen on receptors site in a T-cell plamsma membrane and b. Activation of T-helper cells results after Va and Vb regions react with antigen fragment and MHC molecule on the present on cell surface

The CTLs can destroy target cells in two ways:

- The CD95 pathway, which induces apoptosis or programmed cell death,
- The perforin pathway, which is more frequently employed by CTLs, and uses perforin and granzymes to stimulate both osmotic cytolysis and apoptosis (Fig. 126).

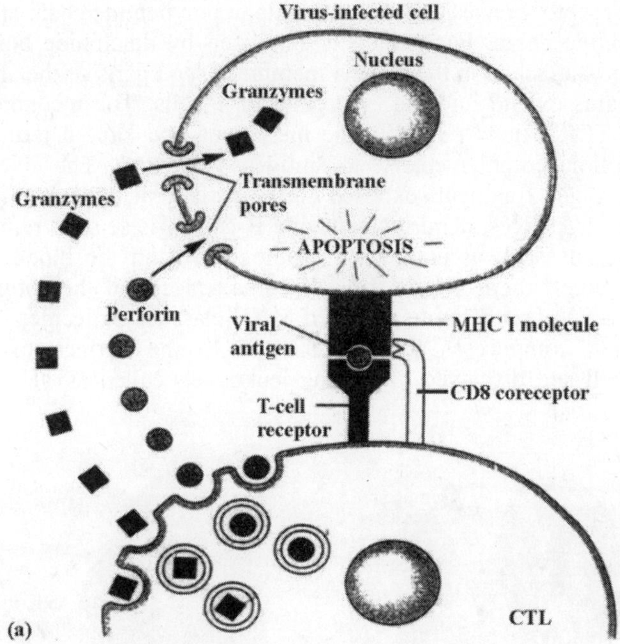


**Fig. 126** Virus-infected target cell by an effector cytotoxic T-cell

## Regulatory T-cells

They control the development of effector T-cells. There are two types – TH-cells (CD4 cells) and T suppressor (Ts) cells. There are three subsets of TH-cells, which are referred as TH1, TH2 and TH0. The TH1 and TH2 produce and secrete a specific mixture of cytokines. The TH1-cell produces IL-2, gamma IFN and TNF-beta that are involved in cellular immunity. These cytokines are responsible for DTH reactions and macrophage activation. The TH2-cells also produce various cytokines such as IL-4, IL-5, IL-10 and IL-13, and are involved in humoral immunity. The TH2-cells are helpers for B-cell antibody reaction and defence against helminthes. The TH0-cells are undifferentiated precursors of TH1-cells and TH2-cells.

Though there is no concrete evidence, most of the immunologists believe in the existence of Ts-cells that can suppress T-cell and B-cell response. Subtypes of Ts-cells specific to a given antigen can be stimulated to proliferate by the IL-2 from activated TH cells. This proliferation occurs at a slow rate to provide negative feedback control for that part of immune response known as acquired immune tolerance.

**Function of T-cells:** Specific TCRs are present on the plasma membrane of surface of T-cells, which help in binding the antigens to T-cells. Each receptor consists of two parts, alpha polypeptide chain and a beta polypeptide chain. Each chain is stabilized by disulphide bonds. The receptor is attached in the plasma membrane, and parts of the alpha and beta chains extend into the cytoplasm of T-cells. The recognition sites of the TCR extend out from the membrane and have a terminal variable section complementary to antigen segments. The T-cells respond to antigen fragments exposed on the surface of APCs, most of which are macrophages, dendritic cells and B-cells. These cells remove foreign materials such as antigens or pathogens, from the blood and tissues, and digest them. Inside the cells, small pieces of the antigens bind to a special class of proteins called MHC class III molecules. The MHC antigen complex is then transported to the surface of the presenting cell and displayed to passing leukocytes called as TH cells (Fig. 127).

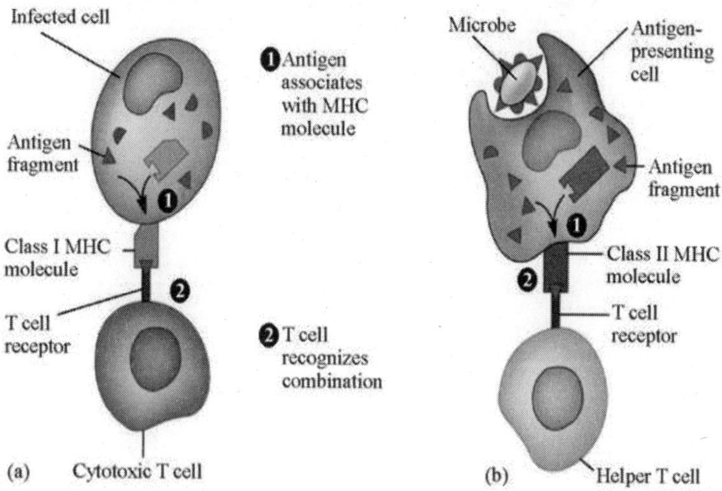


**Fig. 127** Functions of T-cells

## Structure and functions of B-cells

The B-cell precursors called lymphoblasts produced from stem cells are present in the bone marrow. Lymphoblasts mature into B-cells. Mature

B-cells possess cell surface transmembrane immunoglobulin molecules associated with IgA/IgB heterodimer proteins, which are called B-cell Antigen Receptors (BCRs), which can contain IgM and IgD immunoglobulins. However, it is the IgM, which constitutes as the major receptor for its suitable antigen. The B-cells also contain receptor for the Fc part of some immunoglobulin classes and may possess receptors for complement C3b. Each B-cell may possess about fifty thousand of these membrane-bound immunoglobulins. The total number of B-cell population of each individual carrier BCRs specific for many antigens; however, each mature B-cell possesses BCRs specific only for a particular antigenic determinant.

Antibodies are secreted by plasma cells and they bind to the antigens in the solution. The various classes of antibodies such as IgG, IgM, IgA, IgD and IgE are related to the membrane-bound BCRs. In addition to antigen binding, BCRs also stimulate the endocytosis of an antigen leading to it being processed inside the B-cell. A small antigen fragment is presented on the surface of the cell in association with MHC class II molecules. Thus, B-cells have two immunological functions:

1. They proliferate and differentiate into plasma cells, and respond to antigens by making antibodies.
2. They can act as APCs.

**Antigen and Antibody Binding:** Antigen binds to an antibody at the antigen binding site present within the Fab region of the antibody. Specific amino acids contact the antigen epitope or haptome groups at this binding site and multiple non-covalent bonds are formed between the antigen and amino acids of the binding site (Fig. 128).

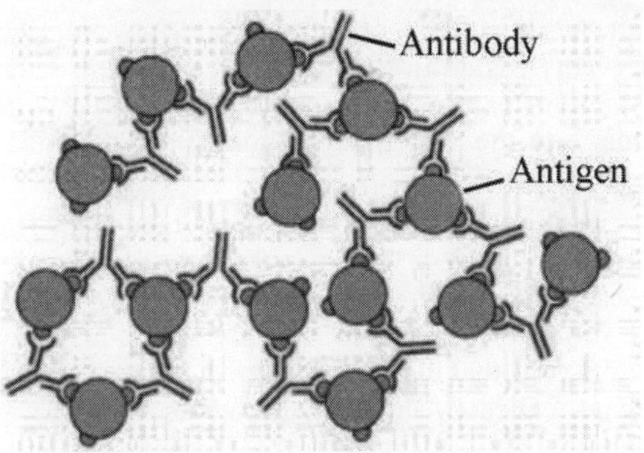


**Fig. 128** Antigen and antibody binding

As binding of antigen and antibody is due to weak non-covalent bonds, such as hydrogen bonds and also electrostatic attractions, the antigens' shape must exactly match with that of the antigen binding site.

If the shape of the epitope and binding site are not exactly complementary, the antibody will not effectively bind the antigen. Though a lock and key mechanism may normally operate, atleast in one case the antigen binding site does change the shape when it complexes with the antigen by induced fit mechanism. Thus, it is clear that antibody specificity results from the nature of antibody and antigen binding.

**B-cell activation:** The mature B-cell, upon activation by a suitable signal, proliferates and differentiates into a plasma cell that secretes antibody. Activation may be antigenic-specific or antigenic non-specific based on the involvement of T-cells. The B-cell activation is of two types:

**T-dependent antigen activation:** Most antigens possess more than one type of epitope or antigenic determinant sites. The B-cells are specific for a given epitope on the antigen, e.g., epitope X often cannot develop into plasma cells capable of secreting antibody e.g., anti X, without the cooperation of TH cells. In other words, binding of the epitope X to the B-cell may be necessary, but it is not sufficient for B-cell activation. Antigens that give a response with the help of TH-cells are called T-dependent antigen activation. Examples include bacteria, foreign red blood cells, certain proteins and hapten carrier combinations.

The basic mechanism of T-dependent antigen activation of a B-cell involves three cells:

a. A macrophage or other APC to process and present the antigen

b. A TH-cell capable of recognizing the antigen and respond to it

c. A B-cell specific for the antigen

When all these cells and antigens are present, the following sequence of events occurs in the activation of B-cells (Fig. 129). The macrophage presents part of the antigen and its own class II MHC to the TH cell (signal-1 to the T-cell). Costimulation is provided by the B7-CD28 interaction (signal-2).

The ILs 1 and 6 stimulate T-cells to divide and make ILs 2 and 4–6. The IL-1 also stimulates the hypothalamus to sense the body temperature, causing a fever, which enhances the activity of T-cells. The IL-2 and 4–6 stimulate TH cells to multiply. These helper cells directly associate with B-cells that display the correct antigen-class II MHC and secrete cytokines called B-cell Growth Factors (BCGFs), which cause B-cells to multiply.

As the number of B-cells increase, TH-cells produce other cytokines called the B-cell Differentiation Factors (BCDFs). These instruct some B-cells to stop replicating, to differentiate into plasma cells and to start producing antibodies.

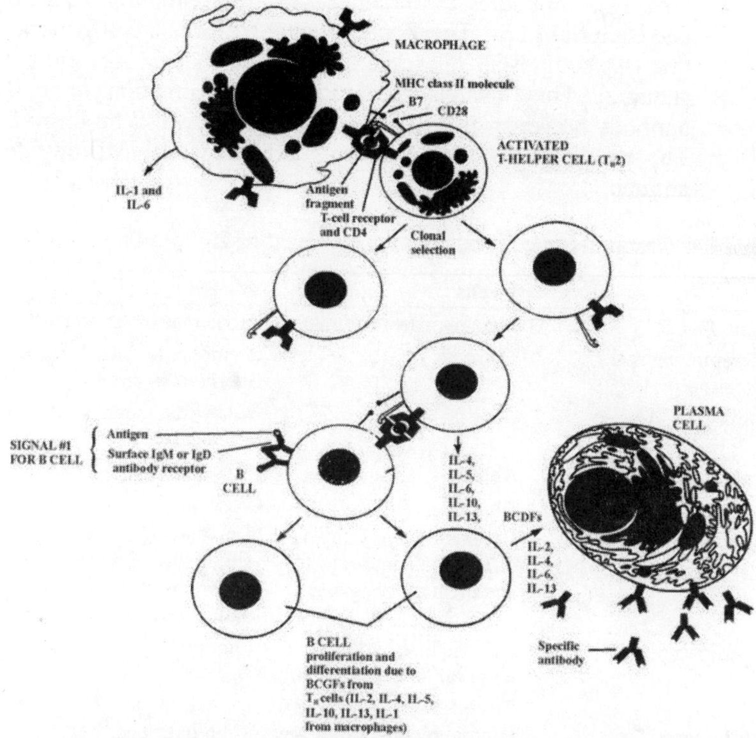


**Fig. 129** T-dependent antigen targeting of a B-cell

B-cell activation involves more than TH cell activity. The B-cell also recognizes the antigen through its surface IgM receptors for the antigen (signal-1 for the B-cell). The B-cell is subsequently triggered to proliferate and differentiate into plasma cells that secrete the antibody. One plasma cell can synthesize more than 10 million antibody molecules per hour. It is noted that B-cells are more effective antigen-presenters than macrophages at low antigen concentrations if the antigen can bind to surface immunoglobulins. The B-cells bind antigen, take it up by receptor-mediated endocytosis and present it to TH-cells to activate them. In this situation, B-cells and TH-cells activate each other. All these immune responses that produce IgG, IgA and IgE involve this T-dependent antigen activation.

1. **T-Independent antigen activation:** T-cell help is not required for all antibodies. There are antigens that activate B-cells to produce immunoglobulins without the help of T-cell. These antigens are called T-independent antigens and the process stimulated by them is called T-independent antigen activation e.g., certain tumor promoting agents, anti-immunoglobulins

(anti Ig), antibodies to certain B-cell differentiation antigens and bacterial LPSs. The T-independent antigens are polymeric, that is composed of repeating polysaccharide or protein subunits. They exclusively help in the formation of IgM antibody however; little quantities of IgG may also be formed. The resulting antibody generally exhibits a low affinity to antigen.

**Table 74:** Comparison of lymphocytes involved in the immune response

| | **T-cells** | **B-cells** |
|---|---|---|
| Origin | Bone marrow in adults | Bone marrow in adults |
| Differentiation and maturation | Thymus | Lymphoid tissue; bursa of fabricius in birds |
| Mobility | Great | Very little (some stages circulate) |
| Complement receptors | Absent | Present |
| Surface Ig | Absent | Present |
| Proliferation | Upon antigenic stimulation, differentiate into effector and memory cells | Upon antigenic stimulation, proliferate and differentiate into plasma and memory cells |
| Immunity type | Cell mediated and humoral (B-cell activation by $T_H$-cells) | Humoral |
| Distribution | High in blood, lymph, and lymphoid tissue | High in spleen, lymph nodes, bone marrow, and other lymphoid tissue; low in blood |
| Secretory product | Cytokines | Antibodies |
| Subset and functions | $T_H$ cell: necessary for B-cell activation by T-dependent antigens and T-effector cells. There are three types of $T_H1$, $T_H2$ and $T_H0$. | Plasma cell: a cell arising from a B-cell that manufactures specific antibodies. |
| | $T_S$-cell: blocks induction and/or activation of $T_H$-cells and B-cells; helps maintain tolerance | Memory cell: a long-lived cell responsible for the anamnestic response. |
| | $T_C$-cell: differentiates into a CTL that lyses cells recognized as nonself and virus or parasite-infected cells | |
| | Memory cells | |

The mechanism for activation of T-independent antigen mostly depends on their polymeric structure. These large molecules present a large group of identical epitopes to a B-cell specific for that determinant. The multivalent nature of the cell receptor, surface IgM, is such that cell activation occurs and IgM is secreted. As there is no T-cell help, the B-cell cannot induce to high-affinity IgG production and no memory B-cells are formed. Summary and comparison of the important properties of lymphocytes are given in the Table 74 and Table 75.

**Table 75:** Antigen recognition by T-cells and B-cells

| | **T-cells** | **B-cells** |
|---|---|---|
| Binds soluble antigen | No | Yes |
| Biochemistry of the antigen | Mostly proteins, but some glycolipids presented on MHC molecules | Proteins, glycolipids, polysaccharides |
| Antigen recognition | Antigens processed internally and presented as linear peptides-bound MHC molecules | Accessible areas of protein structure containting sequential amino acids and non-sequential amino acids |
| | | Immunoglobulin receptor binds antigen in native conformation |
| | Involves three partners: TCR, processed antigen, and MHC molecule | Involve two partners: antigen and membrane immunoglobulin |

## 11.4.2 Monocytes and Macrophages

These cells are highly phagocytic and form monocyte-macrophage system (Fig. 130).

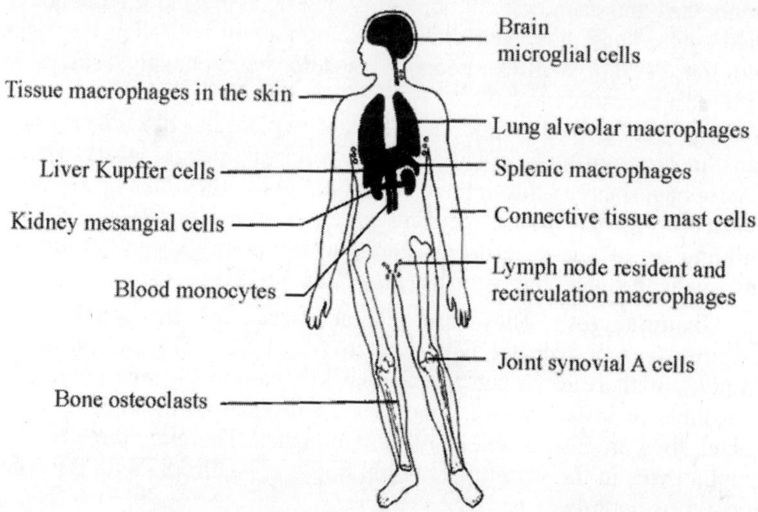


**Fig. 130** The monocyte-macrophage system

**Monocytes:** They are phagocytic leukocytes with only one nucleus in each cell, which is either ovoid or kidney shaped. The cytoplasm contains many granules, which stain gray-blue. They are produced in the bone marrow. After the production, they enter into blood and remain in blood circulation for 8 h, then enlarge, migrate to the tissues, where they mature into macrophages (Fig. 131).

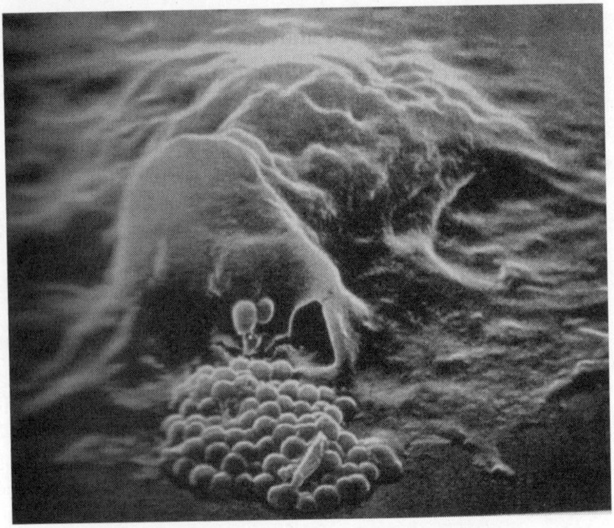

**Fig. 131** Phagocytosis by a macrophage

**Macrophages:** As described above, they are derived from monocytes and each cell contains only one nucleus, and are phagocytic leukocytes. They, however, differ from monocytes with their large size, with more cell organelles, specially lysosomes and phagolysosomes and with plasmamembrane covered with microvilli and suffler. They possess receptors for antibodies and complement, which can bind microorganisms and foreign material, and thereby, enhance phagocytosis. Though macrophages spread throughout the animal body, they take up residence in specific tissues where they are given special names. They help in building up of non-specific resistance to the host body by swallowing pathogens through phagocytosis (Fig. 122).

**Granulocytes:** They are characterized by the presence of irregularly-shaped nuclei with two to five lobes. The cytoplasm has granules with reactive substances that kill pathogenic microorganisms and enhance inflammation. Due to the presence of irregularly-shaped nuclei, they are also called as PMN leukocytes. There are three types of granulocytes in the vertebrate animal body: a. Neutrophils, b. Basophils and c. Eosinophils.

1. **Neutrophils:** They possess a nucleus with 3–5 lobes connected by fine threads of chromatin, and contain inconspicuous primary and secondary granules. They stain readily at neutral pH. They are highly phagocytic because they possess receptors for antibodies and complement proteins like macrophages. They do not reside in healthy tissue, unlike macrophages, but rapidly migrate to the region of tissue damage where they destroy the pathogenic microorganisms or foreign material by phagocytosis and microbicidal activity. The large primary and smaller secondary granules of the cytoplasm of neutrophils contain lytic enzymes and bactericidal substances; the former contain peroxidase, lysozyme and various hydrolytic enzymes while the later possesses collagenase and lysozyme. Neutrophils generate antimicrobial substances and defensins to kill ingested microorganisms by oxygen-dependent and oxygen-independent mechanisms (Fig. 122).
2. **Basophils:** They possess irregularly-shaped nucleus with two lobes and the granules stain blue-black with basic dyes. They are non-phagocytic cells, which perform their function by releasing serotonin, prostaglandins, histamines and leukotrienes from their granules upon suitable stimulation. As these chemical mediators influence the tone and diameter of blood vessels, they are generally called as vasoactive mediators. Basophils possess high-affinity receptors for an immunoglobulin, and thereby, become coated with IgE antibodies. Once coated, antigens can activate the cell to secrete vasoactive mediators. These inflammatory mediators play a major role in certain allergic responses such as eczema, hay fever and asthama (Fig. 122).
3. **Eosinophils:** They have a two-lobed nucleus and the lobes remain connected with fine threads of chromatin, and the cytoplasmic granules stain red with acidic dyes. Unlike basophils, they are mobile cells that can migrate from the blood stream into tissue spaces. They play an important role in conferring defence against protozoan and helminth parasite, mainly releasing cationic proteins and reactive oxygen metabolites into the extracellular fluid to damage the parasites' plasmamembranes (Fig. 122).

**Null cells:** They constitute a small proportion of lymphocytes, which cannot be classified as a category of lymphocytes. Though their nature and functions are not fully understood, it is reported that they kill the target cells. Based on their function, two distinct sub-populations are recognized:

1. **Killer (K) cells:** They possess surface receptors for the Fc part of IgG antibody. They are capable of lysing or killing the target cells sensitized with IgG antibodies. They do not require the help of complement proteins. They are responsible for ADCC in contrast to the action of CTLs, which are independent of antibodies.
2. **NK cells:** They are the large lymphocytes containing azurophilic granules in the cytoplasm and have therefore, been called Large Granular Lymphocytes (LGL). The NK cells are capable of non-specific killing in virally-transformed target cells, and are involved in allograft and tumor rejection. They differ from killer cells in being independent of antibody. Gamma IFN increases the activity of NK cells. Their cytotoxic activity is non-sepcific and can be demonstrated on tumor cell lines or fetal fibroblast cultures. They are present in the spleen and peripheral blood of humans and animals. Though their exact function is not known, they are believed to play an important role in antiviral and antitumor immunity.

**Mast cells:** They are found in connective tissue and are derived from the bone marrow. They contain granules, which possess histamine and pharmacologically-active compounds that contribute to the inflammatory response. Along with basophils, they play an important role in the development of allergies and hypersensitivities.

**Dendritic cells:** They represent white blood cells, and are present in extremely low proportion when compared to other leukocytes, about only 0.2% of the total leukocytes of blood. They are present in the skin and mucous membranes of nose, lungs and intestine. Immature dendritic cells can recognize specific PAMP on microorganisms and play an important role in non-specific resistance. They are capable of recognizing potentially harmful microorganisms from self molecules. After the pathogen is recognized, it binds to the dendritic cells' PRPs and then is swallowed. Some immature dendritic cells can also kill viruses immediately by secreting IFN alpha. After maturation, dendritic cells migrate into the blood stream, lymph nodes and spleen where they interact with the other cells of immune system such as B-cells, which make antibodies and NK cells that attack pathogens and infected cells. As dendritic cells present antigens to T-cells, they also play an important accessory role in the specific immune response (Fig. 122).

## REVIEW QUESTIONS

**I. Answer the following questions:**

1. What is the function of the thymus, lymphnode and spleen?

2. Describe the primary lymphoid organs and tissues involved in immunity.
3. Discuss briefly about the types of immunity.
4. Define and give examples of an antigen. What is an antigenic determinant or epitope?
5. Describe hapten and super antigen.
6. Discuss the biological significance of cluster of differentiation molecule.
7. Discuss about the types and functions of T-cells.
8. Compare and contrast T-dependent and T-independent antigen activation.
9. Describe the structure and functions of B-cells.
10. What are lymphoid organs? List the peripheral lymphoid organs and their functions.
11. What do you understand by primary and secondary lymphoid organs? List the primary organs and their functions in immune response.
12. Describe briefly about the cooperation and regulation between T-cells and B-cells.
13. Describe three types of functions of T-lymphocytes.
14. Explain primary characterstics of a generalized immune response.
15. How does a T-cell recognize an antigen?
16. Explain the role of T-cells and B-cells in immunity.
17. Give diagram of a typical antibody and give an account of heavy chain, light chain and constant variable, and Fc regions.

**II. Write short notes on the following:**

1. Specific immunity and non-specific immunity
2. Monocytes
3. Macrohages
4. Granulocytes
5. Null cells
6. Mast cells
7. Dendritic cells
8. Eosinophils
9. Neutrophils
10. Basophils

## 11.5 FURTHER READING

Alcami A and Kaszinowskii UH 2000 Viral mechanisms of immune evasion, *Immunology Today* **21** 447

Biran CA 1999 Initial and innate responses to viral infections – pattern setting in immunology or disease, *Current Opinion in Microbiology* **2** 374–381

Dajldetti M, Salman H, Bergman M, Dejaletti R, and Bessler H 2002 Phagocytosis the mightly weapon, *Micoscopy Research and Technique* **57** 421–431

Kaunfman SHE 1993 Immunity to interacellar bacteria, *Annual Review of Immunology* **11** 129–163

Mims CA, Nash A, and Stephen J 2000 *Mims' Pathogenesis of Infectious Disease*, London: Academic Press

Patrick S and Larkin M 1995 Immunological and Molecular Aspects of Bacterial Virulence, Chichester, Wiley.

Roy C and Sansonetti P 2004 Host–microbe interaction: Bacteria, Host–pathogen interaction, interpreting the dialogue, *Current Opinion Micrbiology* **7** 1–3

Taussing MJ 1965 *Process in Pathology and Microbiology*, Oxford: Blackwell Scientific

Walkelin D 1996 *Immunity to Parasites*, Cambridge: Cambridge University Press

# 12

# Principles of Immunology

## 12.1 INTRODUCTION

The immune system of a vertebrate host distinguishes between self and non-self through an elaborate recognition process. The proteins and other macromolecules present in the host body remove most of the T-cells specific for self determination prior to birth. Subsequently, self substances are recognized from non-self substances and lymphocytes can produce specific immunologic reaction against non-self substances leading to their removal or death.

### 12.1.1 Antigens

Substances that give immune response and react with products of immune response are called antigens e.g., proteins, NPs, polysaccharides and certain glycolipids. Most antigens are complex and large molecules with higher molecular weight, generally more than 10,000 daltons. The ability of a chemical molecule to function as an antigen depends on its size, structure, chemical nature and degree of foreigness to the host.

Each antigen may have several epitopes or antigenic determinant sites. An epitope is the site or location on the antigen that binds specific antibody or a TCR. Antibodies are readily formed in response to determinant sites that project from the antigen, or a foreign molecule or terminal region of a specific polymer chain. Chemically, determinants include aromatic compounds, organic acids and bases, amino acid side chain or sugars. The number of antigenic determinant sites present on the surface of an antigen is called as its valency, which indicates the number of antibody molecules that can bind with an antigen at a time. If an antigen contains one antigen determinant site, it is called monovalent. However, most antigens are polyvalent as they contain more than one determinant site. Multivalent antigens exhibit stronger immune response than monovalent antigen. In other words, immune response of an antigen is directly proportional to its valency.

### 12.1.2 Haptens

These are the smaller organic molecules without antigenic properties, but become antigenic when they bind to a larger carrier molecule such as a protein. Though they cannot stimulate antibody formation or T-cell response, they can react with antibodies once formed (Fig. 133).

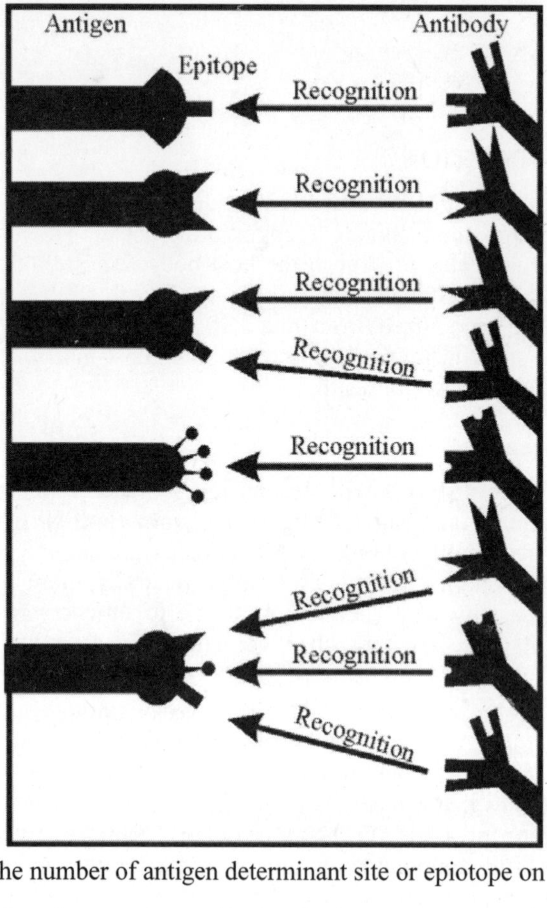


**Fig. 132** The number of antigen determinant site or epiotope on an antigen

When lymphocytes are stimulated, they can react either with the hapten or the larger carrier molecule. Haptens become immunogenic when they conjugate with a carrier protein because the later can be processed and presented to specific T-cells. This results in the formation of hapten-specific and carrier-specificantibodies. For example, penicillin is not an antigen, but when it combines with some serum proteins of sensitive individuals, the resulting molecule can initiate a severe and sometimes, fatal allergic immune reaction. In this case, the hapten is acting as an antigen determinant on the carrier molecule.

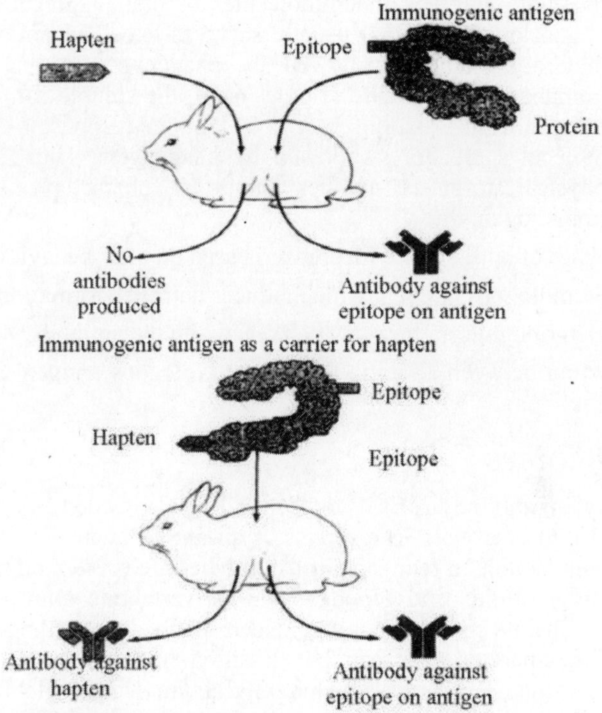


**Fig. 133** Effect of carrier on immunogenicity of hapten

## 12.2 FACTORS AFFECTING ANTIGENICITY

Antigenicity of an antibody is determined by three factors:

1. Molecular size of a substance is an important component of immunogenicity. For example, antigens with low molecular weights i.e., haptens, may bind to antibodies, but cannot induce immune response. However, when bound to larger protein carrier-molecule, haptens become very effective immunogens and initiate a stronger immune response. Most antigens possess a molecular weight greater than 10,000 daltons. Thus, sufficient molecular size is an indication of its potential immunogenicity.
2. Proteins, which are complex and non-repeating polymers, are effective antigens or immunogens. Complex carbohydrates can also function as effective antigens. Nucleic acids and lipids, in contrast, are very poor antigens because they contain repeating monomers. Thus, sufficient molecular complexity also contributes to immunogenicity.

3. Insoluble complex macromolecules or their aggregated form (e.g., clotted proteins) usually serve as excellent antigens, as they are readily swallowed by phagocytes leading to an immune response. On the other hand, the soluble form of the same molecule exhibits very poor immune response as soluble molecules are not swallowed by phagocytes. Thus, suitable physical form is another important character indicating immunogenicity.

Two types of antigens are recognized based on their behaviour:

1. Homologous antigen, which induces antibody formation
2. Heterologous antigens, which react with the antibody.

A reaction between an antibody and heterologous antigen is called as cross-reaction.

## 12.3 ANTIBODIES

Chemically, an antibody is a GP molecule. It is also called as Ig, and is produced by plasma cells in response to an antigen with which it can bind and also be able to recognize it. Antibodies are present on mucosal surfaces, tissue fluids and blood stream of vertebrate animals. It is possible to separate serum GPs by gel electrophoresis according to their charge. The separated GPs can be classified into five types viz, 1. Globulin 2. Alpha-1-globulin 3. Alpha-2-globulin 4. Beta globulin and 5. Gamma globulin. The gamma globulin contains a heterogeneous mixture of Igs named as IgG, IgA, IgM and IgD. They differ from each other in structure, charge, molecular size, carbohydrate content and amino acid composition. Most serum antibodies belong to the IgG class. A fifth Ig called IgE also exists in serum in extremely low concentration and resembles IgD in mobility in gel electrophoresis.

### 12.3.1 Structure of Antibody (Ig)

All Igs are structurally similar and consist of four polypeptide chains, of which two are light chains and the remaining two are heavy chains. They are connected to each other by disulphide bonds. Each light chain has a molecular weight approximately of 25 kilodaltons (KDs) and consists of 220 amino acids. Similarly, each heavy chain has a molecular weight approximately of 50–70 KDs and consists of 440 amino acids. Both light and heavy chains contain two different regions, the constant regions and the variable regions. The constant regions, termed as $C_L$ and $C_H$, possess amino acid sequences, which is almost similar between the antibodies of the same class, whereas the variable regions, termed $V_L$ and $V_H$, exhibit variation in the amino acid sequence among different antibodies. Antigen binding sites are formed due to folding in the variable regions (Fig. 134 and Fig.135).

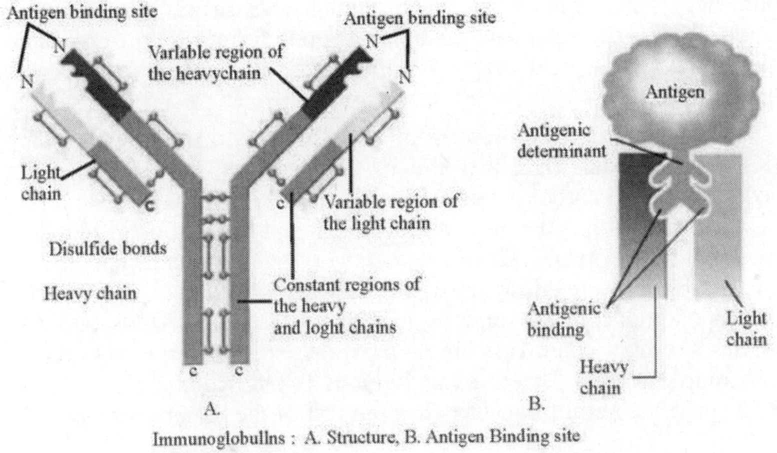


**Fig.134** Immunoglobulin molecules with two similar light and two heavy chains linked together by disulphide bond

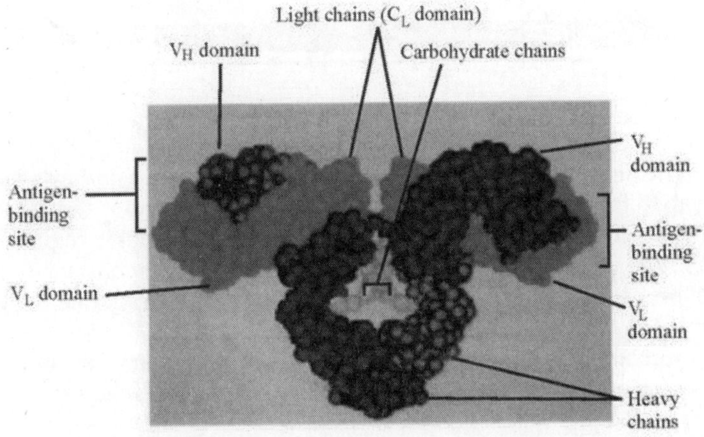


**Fig. 135** Computer-generated model of an immunoglobulin molecule with two light chains and two heavy chains

The two heavy chains and two light chains are arranged in the form of a flexible 'Y' with a distinct hinge region, which makes the antibody molecule to assume a 'T' shape. The stalk of the Y is called as Fc and possesses a site with which an antibody molecule can bind to a cell. There are two antigen binding fragments (Fab) at the top of Y that bind with compatible epitopes or antigen determinant sites. The Fc consists of only constant regions whereas Fab contains both constant and variable regions. There are several homologous units called domains in both heavy and light chains. Each domain consists of 100–110 amino acids. Within each domain, disulphide bonds form a loop of 60 amino acids. Heavy and light chains are also linked by inter-chain disulphide bonds.

The light chain may remain in two distinct forms known as Kappa (K) and Lambda (λ), which can be distinguished by the amino acid sequence of the carboxyl part of the chain. The terminal carboxyl parts of all K chains are identical in human Igs and this region is termed as constant ($C_L$) domain. There are four very similar sequences in the λ chains, which are designated as λ1, λ2, λ3 and λ4, with their corresponding constant regions termed as Cλ1, Cλ2, Cλ3 and Cλ4. Within the light chain variable domain, there are hypervariable regions or Complementary Determining Regions (CDRs) that differ in amino acid sequence more frequently than the rest of the variable domain (Fig. 135).

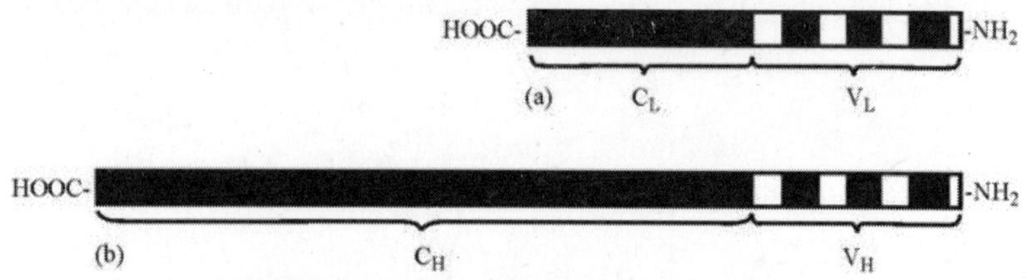


**Fig. 136** Constant and variable domains

The amino ($NH_2$) terminal domain in the heavy chain has a pattern of variability similar to that of $V_K$ and Vλ domains, and is termed as $V_H$ domain. However, this domain contains four hypervariable regions (Fig. 136). The other domains of the heavy chain are called constant domains and are numbered as CH1, CH2, CH3 and CH4, starting with the domain next to the variable domain (Fig. 136). The constant domains of heavy chain form the constant carboxy (COOH) region. The amino acid sequence of this region determines the classes of heavy chains.

In humans, there are five classes of heavy chains designated as gamma (γ), alpha (α), mu (μ), delta (δ) and ipsilon (€). The properties of these five heavy chains determine the five Ig classes i.e., IgG, IgA, IgM, IgD and IgE. Each class of Ig differs from other in general properties, half life, distribution in the body and interaction with other components of the host's defence systems. There are variants within two of the major human Ig classes. These variations are classified as:

1. **Isotypes:** Isotypes are characterized by the variation in the CH regions present in the different classes that are normally present in all individuals.

2. **Allotypes:** Allotypes are genetically-controlled alletic forms of Ig molecules that are not present in all individuals.
3. **Idiotypes:** Idiotypes are the individual specific Ig molecules that differ in the hypervariable region of the Fab portion.

The vast variation in the Ig structure reflects the diversity of antibodies generated by the immune response.

There is more than one antigen combining site in most of the antibodies. For example, most human antibodies possess two combining sites and they are called bivalents. Some bivalent antibodies sometimes combine to form multimeric antibodies that may have upto ten combining sites (Fig. 137).

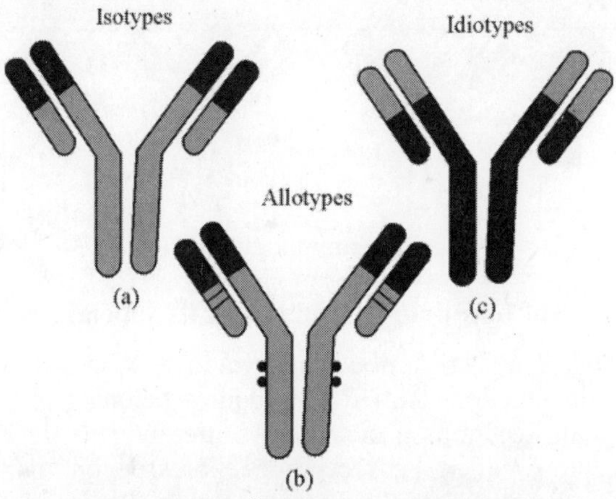


**Fig. 137** Variants of immunoglobulins

### 12.3.2 Classes of Ig

There are five classes of Ig or antibodies, which are termed as IgG, IgM, IgA, IgD and IgE.

1. **Immunoglobulin G (IgG):** It is present in blood plasma and tissue fluids. It is the major Ig in human serum accounting for 80% of the total Ig pool. It acts against bacteria and viruses by opsonizing them and neutralizing toxins. It is one of the two Igclasses that activate complement by classical pathway. It is the only Ig molecule capable of crossing the placenta and providing natural immunity to the newborn at birth. It has a molecular weight of about 1,50,000 daltons.

   There are four human IgG subclasses, which are termed as IgG1, IgG2, IgG3 and IgG4. They differ from one another in their heavy chain composition, and the number, arrangement, inter-chain disulphide bonds (Fig. 138) and biological function. The IgG1 constitutes about 65% of the total serum and IgG2 23%, and IgG3 and IgG4 are present in traces.

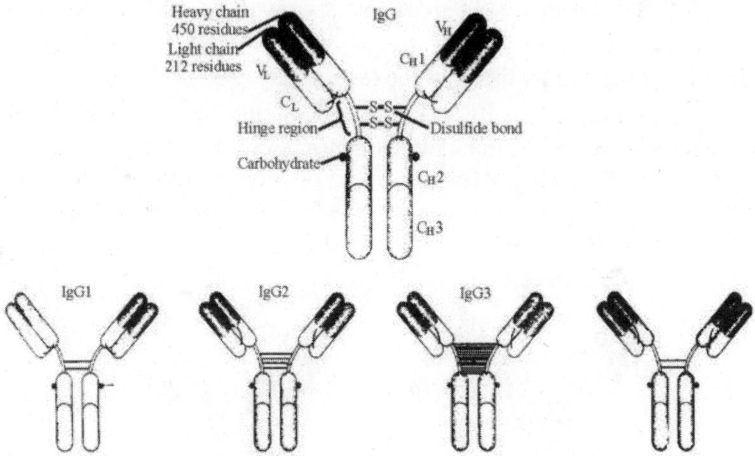


**Fig. 138** Immunoglobulin G

**Basic structure of Immunoglobulin G and its subclasses**

**Functions:** The IgG antibodies develop in response to anti-toxin and opsonic in function. Anti Rh antibodies belong to IgG1 or IgG3 subclasses; upon recognition of their specific antigens, IgG1 and IgG3 bind to receptors called Fc receptors, expressed on monocytes and macrophages, and stimulate them to phagocytes, the invader. The IgG4 antibodies function as skin sensitizing Ig.

1. **Immunoglobulin M (IgM):** The IgM is the first Ig made at the time of B-cell maturation and is expressed as membrane-bound antibody on B-cells. It is secreted into blood serum during a primary antibody response. It accounts for about 10% of the serum Ig pool. It is generally a polymer of five monomeric Ig molecules and called as pentamer. The pentamer has a molecular weight of about 9,70,000 daltons. Each monomer consists of two heavy chains and two light chains. The monomers are arranged in a pinwheel manner with Fc ends in the centre, and are held together by disulphide bonds and a special joining (J) chain (Fig. 139).

Though most of the IgM remain in pentameric form, about 5% or even less, human blood-serum IgM exists in hexamine form due to combination of six monomeric units. However, the hexamine form lacks J chain. It is believed that hexameric form activates complement upto twenty-fold more effectively than the normal pentameric form. It has been reported that bacterial cell wall antigens such as polysaccharides of gram-negative bacteria, may directly stimulate B-cells to form hexameric IgM without J chain. In such cases, the Ig formed during primary immune response remains less homogeneous.

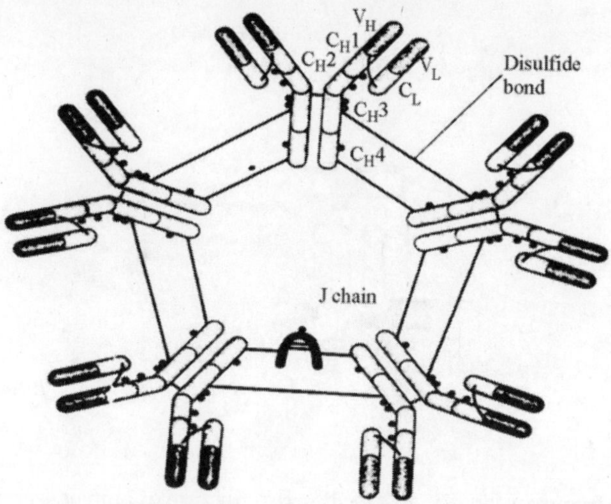


**Fig. 139** Immunoglobulin M: The pentameric structure of human IgM

**Functions:** As IgM is very large, it does not leave the blood stream and is unable to cross the placenta. It agglutinates bacteria, enhances the ingestion of pathogens by phagocytic cells and activates complement by classical pathway.

2. **IgA:** The IgA has a molecular weight of about 1,50,000 daltons. It accounts for about 15% of the serum Ig pool. It is present in saliva, tears, breast milk, colostrum, gastrointestinal secretions and mucosal secretions of respiratory and gastrointestinal tract. In serum, it is present in monomeric form with two heavy and light chains. However, most of the IgA occurs in mucosal secretions in dimeric form, which are bound together by a J chain. When it is transported from MALT to mucosal surfaces, it acquires a protein called the secretory component and then it is called as sIgA (Fig. 140).

**Functions:** IgA plays an important role in protecting surface tissues against infectious microorganisms by the formation of immune barrier.

- In breast milk, it helps to protect nursing newborns against pathogens. It attaches viruses, bacteria and protozoan parasite such as *Entamoeba histolytica,* which prevents adherence of pathogens to mucosal surfaces and invasion of host tissues. This phenomenon is called as exclusion.

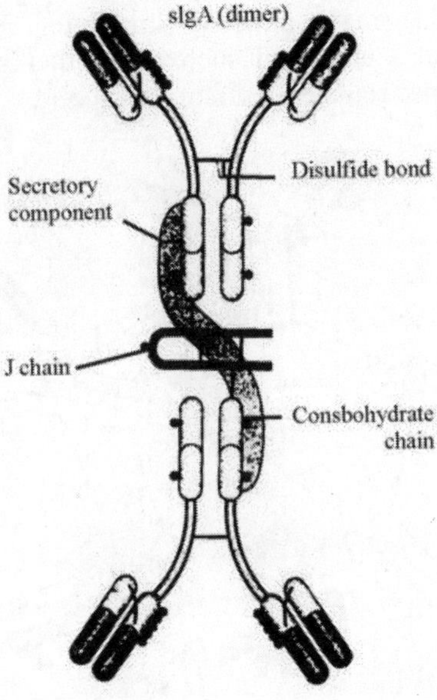


**Fig. 140** Immunoglobulin A: The dimeric structure of human secretary IgA

- It binds to antigens within the mucosal layer of small intestine, forming a complex that is excreted through the adjacent epithelium into the lumen of gut from where they are expelled out through faeces. It also helps the alternative complement pathway.

3. **Immunoglobulin D (IgD):** It has a molecular weight of about 1,80,000 daltons. It is present in traces in the blood stream. It has a monomeric structure similar to IgG (Fig. 141).

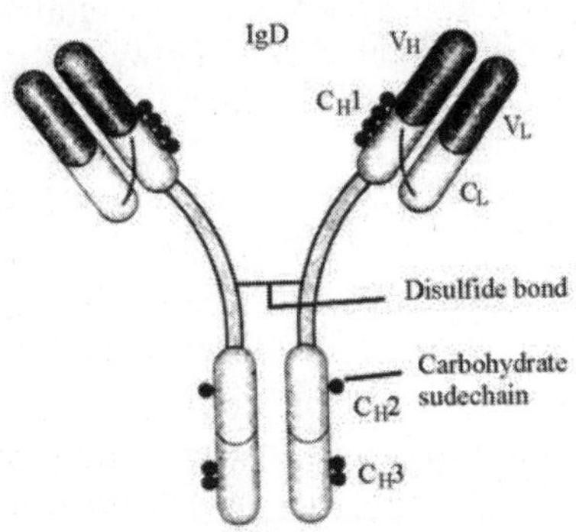


**Fig. 141** Immunoglobulin D: The structure of human IgD

**Functions:** The IgD antibodies neither cross the placenta nor fix a complement. In combination with IgM, they are abundant on the surface of B-cells and bind antigens, which stimulate the B-cells to produce antibodies.

4. **Immunoglobulin E (IgE):** The molecular weight of IgE is 1,90,000 daltons, which is significantly higher than the other classes of Ig. It is present in blood serum in extremely small amounts (about 1 in every 50,000 Igs). Despite its low concentration, it is important because immediate-type hypersensitivities are mediated by IgE. The IgE molecules have four constant region domains on their heavy chains, designated as $C_E1$, $C_E2$, $C_E3$ and $C_E4$. The Fc part of the $C_E4$ chain can bind to special FcE receptors on mast cells and basophils (Fig. 142).

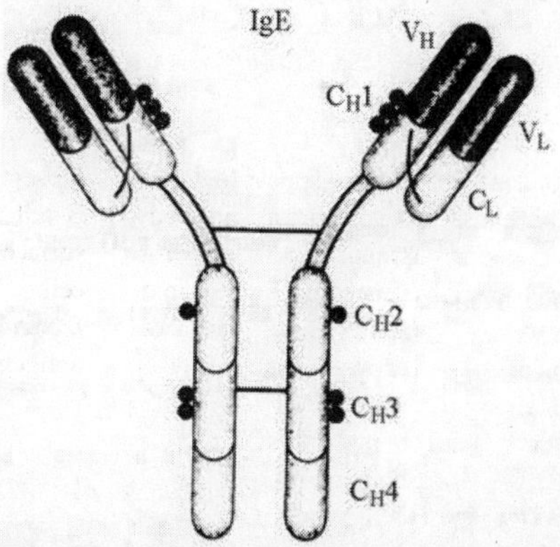


**Fig. 142** Immunoglobulin E: The structure of human IgE

**Functions:** When two IgE molecules on the surface of a mast cell or basophil are cross-linked by binding to the same antigen, the cells get degranulated. This degranulation releases histamins and other pharmacological mediators of anaphylaxis. It also stimulates eosinophilia and increased rate of movement of the intestinal contents (hypermotility) that help in the elimination of helminth parasites. Certain physicochemical properties of human Ig classes are given in Table 76.

**Table 76:** Physicochemial properties of Ig classes

| | Immunoglobulin Classes | | | | |
|---|---|---|---|---|---|
| Property | IgG | IgM | IgA | IgD | IgE |
| Heavy chain | Γ | μ | α | δ | € |
| Mean serum concentration (mg/ml) | 9 | 1.5 | 3.0 | 0.03 | 0.00005 |

| | Immunoglobulin Classes | | | | |
|---|---|---|---|---|---|
| Percent of total antibody | 80–85 | 5–10 | 5–15 | <1 | 0.002–0.05 |
| Valency | 2 | 5(10) | (2–4) | 2 | 2 |
| Mass of heavy chain (kDa) | 51 | 65 | 56 | 70 | 72 |
| Mass of entire molecule (kDa) | 146 | 970 | 160[2] | 184 | 188 |
| Placental transfer | + | - | - | - | - |
| Half-life in serum (days)[d] | 23 | 5 | 6 | 3 | 2 |
| Complement activation | | | | | |
| Classical pathway | ++ | +++ | - | - | - |
| Alternative pathway | - | - | + | - | - |
| Induces mast cell degranulation | - | - | - | - | - |
| Major characteristics | Most abundant Ig in body fluids; neutralizes toxins, opsonizes bacteria, activates complement, transplantal antibody | First to appear after antigen stimulation; very effective agglutinator; expressed as membrane-bound anti-body on B-cells | Secretory antibody; protects external surfaces | Present on B-cell surface, B-cell recognition of antigen | Anaphylactic -mediating antibody; resistant to helminthes |
| % carbohydrate | 3 | 7–10 | 7 | 12 | 11 |

**Table 77:** Some autoimmune diseases in humans

| Disease | Autoantigen | Pathophysiology |
|---|---|---|
| Type II Hypersensitivity: Antibody to Cell-Surface or Cell Matrix Antigens (Cytotoxic) | | |
| Acute rheumatic fever | Streptococcal cell wall antigen; antibodies cross-react with cardiomyocytes | Arthritis, scarring of heart valves, myocarditis |
| Autoimmune haemolytic anaemia | Rh blood group, I antigen | RBCs are destroyed by complement and phagocytosis, anaemia |
| Autoimmune thrombocytopenia purpura | Platelet integrin | Perfuse bleeding |
| Goodpasture's syndrome | Basement membrane collagen | Glomerulonephritis, pulmonary haemorrhage |
| Graves' disease | Thyroid-stimulating hormone receptor | Hyperthyroidism |

| Disease | Autoantigen | Pathophysiology |
|---|---|---|
| Myasthenia gravis | Acetycholine receptor | Progressive muscular weakness |
| Pemphigus vulgaris | Cadherin in epidermis | Skin blisters |
| Type III Hypersensitivity: Immune Complex Disease | | |
| Systemic lupus erythematosus | DNA, histone, ribosome | Arthritis, glomerulonephritis, vasculitis, rash |
| Type IV Hypersensitivity: T-cell-Mediated Disease | | Rejection of transmitted tissues, contact dermatitis |
| Type 1 diabetes mellitus | Pancreatic beta-cell antigen | Beta cell destruction |
| Multiple sclerosis | Myelin basic protein | Demyelination of axons |
| Rheumatoid arthritis | Unknown synovial joint antigen | Joint inflammation and destruction |

Components of complement activation, antigen-antibody reaction and antibody basic techniques are discussed in **Chapter 11**.

## REVIEW QUESTIONS

### I. Answer the following questions

1. Describe factors affecting antigenicity.
2. Explain the general structural characters of an immunoglobulin.
3. Explain five classes of heavy chain and their variants.
4. Describe the structure and functions of IgG.
5. Explain the structure and functions of IgM.

### II. Write short notes on the following

1. Antigen
2. Hapten
3. Antigen determinant site
4. Monovalent
5. Polyvalent
6. Kappa and Lambda
7. IgA
8. IgD
9. IgE

## 12.4 FURTHER READING

Immunology Today: Issues contains understandable Reviews on current topics in immunology

Janeway C 2004 *Immunology*, New York: Garland Science

Janeway CA and Travers P 1996 *Immunology*, Garaland science, New York

Linston PJ and Dorsland K 2004 Age related changes in lymphocytes development and function, *Nature Immunology* **5** 133–139

Male D et al 1996 *Advanced Immunology*, St. Louis, Mosby

Roitt I, Brostoff J, and Male D 1996 *Immunology,* St. Louis, Mosby.

13

# Polyclonal and Monoclonal Antibodies

## 13.1 PRODUCTION AND APPLICATION

### 13.1.1 Polyclonal Antibodies

The immune response to an antigen results in the production of Ig molecules or antibodies, which are directed at the numerous determinants present on the antigen. Only a few of the many Igs are directed towards each antigen determinant. The resulting antiserum with a mixture of different antibodies is known as polyclonal antiserum and the mixture of different antibodies is known as polyclonal antibodies. Therefore, polyclonal antibodies are a mixture of different antibodies produced by many different B-cell clones.

Polyclonal antisera consist of antibody populations i.e., polyclonal antibodies that provide sufficient immune protection to the host, but are usually specific for a variety of determinants. These antisera are not reproducible because they are the sum of the antibody responses produced by an individual animal at a single time. Thus, polyclonal antisera with polyclonal antibodies, though very potent, are extremely difficult to standardize for immune diagnostic procedures.

### 13.1.2 Monoclonal Antibodies

Generally each Ig is produced by a single B-cell. A B-cell cloned in the laboratory is capable of producing limitless number of single, monospecific Igs. Antibodies produced by a single cloned B-cell are called monoclonal antibodies. The B-cell clones can be stored as frozen cell lines and later reconstituted providing a reproducible source of specific antibodies. As a result, monoclonal antibody technology has substituted standard polyclonal antibody techniques for many immunological applications. Comparative account of properties of polyclonal antibodies with that of monoclonal antibodies is given in Table 78.

**Table 78:** Properties of monoclonal and polyclonal antibodies

| Monoclonal | Polyclonal |
|---|---|
| • Contains a single antibody recognizing only a single determinant | • Contains many antibodies recognizing many determinants on an antigen |
| • Single class of antibody produced | • Various classes of antibodies are present, IgG, IgM etc. |
| • Can make a specific antibody using an impure antigen | • Can make a specific antibody using only a highly-purified antigen |
| • Highly reproducible and standardizable | • Reproducibility and standardization difficult |

## 13.2 MONOCLONAL ANTIBODIES AND HYBRIDOMAS

Antibody producing B-lymphocytes generally die after several weeks in cell culture *in vitro*. Therefore, antibody producing B-lymphocytes must be modified to enable them to live in cell culture for longer time. This is possible when antibody producing B-lymphocytes are fused with B-cell tumors called myelomas. Myeloma cells grow and divide indefinitely *in vitro*. The important cell lines that result from the B-cell-myeloma fusion are hybrid cell lines called hybridomas. The hybridoma cell lines share the properties of B-lymphocytes and myeloma cells. They grow indefinitely *in vitro* and produce antibodies.

A mouse is immunized with the antigen of interest to produce a monoclonal antibody. During the next several weeks, antigen-specific B-cells proliferate and begin producing antibody in the mouse. The spleen tissue, rich in B-lymphocytes, is then removed from the mouse and the B-cells are fused with myeloma cells. However, viable hybridoma cell fusions are only a small fraction of the remaining cell population. The hybridoma cells are selected from other cells by addition of hypoxanthine, aminopterin and thymidine to the *in vitro* cell culture medium i.e., Hypoxanthine–Aminopterin–Thymidine (HAT) medium. The HAT medium stops the growth of unfused myeloma cells because these cells, though able to grow indefinitely in cell culture, are unable to use the metabolites hypoxanthine and thymidine to bypass a metabolic block caused by aminopterin, which is a cell poison. In contrast, fused hybrid cells can use hypoxanthine and thymidine to bypass the aminopterin block and grow normally in HAT medium because they receive the genetic information for use of hypoxanthine and thymidine from the B-cell fusion partner. Unfused B-cells die in a few days because they cannot grow in culture for more than few cell divisions. Following fusion, the antibody producing clones must be identified (Fig. 143).

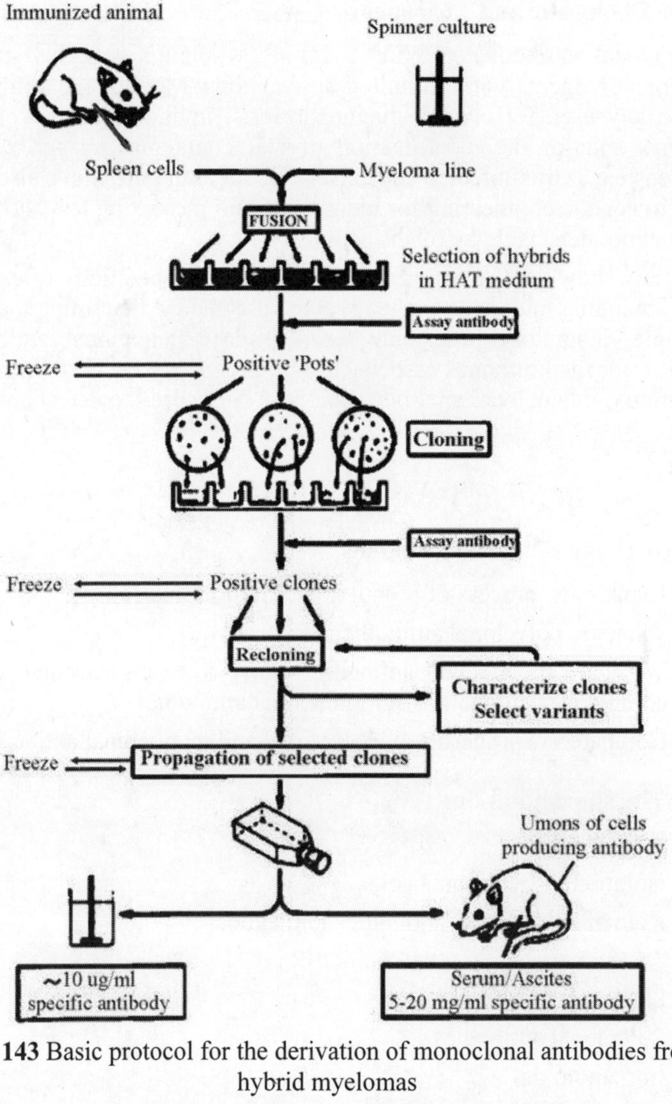


**Fig. 143** Basic protocol for the derivation of monoclonal antibodies from hybrid myelomas

A sensitive ELISA test can be used to identify hybridomas that produce monoclonal antibodies. Several distinct clones are isolated from a typical fusion, each making monoclonal antibody to a different determinant on the antigen. Once the clones of interest are identified, they can be grown in cell culture and antibody can be harvested from the culture supernatant. Hybridomas can grow indefinitely or can be stored as frozen cells. Hybridomas can then be thawed and grown in tissue culture to produce more of the specific monoclonal antibody.

### 13.2.1 Diagnostic and Therapeutic Uses

Monoclonal antibodies provide a limitless supply of highly-specific biological reagents with unlimited applications. Monoclonal antibodies are widely used for clinical diagnostic tests, immunological typing of bacteria and for the identification of cell containing foreign surface antigens e.g., virus-infected cells. Monoclonal antibodies have also been used in genetic engineering for identifying and measuring levels of gene product not detectable by other methods.

They show greater promise for increasing the specificity of existing tests including blood typing and even pregnancy determination. For example, laboratory pregnancy tests employ monoclonal antibodies specific for the hormones associated with pregnancy HCG. Due to their specificity, monoclonal antibodies are now being used to detect and treat cancer.

## REVIEW QUESTIONS

**I. Answer the following questions**

1. Explain the process of production of hybridoma cells.
2. What are polyclonal antibodies? Explain.
3. What are monoclonal antibodies? Give a concise account of the commercial production of monoclonal antibodies.
4. Compare the properties of polyclonal and monoclonal antibodies.

**II. Write short notes on:**

1. HAT medium
2. Isolation of hybridoma cells
3. Role of monoclonal antibodies in diagnosis
4. Aminopterin
5. Koeler and Milstein
6. Polyclonal antibodies
7. Immunotoxin
8. Immunopurification

## 13.3 FURTHER READING

Birch JR and Lennox ES 1995 *Monoclonal Antibodies: Principles and Applications*, New York: John Wiley

Goldsay RA, Kndt TJ, and Osborne BA 2000 *Kuby Immunology*, New York: W.H. Freeman Company

King DJ 1998 Applications and Engineering of Monoclonal Antibodies, London: Taylor and Francis

Kondilis HD, Kor B, Steckman B, and Kargel M 2005 Regulation of T-cell receptor B allelic exclusion at a level beyond accessibility, *Nature Immunology* **8** 189–197

# 14

# Immune Disorders – Hypersensitivity

## 14.1 IMMUNE DISORDERS

Though the vertebrate immune system is oriented in such a way that it protects the host against the continuous attack of microbial invaders and cancer cells, disorders also occur in the immune system. They can be classified as:

1. Hypersensitivities
2. Autoimmune diseases
3. Immunodeficiencies
4. Transplantation or rejection or tissue rejection

### 14.1.1 Hypersensitivity

Hypersensitivity can be defined as an abnormal immune response, which leads to tissue damage. It may result in an individual on first, second or subsequent exposure to an antigen. Hypersensitivity reaction may occur immediately or delayed. The former appear faster than the later. Perter Gell and Robert Coombs (1963) classified hypersensitivities into four types depending upon the nature of immune response on the antigen.

**1. Type I Hypersensitivity:** An allergy is one kind of type I hypersensitivity reaction. An individual develops allergy that has produced IgE antibody in response to exposure to a harmless antigen or allergen and develops same allergy on subsequent exposure.

The B-cells are stimulated to differentiate into plasma cells upon initial exposure to a soluble antigen. The plasma cells produce IgE antibody with the help of T-cells. The IgE is also called as reagin, which is also produced as a manifestation of heredity in some individuals. The IgE, thus formed, sensitizes the mast cells by binding to their Fc receptors, making the individual allergic to the allergen. On second exposure of an individual with the same allergen, it binds to the surface of IgE present on the surface of the sensitized mass cells, which leads to degranulation. Degranulation leads to release of physiological mediators such as histamine leukotrienes, heparin, prostaglandins, Platelet Activation Factor (PAF), Eosinophil Chemotactic Factor of Anaphylaxis (ECF-A) and proteolytic enzymes. These substances stimulate contraction of smooth muscles, vasodilation, increased vascular permeability and mucous secretion. All these phenomena are together called as anaphylaxis, which is of two types – systemic and localized (Fig. 144).

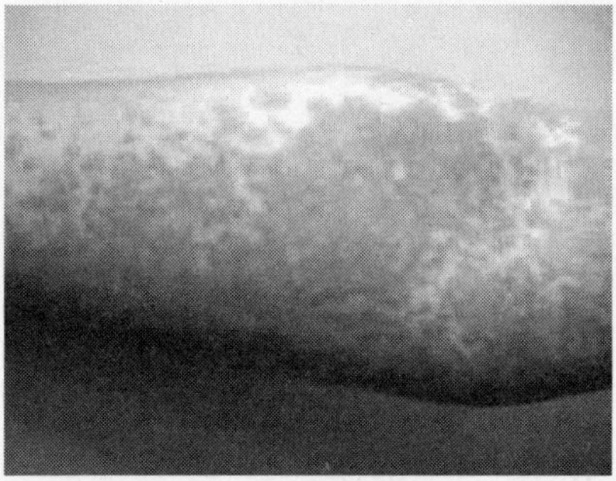

**Fig. 144** The anaphylactic response to bee venom

Systematic anaphylaxis: It is a generalized reaction, which occurs in an individual sensitized to an allergen that receives it again due to second or subsequent exposure. The reaction is immediate due to the release of large amount of physiological mediators within a short time. These changes lead to impairment of respiratory system caused by the contraction of smooth muscles of bronchioles, arterioles dilatation leading to drop in blood pressure and capillary permeability with rapid loss of fluid into the tissue spaces. Due to these abnormal reactions, the individual may die within few minutes from reduced blood pressure, circulatory shock or reduced venous-blood return (Fig. 145). Penicillin drugs, peanuts, insect venoms from the stings or bites of wasps, bees or hornets and passively administered antisera are the examples of allergies, which cause this type of anaphylaxis.

**Localized anaphylaxis:** It is also called as atopic allergy. The type of allergy and its symptoms depend mainly on the channel through which allergen enters the body. A good example of atopic allergy is the allergic rhinitis or hay fever, which involves the organs of upper respiratory tract. Most cells present in the mucous membranes get sensitized due to initial exposure to air-borne allergens such as fungal spores, plant pollen, house dust, mites etc. Re-exposure to these allergies results in the typical localized anaphylactic response characterized by itching and tearing eyes, congested nasal passages resulting in coughing and sneezing. Antihistamines are generally used for relieving hay fever and their disorders.

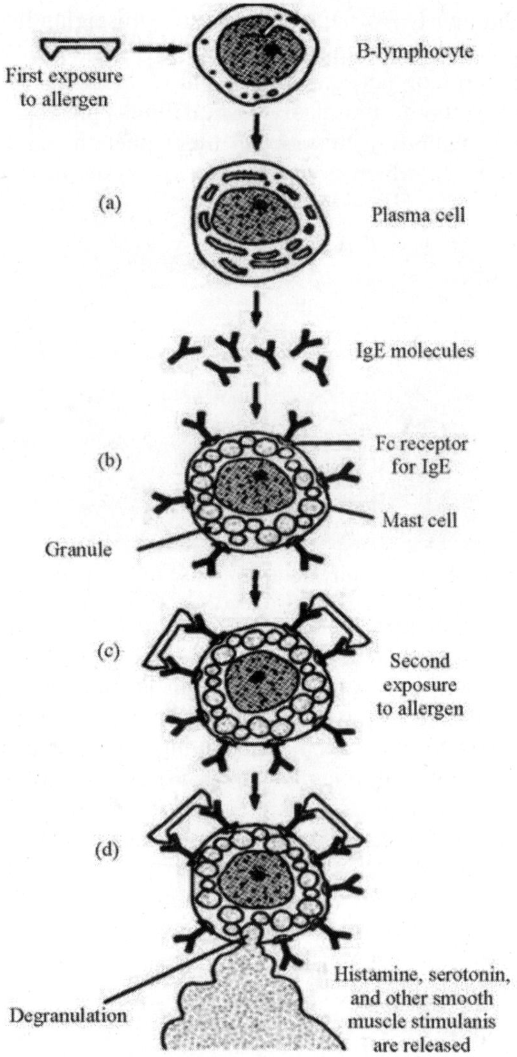


**Fig. 145** Type I hypersensitivity

Another example for atopic allergy is the bronchial asthama, which involves organs of lower respiratory tract. During bronchial asthama, alveoli or air sacs of lungs swell abnormally due to accumulation of fluid and mucous, and narrowing of bronchi due to constriction of their smooth muscles occurs. Bronchial constriction leads to wheezing and whistling sound during exhalation. The allergies, which cause hay fever, are also responsible for the production of bronchial asthma. Relief can be obtained by bronchodilators, which relax constricted muscles, and by expectorants and liquefacients, which dissolve the fluid and mucous present in the alveoli.

It is possible to identify the specific allergen responsible for allergy using skin tests. In this test, small amount of suspected allergen is inoculated onto the skin. Sensitivity to the injected allergen is manifested in the form of rapid inflammatory reaction marked by redness, swelling and itching in the place of inoculation. Once the allergen is identified, the individual should avoid its contact. However, this is not possible in many instances. Therefore, desensitization is generally followed. In this process, a series of allergen doses are injected beneath the skin to stimulate the production of the antibody IgG rather than IgE. These antibodies act as blocking antibodies by intercepting and neutralizing the allergen before they react with vast cell bound IgE antibodies. This process is 65–75% effective in those individuals who develop allergies due to inhaled allergens.

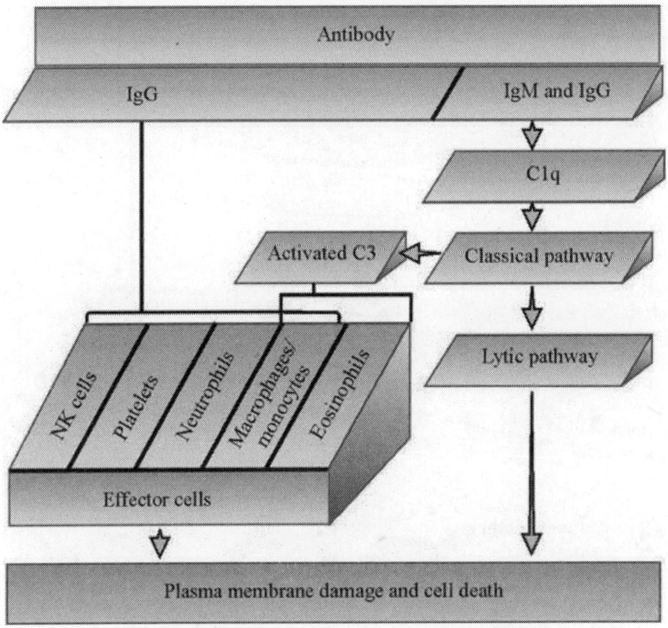


**Fig. 146** Type II hypersensitivity

**1. Type II Hypersensitivity:** It is also called as cytotoxic or lytolytic reactions as it results in the destruction of host cells due to lysis or toxic mediators. In this hypersensitivity, the antibodies IgG or IgM act against cell surface antigens or tissue-associated antigens. They normally stimulate complement pathway and different effector cells such as eosinophiles, neutrophiles, platelets, macrophages / monocytes and NK cells. The antibodies interact with complement and the effector cells through their Fc regions. The destructive mechanisms are reflection of the abnormal physiological processes involved in interaction of the immune system with pathogen (Fig. 146). A good example of this type of hypersensitivity is when a person receives a transfusion of blood from a donor with a different blood group.

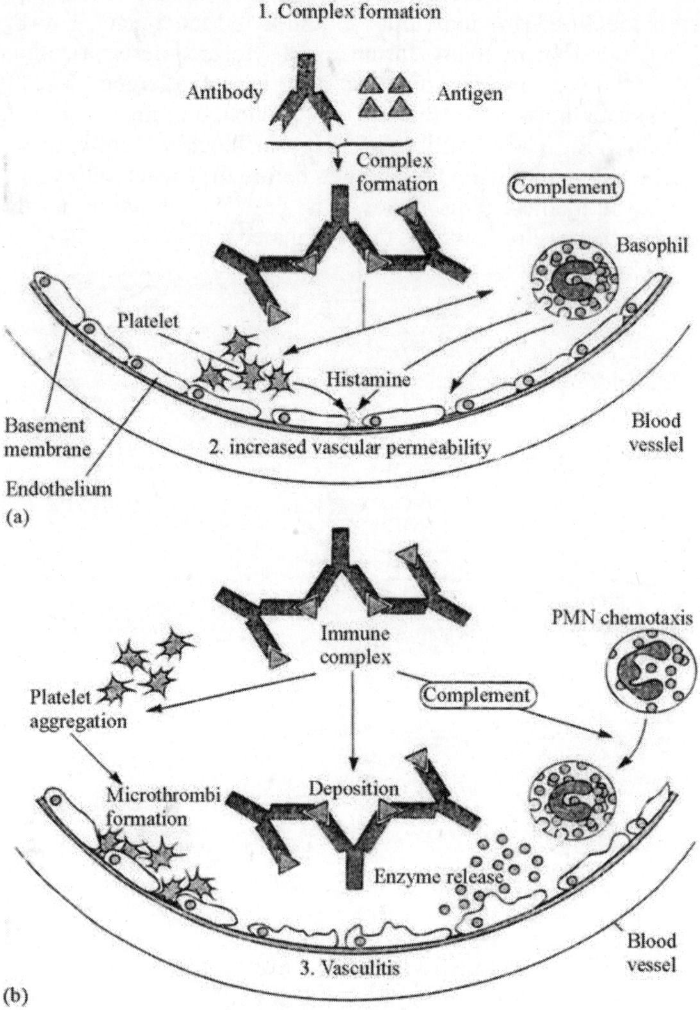


**Fig. 147** Type III hypersensitivity

**2. Type III Hypersensitivity:** It results due to the formation of immune complexes, which are usually removed by the fixed monocytes and macrophages of monocyte-macrophage system. These immune complexes are formed due to the presence of excess amounts of soluble antigen due to which the antibody and antigen complexes may not be removed. The accumulation of these complexes stimulates complement that triggers a variety of inflammatory processes, which may lead to damage of blood vessels, membranes of kidneys (glomerulonephritis), arthritis of joints and skin (Fig. 147). The human diseases occurring due to this type of hypersensitivity can be grouped into three types:

a. A prolonged presence of bacterial, viral and protozoan infection along with presence of weak antibody reaction leads to enormous immune complex formation, which leads to their deposition in the host tissue
b. During autoimmune diseases, there is continuous production of autoantibody to self antigen, which leads to prolonged immune complex formation. This leads to overload of the monocyte-macrophage system, which results in the deposition of complexes.
c. Formation of immune complexes at the body surfaces such as lungs following repeated inhalation of allergens from plants (such as pollen), molds (spores conidia) or animals. For example, in farmer's lung disease, inflammation of alveoli of lungs occurs due to formation of immune complexes due to circulation of antibody (IgG) to fungal spores. In humans, immunologically-mediated glomerulonephritis is produced because of streptococcal infection. It results due to deposition of immune complexes consisting of streptococcal antigen and its antibody in the glomerulus of kidneys.

**3. Type IV Hypersensitivity:** It is called as delayed T-cell-mediated immune reaction, because certain time, usually day or more, is required for the migration and accumulation of a type of TH-cell (designated as delayed-type hypersensitivity cell T DTH) around the antigen.

This type of hypersensitivity occurs when antigen, especially, those binding to tissue cells, are phagocytosed by macrophages and then presented to receptors in the TH-cell leads to proliferation and release of cytokines, which attracts basophils, macrophages and lymphocytes to the affected tissue. This may result in excessive tissue damage. Tuberculin hypersensitivity test (the TB skin test), allergic contact dermatitis, some autoimmune diseases, transplantation-rejection and killing of cancer cells are some of the examples of this type of hypersensitivity (Fig. 148). In the following lines, the TB skin test and allergic contact dermatitis are described in detail.

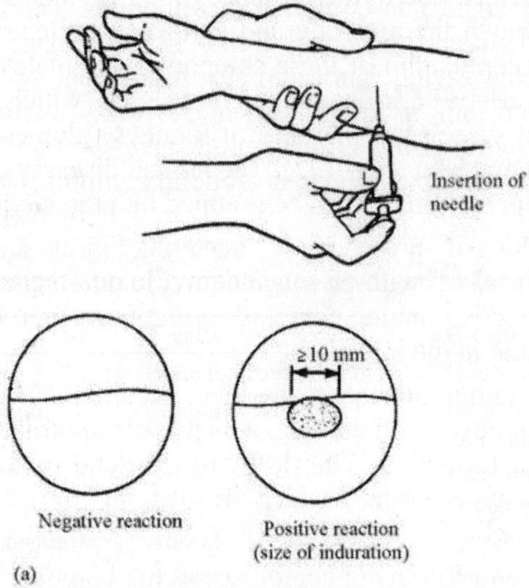


**Fig. 148** Mechanism of type IV hypersensitivity

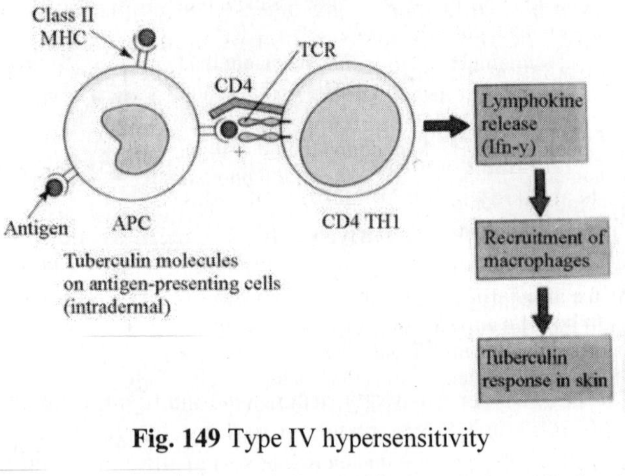


**Fig. 149** Type IV hypersensitivity

In the TB skin test, a partially-purified protein called tuberculin, extracted from the bacterium (*Mycobacterium tuberculosis*), which causes the disease tuberculosis, is injected into the skin of forearm. Hypersensitivity reaction in tuberculin positive individuals starts in about 8 h and is indicated by the formation of red coloured induration at the site of injection. The $TH^1$-cells that migrate to the site of injection are responsible for the formation of this indurated structure. The reaction reaches its peak in about 48 h and then subsides. The size of induration depends upon the amount of tuberculin injected into the skin and the degree of hypersensitivity of the tested individual (Fig. 149).

Allergic contact dermatitis is caused by the allergen formed by a reaction between the haptens and skin protein. This allergen causes immune response. The haptens are the antigenic determinants and the skin proteins act as carriers to the haptens. Topical chemotherapeutic agents, plant material such as catechol molecules from poison Ivy and Oak, jewellry (especially jewellry containing nickel) and metals are some of the examples, which act as haptens (Fig. 150).

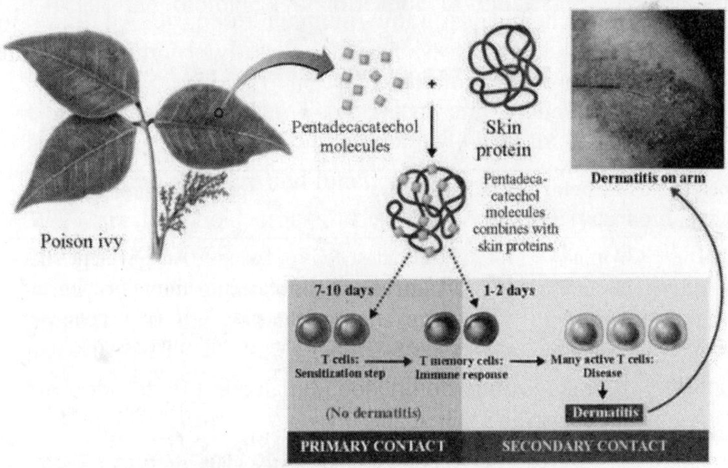


**Fig. 150** Allergic contact dermatitis

This type of hypersensitivity leads to several chronic diseases, which involve tissue damage. These diseases are caused by several types of microorganisms such as bacteria, viruses, fungi, protozoa and others. Important diseases include leprosy, tuberculosis, herpes complex lesions, candidiasis and leishmaniasis.

### 14.1.2. Autòimmunity and its Significance

The body of a vertebrate animal such as human body is normally able to distinguish its own self antigens from foreign non-self antigens and does not mount an immunologic attack against self antigens. The phenomenon is called immune tolerance. Sometimes the body looses tolerance and mounts an abnormal immune attack either with antibodies or T-cell against a person's ownself tissue antigens. This immune attack is called as autoimmunity.

Autoimmunity is often benign where as autoimmune disorder or disease is often fatal. Autoimmunity is characterized only by the presence of serum autoantibodies. It is induced by aging or infectious agents or organisms or drugs. It is often reversible because it disappears when the inducing agent is removed or eradicated. Autoimmune disease results from the activation of self-reactive T-cells and B-cells that causes actual tissue damage following stimulation by genetic or environment triggers.

Four factors influence the development of autoimmune diseases. Two major factors are genetic and viral. The third factor is endocrine. The hormone estrogen promotes autoimmune disease and androgen acts as natural immunosuppressive agents. The fourth is by the influence of stress and neurochemicals on the immune response. All these four factors can effect gene expression, which directly or indirectly interfere with important immune regulatory function (Table 79).

**Table 79:** Factors influencing autoimmune diseases

| Type of reactions | Time before clinical signs | Characterstics | Examples |
|---|---|---|---|
| Type I | <30 min | IgE binds to mast cells or basophils and it causes degranulation of mast cells or basophils, and release of reactive substances such as histamine occurs. | Anaphylactic shock from injection and insect venom, common allergic conditions such as hey fever, asthama |
| Type II | 5–12 h | Antigen causes formation of IgM and IgG antibodies that bind to target cell; when combined with action of complement, destroys the target cell | Transfusion reactions, Rh incompatibility |
| Type III | 3–8 h | Antibodies and antigens form complexes that cause damaging inflammation | Arthrites reaction, serum sickness |
| Type IV | 24–48 h | Antigens activate Tc-cells that kill the target cells | Rejection of transplanted tissues, contact dermatitis, chronic diseases such as tuberculosis |

These diseases, although their causal mechanism is not clear, are more common in older people and may involve viral or bacterial infections. It is reported that the release of abnormally large quantities of antigens may occur when the infectious agents cause tissue damage. The same agent may also cause body production or T-cell activation. Simultaneously, the activity of Ts-cells, which normally limits this type of reaction, seems to be repressed. Many autoimmune diseases have genetic component. For example, there is well known association between an individual's susceptibility to Graver disease and multiple sclerosis, and specific determinants on the MHC.

## REVIEW QUESTIONS

### I. Answer the following questions

1. Describe type I hypersensitivity with suitable examples.
2. Write critical notes on Type II hypersensitivity reactions.
3. Explain Type III hypersensitivity.
4. TB skin test belongs to which type of hypersensitivity? Explain.

### II. Write short notes on the following

1. Immune disorder
2. Asthama
3. Immune complexes
4. Significance of autoimmunity

## 14.2 FURTHER READING

Abbas AK and Litchman AH 2005 *Cellular and Molecular Immunology*, Philadelphia, Penngulvania, USA: Elsevier Saunders

Ananthanarayan R and Jayaram Panikar CK 2003 *Text Book of Microbiology*, Chennai: Orient Longman

Greenwood D, Slack RCB, and Peutherer JF 2006 *Medical Microbiology*, New Delhi: Reed Elsevier, India Pvt. Ltd.

Irving W, Boswell T, and Dlawer AA 2009 *Medical Microbiology*, New Delhi: Viva Books Pvt. Ltd.

Madigan MT, Martinko JM, and Parker J 2003 *Brock Biology of Microorganisms*, New Jersey: Prentice Hall

Murray PR, Rosenthal KS, Kobayashi GS, and P Faller MA 1998 *Medical Microbiology*, London: Mosby

Nagoba BS 2005 *Clinical Microbiology*, New Delhi: B I Publications

Prescott M, Harley JP, and Klein DA 2005 *Microbiology*, New York: Mc Graw Hill Higher Education

Ranjan Kumar 2007 *Diagnostic Microbiology*, New Delhi: Jaypee Brothers Publishers Pvt. Ltd.

Ranjan Kumar 2007, *Clinical Microbiology*, New Delhi: B I Publications

Thomas PA 2007 *Clinical Microbiology*, Chennai: Orient Longman Pvt. Ltd.

Wetz JAE, Melmick JL, and Adelberg EA 1987 *Review of Medical Microbiology*, New Jersey: Prentice Hall International.